Molecularly Imprinted Polymers for Environmental Monitoring

Fundamentals and applications

Online at: https://doi.org/10.1088/978-0-7503-4962-8

Molecularly Imprinted Polymers for Environmental Monitoring

Fundamentals and applications

Edited by

Raju Khan

CSIR-Advanced Materials and Processes Research Institute (AMPRI), Bhopal, India

Ayushi Singhal

CSIR-Advanced Materials and Processes Research Institute (AMPRI), Bhopal, India

IOP Publishing, Bristol, UK

ISBN   978-0-7503-4962-8 (ebook)
ISBN   978-0-7503-4960-4 (print)
ISBN   978-0-7503-4963-5 (myPrint)
ISBN   978-0-7503-4961-1 (mobi)

DOI   10.1088/978-0-7503-4962-8

Version: 20230801

IOP ebooks

British Library Cataloguing-in-Publication Data: A catalogue record for this book is available from the British Library.

Published by IOP Publishing, wholly owned by The Institute of Physics, London

IOP Publishing, No.2 The Distillery, Glassfields, Avon Street, Bristol, BS2 0GR, UK

US Office: IOP Publishing, Inc., 190 North Independence Mall West, Suite 601, Philadelphia, PA 19106, USA

One individual may die for an idea, but that idea will, after his death, incarnate itself in a thousand lives.

—Subhas Chandra Bose

Contents

10 Molecularly imprinted polymers for the detection of environmental estrogens   10-1

Melkamu Biyana Regasa and Tebello Nyokong

11 Molecularly imprinted polymer based detection of pesticides   11-1

Priya Chauhan and Annu Pandey

14 Future perspectives on molecularly imprinted polymers for   **14-1**
environmental monitoring
Sadhna Chaturvedi

Preface

The field of molecular imprinting is growing rapidly and has become broad during the past few years. Molecularly imprinted polymers (MIPs) are artificial templates for the binding of a specific analyte. An MIP operates on the principle of a 'lock and key' mechanism to bind specifically to molecules such as amino acids, metal ions, antibiotics, drug additives, food additives, pesticides, and various environmental pollutants with which they were imprinted during the process of production. MIPs can complement natural antibodies. Natural antibodies are functional polypeptide chains and are a biological material. They exhibit several drawbacks, for example, they are unstable if exposed to extreme conditions, it is difficult to obtain them in large batches, and the synthesis process is expensive. Alternatively, MIPs are synthetic materials and overcome the limitations of antibodies as they are cheap, physically and chemically stable, highly specific and selective for the target analytes, and also possess high durability. Due to their advantages, MIPs have been widely used as potential materials in sensor fabrication, binding assays, and antibody mimics, and as a stationary phase in chromatography. The most common applications of MIPs include immunoaffinity separation, biosensors, bioimaging, food analysis, explosive detection, pathogen detection, chiral molecule detection, pharmaceutics detection, and clinical treatment as drug delivery agents. MIPs are artificial materials with imprinted recognition sites that are able to bind specifically and selectively to an analyte. These MIPs are obtained by the polymerization of various monomeric units in the presence of crosslinkers. The monomers are selected based on their capability to interact with functional groups present on the surface of the template. There are many advantages to using MIPs, as they are highly selective, stable against extreme conditions of pH, heat, mechanical stress, economical, have a long storage life, are functional for several uses, and are chemically inert in acids, bases, and other organic solvents. However, the imprinting of long-chain peptides, proteins, etc, molecules of high molecular weight is still a major concern. MIPs are used extensively for the separation and detection of specific analytes such as amino acids, metal ions, antibiotics, drug additives, food additives, pesticides, and various environmental pollutants from matrix materials such as water, soil, food, biological samples, etc.

Environmental monitoring and analysis to examine various pollutants such as gases, heavy metals, pharmaceutical waste, adulteration chemicals, pesticides, herbicides, etc, are the prime requirements. To minimize the adverse effects of these pollutants and maintain the environment, various sensors have been developed.

MIP-based sensors have gained much attention due to their several advantages. Overall, the advantages of the MIP-based sensors make them suitable candidates for not only the analysis of the environmental pollutants but also the detection of biological species as in the diagnosis of numerous disease biomarkers. In the future, they could be miniaturized into portable devices and be commercially available, which would extend their future applicability in sensing applications. This book fills a gap in the literature and provides a compiled literature review of the fabrication and development of MIP-based devices for various sensing applications. This book is intended to provide users with concise and well-framed information about MIP-based biosensors which can be exploited by researchers and device manufacturers for the development of novel sensors and their commercialization for environmental monitoring.

This is a multidisciplinary book which connects and interfaces between nano-biotechnology, environmental science, environmental chemistry, material science, chemistry, physics, and the healthcare industry. It covers the approaches for designing novel, cost-effective MIP-based sensors for environmental monitoring and analysis and sheds light on the properties of nanomaterials with the specific focus of MIPs for sensitive and specific detection of environmental pollutants.

Acknowledgments

The editors are sincerely grateful to every contributor to this book. Their sincere efforts, hard work, and analytical approach are greatly acknowledged. At the same time, the editors would like to acknowledge the publisher and the associated team for continuous support, guidance, and motivation which constantly pushed us forward to complete this book.

The valuable advice and guidance provided by Dr Avanish Kumar Srivastava, Director, CSIR-Advanced Materials and Processes Research Institute, Bhopal, MP, India, is duly acknowledged.

Raju Khan expresses his special thanks to his parents, his wife Shazia M Siddiqui, his daughter Sara Khan, and his son Aayan Khan for their everlasting love, enthusiasm for science, and encouragement to pursue every task successfully. Ayushi Singhal expresses her special gratitude to her inspiration Dr Pragya Bharti, her father Rajendra Singhal, her mother Rashmi Singhal, and all her family members for their blessings and constant support.

Further, the editors are sincerely grateful to all who have directly or indirectly rendered valuable input to this book.

Raju Khan
Ayushi Singhal

Editor biographies

Raju Khan

Raju Khan is currently working as a Principal Scientist and Associate Professor at the CSIR-Advanced Materials and Processes Research Institute (AMPRI), Bhopal, India. He received his MSc in chemistry in 2002 and PhD in 2005 from Jamia Milia Islamia, Central University, New Delhi, India. Soon after, he joined the Sensor Laboratory, University of the Western Cape, Cape Town, South Africa, as a Post-Doctoral researcher from 2005 to 2006. From there he joined the Biomedical Instrumentation section, National Physical Laboratory, New Delhi, India as a Fast Track Young Scientist from 2007 to 2008. He was awarded the reputed BOYSCAST fellowship under the scheme of the BOYSCAST fellowship from the Department of Science and Technology and worked as a visiting scientist for one year at the University of Texas at San Antonio, USA. He has more than 15 years of experience in electrochemistry, exploring the electrochemical properties of nanostructures to design and develop efficient biosensor devices for healthcare monitoring. He has several ongoing projects and has completed international scientific collaborations with the Czech Republic and Russia. His research is highly interdisciplinary and overlaps with materials science and biomedical engineering. His current research is on electrochemical biosensors integrated with microfluidics for clinical applications.

Ayushi Singhal

Ayushi Singhal is currently pursuing her doctorate from CSIR-Advanced Materials and Processes Research Institute (AMPRI), Bhopal. She was a gold medalist in MSc (Environmental Chemistry) at Jiwaji University. Her research interests are towards the synthesis of molecularly imprinted polymers and polymer nanocomposites, the fabrication of biosensors, and the applications of biosensors in the analysis and monitoring of environmental pollutants in different matrices.

List of contributors

G K Athira
Institute of Chemical Technology, Mumbai–Marathwada Campus, Jalna, Maharashtra 431203, India

Yasmin Bano
Department of Molecular and Human Genetics, Jiwaji University, Gwalior, India

Department of SOS in Biochemistry, Jiwaji University, Gwalior, India
Department of Biotechnology, Cancer Hospital and Research Institute, Gwalior, India

Sadhna Chaturvedi
Department of Biotechnology, School of Sciences, ITM University, Gwalior, India

Jyoti Singh Chauhan
Council of Scientific and Industrial Research-National Botanical Research Institute, Prem Nagar, Hazratganj, Lucknow 226001, UP, India

Priya Chauhan
SOS in Environmental Chemistry, Jiwaji University, Gwalior 474011, MP, India

Rashmi Dahake
Academy of Scientific and Innovative Research (AcSIR), Ghaziabad, Uttar Pradesh 201002, India
CSIR-National Environmental Engineering Research Institute (NEERI), Nagpur, Maharashtra 440020, India

K B Divya
Dairy Technology Department, Verghese Kurien Institute of Dairy and Food Technology, Mannuthy, Kerala, India

M P Divya
Dairy Chemistry Department, Verghese Kurien Institute of Dairy and Food Technology, Mannuthy, Kerala, India

Cem Erkmen
Ankara University, Faculty of Pharmacy, Department of Analytical Chemistry, 06560 Ankara, Turkey

Sarang Gumfekar
Department of Chemical Engineering, Indian Institute of Technology Ropar, Rupnagar, Punjab 140001, India

Vaishnavi Hada
Council of Scientific and Industrial Research-Advanced Materials and Processes Research Institute, Hoshangabad Road, Bhopal 462026, MP, India

L R Hemanth
Scientist, Research and Development, Samco Inc., Waraya-cho, Takeda, Fushimi-Ku, Kyoto 612 8443, Japan

Shiv Kumar Jayant
Department of SOS in Biochemistry, Jiwaji University, Gwalior, India

Ashish Kapoor
Department of Chemical Engineering, Harcourt Butler Technical University, Kanpur, Uttar Pradesh 208002, India

Priya Kumari
School of Bioscience, Indian Institute of Technology, Kharagpur, West Bengal 721302, India

Sapna Kumari
Sant Hirdaram Girls College, Lake Road Sant Hirdaram Nagar, Bairagarh, Bhopal, Madhya Pradesh 462030, India

Hitesh Malviya
Medilux Laboratories, Indore 474011, MP, India

Ritesh Mishra
Medilux Laboratories, Indore 474011, MP, India

Rafia Nimal
Department of Chemistry, Quaid-i-Azam University, 45320 Islamabad, Pakistan

Tebello Nyokong
Institute for Nanotechnology Innovation, Rhodes University, Makhanda 6140, South Africa

Annu Pandey
Department of Chemistry, Institute of Science, Chandigarh University, Ajitgarh 140413, India

Suman Pawar
Department of Chemical Engineering, Siddaganga Institute of Technology, B H Road, Tumakuru 572 103, Karnataka, India

Antony Nitin Raja
Medilux Laboratories, Indore 474011, MP, India

Y S Rajput
Ex-Emeritus Scientist, Animal Biochemistry Division, ICAR-National Dairy Research Institute, Karnal 132001, India

Melkamu Biyana Regasa
Institute for Nanotechnology Innovation, Rhodes University, Makhanda 6140, South Africa

Manish S Sengar
School of Medical Science and Technology, Indian Institute of Technology, Kharagpur, West Bengal 721302, India

Neha Sengar
Department of Chemistry, Delhi University, Delhi 110007, India

Afzal Shah
Department of Chemistry, Quaid-i-Azam University, 45320 Islamabad, Pakistan

Rajan Sharma
Dairy Chemistry Division, ICAR-National Dairy Research Institute, Karnal 132001, India

Abhinav Shrivastava
Department of Biotechnology, Cancer Hospital and Research Institute, Gwalior, India

Apoorva Shrivastava
Dr D Y Patil Biotechnology and Bioinformatics Institute, Dr D Y Patil Vidyapeeth, Sr No. 87–88, Mumbai–Bangalore Highway, Tathawade, Pune, Maharashtra 411033, India

Piyush Shukla
Laboratory of Natural Products, Department of Rural Technology and Social Development, Guru Ghasidas University, Bilaspur, India

Muhammad Siddiq
Department of Chemistry, Quaid-i-Azam University, 45320 Islamabad, Pakistan

Ayushi Singhal
Industrial Waste Utilization, Nano and Biomaterials, CSIR-Advanced Materials and Processes Research Institute (AMPRI), Hoshangabad Road, Bhopal - 462026, MP, India
Academy of Scientific and Innovative Research (AcSIR), Ghaziabad-201002, India

Mittali Tyagi
Dr B Lal Institute of Biotechnology, University of Rajasthan, Jaipur, Rajasthan 302017, India

K Uma
Department of Chemistry, Siddaganga Institute of Technology, B H Road, Tumakuru 572 103, Karnataka, India

Bengi Uslu
Department of Analytical Chemistry, Faculty of Pharmacy, Ankara University, 06560 Ankara, Turkey

Vijay Vaishampayan
Department of Chemical Engineering, Indian Institute of Technology Ropar, Rupnagar, Punjab 140001, India

IOP Publishing

Molecularly Imprinted Polymers for Environmental Monitoring
Fundamentals and applications
Raju Khan and Ayushi Singhal

Chapter 1

An introduction to molecularly imprinted polymers: the plastic antibodies

Antony Nitin Raja, Ayushi Singhal, Ritesh Mishra and Raju Khan

The molecular imprinting technique is an attractive approach in which templates are formed for specific recognition sites in a polymer. These artificial receptors are imbued with some astonishing properties, such as robustness, high affinity, specificity, and low-cost production. Molecular imprinted polymers (MIPs) have the ability to structurally mimic natural antibodies. As an alternative to antibodies, they are also called plastic antibodies. MIPs have gained enormous attention; researchers are showing significant interest with innovative findings and a large number of research papers related to MIP sensors being published. Researchers have applied this technique in various fields, such as the determination of biomolecules, drugs, and explosives, and further taken this technology toward environmental, medical and forensic diagnostics.

1.1 Introduction

Molecular recognition is a fundamental aspect of many biological processes. This recognition is very precise so molecules fit only to their natural and selected targets. For example, antibodies have specific sites to bind specific antigens and enzymes to their substrates, hormone receptors to their hormones, etc. These bio-macromolecules also serve as a molecular recognition tools for modern biomedical and analytical chemistry. They are used as recognition systems in affinity technology with numerous applications in various fields [1, 2]. Molecularly imprinted polymers (MIPs) are best described as man-made structures homologous to the natural biological antibody–antigen systems. The antigen–antibody interaction is based on a lock and key system [3] and in a similar way MIPs are specific to their targets. They bind very selectively to the molecule with which they were templated during production [4]. The natural antibodies have certain drawbacks, for example under conditions targeting structures of low immunogenicity, natural antibodies do not perform well. MIPs have been suggested to replace such antibodies and are therefore also called molecularly imprinted synthetic antibodies and plastic antibodies. Plastic antibodies can recognize and neutralize specific bio-macromolecules with effectiveness comparable to antibodies and are of significant interest as an abiotic alternatives to antibodies [5, 6]. The key sensing feature of MIPs is specific recognition sites for target analytes. These specific sites are called cavities. These cavities are embedded in the polymer matrix. In MIPs cavities are formed in three steps. In the first step functional monomer molecules are prearranged around the template molecule in solution, the second step involves polymerization of the resulting complex in the presence of cross-linking monomers and porogen (e.g. suitable solvents or ionic liquids), and in the last step the template is removed from the synthesized MIPs [7, 8]. To accomplish any significant endeavor a strong base is required. The first step is vital and acts as a base for generating highly selective molecular cavities. The prearrangement of functional monomers can be achieved either by forming covalent bonds or self-assembly with non-covalent bonds. In this process functional and cross-linking monomers are polymerized together in the presence of a target molecule or a derivative. The target molecule acts as a molecular template. The functional monomers attach to the imprint molecule and form a complex. This step is followed by polymerization; here, the highly cross-linked polymeric structure keeps the functional group in a fixed position [9–11]. In the final step, removal of the imprinted molecule reveals binding sites that are complementary in size and shape to the template. The chosen MIP synthesis approach depends mainly on the application and required morphology. MIPs can be synthesized by using different polymerization techniques such as bulk polymerization, emulsion polymerization, electropolymerization, and controlled polymerization. These techniques are specific and selective to the biological receptors. MIPs have numerous advantages such as durability with respect to environmental conditions and economical cost [12, 13]. Currently, various sophisticated and powerful analytical instrumentation and analytical methods, including high performance liquid chromatography and liquid chromatography with mass spectrophotometry, are used. Even though these chromatographic and spectrophotometric procedures have been employed systematically for many analyses with high sensitivity and good selectivity, these methods suffer from disadvantages such as expensive instrumentation, a requirement for hazardous organic solvents in large quantities, and time-consuming pre-treatment. Hybrid chromatographic techniques coupled to common detectors (UV, fluorescence, PDA) or, more recently, mass spectrometry (MS) and gas chromatography are coupled and used routinely in analytical laboratories for the determination of target analytes. However, some of the samples are crude extracts and direct injection onto the column is not possible even when using the selective detection provided by MS. Large molecules can affect the ionization process and due to incomplete ionization accurate quantification may not be possible compared to other recognition systems. In contrast, MIPs possess three major unique features: structure predictability, recognition specificity, and application universality. They have received widespread attention and become attractive in many fields, such as the environmental and medical fields, and in techniques such as separation, chemo/

biosensing, artificial antibodies, drug delivery, catalysis, and degradation. The capacity to compare with and be integrated with existing systems, which gives high recognition ability, is the most crucial component of MIP-enabled biosensors. Depending on the sorts of samples to be analyzed, MIPs can be employed in a variety of point-of-care testing (POCT) applications owing to their stability as shown in figure 1.1. MIPs have high physical stability, straightforward preparation methods, and are robust and economical [14–20].

Despite having some advantages over other analytical techniques, MIPs have some drawbacks as well. One of the major causes is that the functional monomers used in molecular imprinting are inadequate. This downside of MIPs restricts their selectivity and due to this the applications of MIPs are also limited to some extent. The other restriction is that molecular imprinting technology for the preparation of MIPs has been neglected to some extent, and this hampers their application in various fields. The fast development of molecular imprinting can be enhanced by paying more attention to the molecular imprinting technology [21–23]. Molecular imprinting technology is a multidisciplinary technology and it should be updated continuously with respect to other technologies, such as polymer technology, nanotechnology, analytical chemistry, environmental science, biotechnology, and so on [24, 25]. For the development of molecular imprinting technology, it must borrow and integrate aspects from other related technologies/strategies to obtain significant breakthroughs and acceleration. For example, the qualitative and

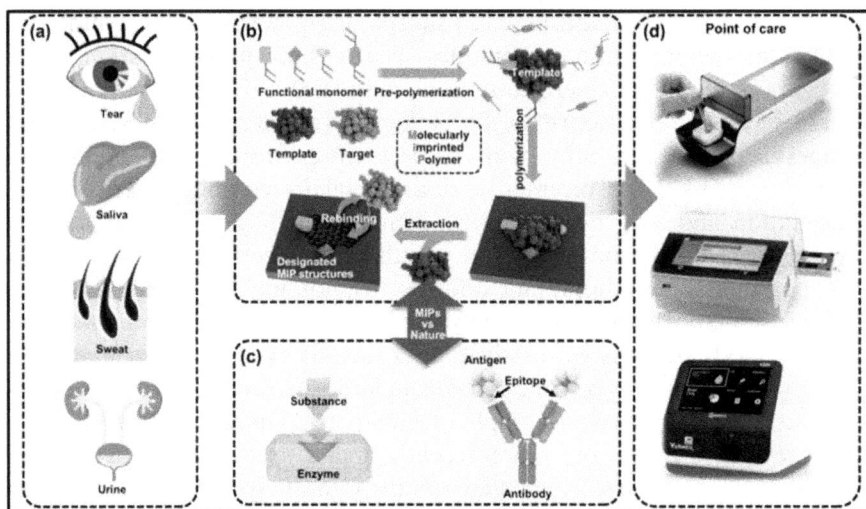

Figure 1.1. Brand-new types of benchtop POCT devices that used MIP-based biosensors for high-precision diagnostics to find biomarkers in biofluids. (a) Tears, saliva, perspiration, and urine are the four forms of representative human biofluid that indicate different health situations. (b) A schematic illustration for creating a molecular imprinting system with biorecognition sites and (c) an illustration of a natural biorecognition system using an enzyme–substrate complex on the left and an antigen–antibody reaction on the right. The MIP biosensing system's biomimetic functional similarity is comparable to that of natural antibodies. (d) A variety of benchtop-scale POCT devices based on immunoassays. (Reproduced from [21]. Copyright 2022 the authors. CC BY 4.0.)

quantitative aspects of the technology can be enhanced by concentrating on sample preparation. A clean sample is generally recommended to improve separation and detection. Typically, several sample treatments are necessary for the best results. In one aspect molecularly imprinted polymers are entirely different from antibodies but on other hand, they have shown many similarities with them. In terms of configuration, MIPs are larger than antibodies and are rigid and insoluble, but they do share the most important feature of the natural receptor molecules: the capability to bind specifically to a target molecule. For the time being, MIPs cannot compete with antibodies for use in techniques in which antibodies are used in their soluble form, such as immunodiffusion, immunoelectrophoretic, immunoblotting, and tissue immunofluorescence [26–28].

MIP techniques have shown great adaptability. In the early period, syntheses of MIPs using various polymerization techniques for bulk micro-particles were achieved. Due to the large particle size, it was found to be very difficult to integrate these into sensor, medical imaging, and drug delivery approaches [29, 30]. After the introduction of MIP nanoparticles (MIP-NPs), this technique has been redeveloped and has greater potential. Nanomaterials have well known properties such as strong adsorption efficiency, a high surface to volume ratio, high surface reactivity, high solubility, better diffusion, and ease of immobilization. In recent years MIP-NPs have become a fast emerging field of research. The number of publications related to the applications of MIP-NPs has increased in recent years, particularly in the field of analytical chemistry, such as applications for solid-phase extraction, liquid chromatography, drug delivery systems, capillary electro-chromatography, enzyme-like catalysis, and sensors. However, despite MIPs' many merits, there is still room for improvement in their integration into sensors, in particular regarding sensitivity.

Areas for further improvement that can be foreseen in MIP development are the reduction of non-specific binding for higher selectivity of the target and reproducible high-quality large-scale production [31–33].

1.2 The early history of MIPs

German physician Paul Ehrlich first used the term antibody in 1891. It refers to proteins present in extracellular fluids that play an important role in specific immune responses. Antibodies are also called immunoglobulins and are made up of glycoproteins. The key function of an antibody is to recognize foreign substances such as viruses, bacteria, chemicals, pollens, etc, called antigens. Antibodies directly neutralize or bind to antigens for successive processing by other components of the immune system. These are the macromolecules, originating from B lymphocytes. Antibodies are Y-shaped globular proteins. They consists of two light chains and two heavy chains. Disulfide bonds hold the chains together. A schematic representation of antibody structure is shown in figure 1.2 [6, 34–36].

In 1952 the antigen–antibody interaction was discovered by Richard J Goldberg. This interaction was found to have specificity, i.e. each antibody is only capable of binding to a specific antigen. The interaction is carried out between the epitope, a specific part of the antigen, and the paratope, a specific part of the antibody.

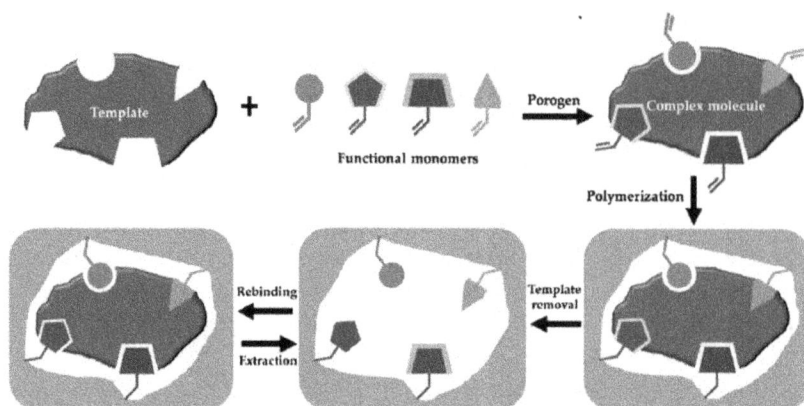

Figure 1.2. Molecularly imprinted polymer-based sensors for priority pollutants. (Reproduced from [34]. Copyright 2021 the authors. CC BY 4.0.)

The antigen–antibody interaction is supported by weak and non-covalent interactions, such as van der Waals forces, hydrogen bonds, and hydrophobic and electrostatic interactions. Agglutination is the term use to describe the antigen–antibody reaction. The main characteristics of this reaction are high specificity and affinity, resulting in the formation of an antigen–antibody complex. As the reaction is reversible, there is a dynamic equilibrium between the dissociated and undissociated forms [37–39].

In 1970 the era of modern molecular imprinting technology began. Takagishi, Klotz, Wulff, Sarhan, and Tahr reported that synthetic polymers could be used to prepare chemical memory. Later they were used for enantiomer recognition. The covalent approach to molecular imprinting was introduced Wulff's group who presented a series of papers on the subject. In these system molecules are recognized based on reversible covalent interaction. In the early 1980s the non-covalent approach to molecular imprinting was introduced by Mosbach's group. The most general forms of interaction are hydrogen bonding, ionic interaction, hydrophobic interaction, etc, through which molecules are recognized. A new approach has been introduced, i.e. metal ion chelation between the template and polymer, which has gain lot of interest in the research field [40–42].

In 1949 the concepts of molecular imprinting were introduced by Dickey. He used organic dyes to precipitate silica gel. This imprinted silica was reported to have high selectivity towards the template dye. Martin and Synge used this dye imprinted silica for the separation of amino acid derivatives. Later these silica gels were used for the crude purification of camosulphonic acid and mandelic acid enantiomers. These silica gels have the ability to separate enantiomers. Unfortunately, the technique developed using these types of experiments had limited application at that time. The uses of silica gels were limited at the time, but some researchers showed tremendous interest and carried out the research on the material. Patrikeev's group used imprinted silica for thin layer chromatography and in column chromatography. The structural elucidation and configurational studies of organic molecules can be performed using imprinted silica. In 1972 a significant study was performed in the field of molecular imprinted technology. Wulff had introduced a functional group into the imprinted cavity through the use of covalent bonds for molecular recognition. This study in fact developed the covalent method of preparing imprinted polymers. Short after that, Mosbach proved that it is achievable to introduce functional groups into the binding cavity through non-covalent interactions. This non-covalent imprinting became the basis for the preparation of most imprinted polymers. In recent time, significant work has been in progress regarding the molecular imprinting techniques for various purposes [43–46].

1.3 MIPs compared to other biorecognition elements

A biomarker shows a specific physical trait or a measurable biologically produced change in the body that is linked to a disease or a health condition. Delays in results are still responsible for a significant amount of death and morbidity, and hence there is a need for faster, cost-effective, and sensitive analytical methods [47]. MIPs could help solve these issues, as MIPs are synthetic materials that can mimic biological recognition well enough to be called 'plastic antibodies'. In the past decade, the number of studies of using MIPs as a biomarker has increased, as these materials hold the potential to provide a sensitive, robust, rapid, and low-cost diagnostic tool [48]. The detection of serum albumin, hemoglobin, ferritin, and avidin, as well as studies of infectious diseases, bone loss, cardiovascular diseases, and various cancers are some areas that are of high interest for the use of MIPs as biomarkers [49]. MIPs bind specifically to various analytes, unlike the natural biorecognition elements (BREs). MIPs are made of synthetic materials that can be synthesized artificially and can be obtained in large batches. The natural BREs cannot be produced in large batches and are very unstable. Natural BREs need to be stored under specific temperatures and storage conditions, and are less durable. MIPs are synthetic, durable, cheap, stable, and are very specific and selective [50–52]. Different target molecules or templates used in MIPs are shown in figure 1.3.

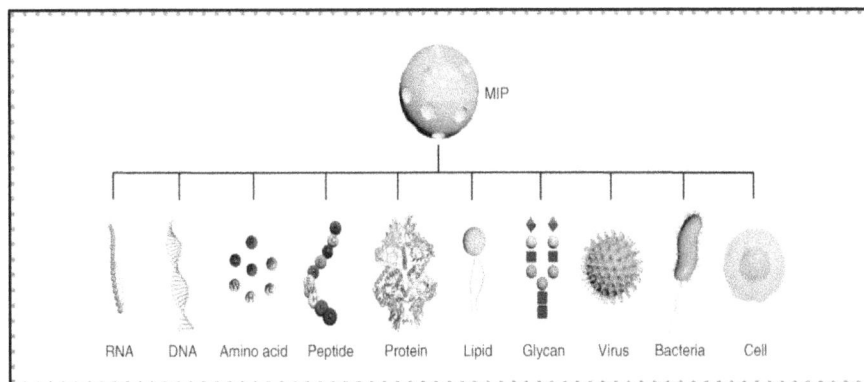

Figure 1.3. Different templates for molecularly imprinted polymers. (Reproduced from [53]. Copyright 2023 Future Science Group. Reproduced with permission from Biotechniques as agreed by Future Science Ltd.)

MIPs target small molecules ranging approximately between ~0.1 and ~10 nm, whereas surface-imprinted polymers (SIPs) are applied for larger biomarkers. SIPs can form cavities directly on the surface of polymers, which enhances the removal of the template, providing better use for larger biomarkers such as viruses, bacteria, and cells [53].

1.4 The properties of MIPs

In contrast to monoclonal antibodies and synthetic receptors such as crown ethers, molecularly imprinted binding sites differ in the precise spatial arrangement of the functional groups, and the access to the site depends on the polarity of the environment. This property of heterogeneity distinguishes them from other binding sites. They resemble polyclonal antibodies and the varying sequences result in various structures at the antigen binding sites [54].

In the case of non-covalently imprinted materials, a large degree of heterogeneity results as the interactions between the monomer and the template are controlled by equilibrium. This results in the presence of a variety of monomer–template complexes as well as not as complexed monomers in the pre-polymerization mixture. The diversity is maintained in the macromolecular material after polymerization [55]. By the impairment of binding sites, causing sites to disrupt on template removal, and causing locally variable impairment of the polymer in a different solvent, processing the material (such as by grinding and sieving), removing the template, and switching the polymerization solvent with a different solvent to study the binding properties can all result in further heterogeneity. 'Specific' sites are those that result from any type of monomer–template complex in the pre-polymerization mixture. These are anticipated to have a higher affinity for the template (and similar structures) and to be more selective. This is due to the precise arrangement of the functional groups and shape selectivity. To put it simply, sites resulting from a non-complexed monomer in the pre-polymerization mixture are proposed to give 'non-specific' sites; these are expected to have lower affinity for the template and to bind other species without discrimination. It is how a polymer with functional monomers arranged randomly would be expected to behave when prepared without a template. Although the dichotomy between 'specific sites' and 'non-specific sites' (or 'imprinted sites' and 'non-imprinted sites') may be useful, it is clear that it does not fully depict the situation, which is one of a continuous spectrum of sites ranging from weaker binding, less selective, to stronger binding, more selective [56–58].

The pre-polymerization mixture will have a higher variety of species than has been predicted. Suppose there are several types of monomers present and the monomer or template can interact with the cross-linker, or there are 'clusters' of the template. The idea that the pre-polymerization species are exactly replicated in the polymerized material is likely naive; studies have shown that these structures change during the polymerization process. However, the general idea that diversity is preserved or enhanced is unquestionably true, for instance, because the polymeric chains are folded in different ways around different sites, the 'outer sphere' interactions for each site will differ [59].

1.5 The applications of MIPs in various fields

MIPs are highly sensitive and selective, and they also have good physiochemical stability and long half-lives. They are exceptionally resistant to hostile environments, including extremely basic and acidic pH, mechanical stresses, and heat, making them suitable for a variety of uses. They can also be modified easily with fluorescent dyes, quantum dots, or other fluorescent materials such as carbon dots and upconversion nanoparticles. They provide a flexible platform for a variety of biomedical applications, such as separation technologies, antibody replacement in diagnostics, sensing, and drug delivery. Some of the major advantages of MIPs are shown in figure 1.4.

1.5.1 Environmental applications

In our fast-developing world, many hazardous materials, so called micropollutants, are releases into the environment. The wastewater discharges from industries such as pharmaceutics, chemical plants, agriculture, oil refineries, etc, contaminate the groundwater and surface waters that are used as drinking water resources. The concentrations of these contaminants are extremely low, typically at levels of parts per billion or parts per trillion, which makes their treatment difficult. This has stimulated studies to develop and adopt novel techniques which consider the removal of pollutants with a premium on economic feasibility, simple instrumentation, and high performance. In the treatment of water, the removal of trace concentration organic compounds and numerous other polluted water effluents is difficult due to the limited affinity of trace compound ions to ion exchange resins [60–63].

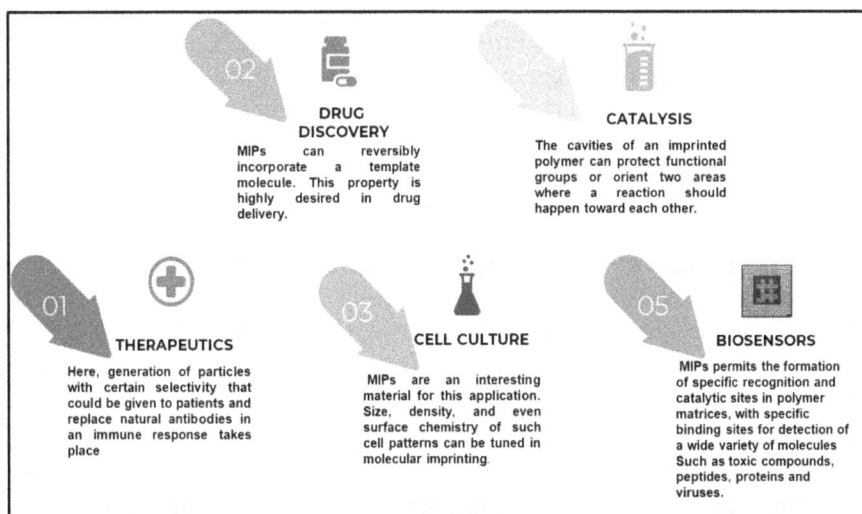

Figure 1.4. Various applications of molecularly imprinted polymers.

The increase of contaminants in environmental water is an alarming issue. Environmental monitoring can be carried out on environmentally hazardous materials such as dyes, persistent organic pollutants, pesticide residues, mycotoxins, heavy metals, and antibiotics. The presence of these contaminants can cause harmful effects on public health, either directly or indirectly. Environmental matrices that can be polluted by pollutants include air, soil, the atmosphere, sediments, and flora and fauna. Pre-treatment samples aim to eliminate matrix interference so that only the target analyte is obtained. MIPs are commonly used as the sorbent in pre-treatment samples with complex matrices because of their advantageous properties compared to conventional sorbents [64–67].

1.5.2 Food analysis

Contaminants from the environment can enter the body through food or drinking water and have the potential to cause harmful effects on health. For example, insecticide and herbicide contaminants used in agriculture can be found in fruits, vegetables, and cereals. Effective analytical methods and technologies are needed to ensure food safety. Food is a complex matrix sample, so MIPs can be used as a sorbent in the analysis process [68–71].

1.5.3 Biomedical diagnostics

MIPs as biomedical diagnostics have shown great suitability. They have the ability to replicate natural receptors, and can replace natural receptor and identify cells. MIPs as an alternative to antibodies are often used as an analytical method for diagnostic purposes, such as breast cancer diagnostics, cardiovascular disease, and dengue fever [72, 73]. Any abnormality in the body can be sensed using biomarkers, and MIPs have valuable characteristics and acts as biomarkers. Compared to natural receptors, MIP has advantages such as high affinity and selectivity, better temperature stability, rapid preparation, and low cost. MIPs are designed to have specific recognition sites for target molecules such as antibodies and enzymes. In the immune system, antibodies must be specific in recognizing certain antigens [74, 75].

1.5.4 Drug delivery

The pharmacological bioavailability of a drug can be increased by direct delivery to the desired place. This method enhances the therapeutic effects by delivering the drug before drug release and absorption. The drug delivery can be more productive if one can control the amount and speed of drug release. When we consider MIPs as drug delivery systems, they give us many advantages, namely a long shelf life, easy preparation, high chemical and physical stability, and low cost. Sometimes the usual practice of drug delivery can generate toxicity in the system and reduces the bioavailability [76, 77]. Another drawback is non-specific drug delivery and rapid clearance from the system. More recently MIP-based drug delivery research has been more focused on how to overcome these limitations. This has bought new evolution in the DDSs. Although the development of methods and processes in clinical practice using MIPs is still in its infancy, research and development will most likely lead to the creation of innovative drug delivery vehicles with commercial value [78, 79].

1.6 Challenges

Molecular imprinting has seen significant evolution in the past decade, both in terms of materials and applications. However, there is still a lot of room for further development. Research teams from all around the world are aggressively tackling the issues that are now seen as major obstacles, particularly the homogeneity of MIPs' binding sites, their compatibility with water, and the potential to synthesize MIPs tailored to particular proteins. Proper monomer selection, washing technique/ template removal, quantification of the rebinding, and reproducibility are the primary areas of concern. The use of charged monomers can result in both undesirable high aspecific binding and strong electrostatic interactions between the monomers and template. Template rebinding is frequently quantified inexactly, results are not critically assessed, and statistical analysis is lacking. Also, biological macromolecules, which are insoluble in the organic solvents often employed in molecular imprinting, present a challenge for imprinting. Template leakage (bleeding), which interferes with MIP particle binding - it is a hindrance in analytical applications. The use of MIPs as biorecognition materials in assays and sensors is not practical if there is a chance that a compromised clinical or forensic analysis could result from a leaky template. The final results are also greatly complicated by variables such as pH and ionic strength. The use of MIPs in diagnostic, pharma-ceutical, and separation applications is constrained by high degrees of non-specific binding, except for a small number of unique rare instances when there is no other option.

1.7 Conclusion and future prospects

It is very important to carry out a complete survey of binding studies. After reviewing the research in the field of molecular imprinted technology, we have found a plethora of papers missing. Most of the areas of research with a low rate of interest or which are underrated are binding studies, binding kinetics, binding isotherms, and binding selectivity [3, 80]. A developed method or technology can be validated if the binding studies are performed. The binding study is the benchmark through which one can verify the applicability of the method and use as a standard for their effective synthesis. Binding capacity is one of the important parameters in molecular imprinting technology, and to an extent it has been overshadowed by other parameters. Comparison to non-imprinted polymers (NIPs) in terms of binding affinity should be presented in the optimization graphs and binding test results to indicate the higher selectivity and usefulness of MIPs. Moreover, the selectivity of MIPs could be determined using cross-reactivity studies, where researchers choose analytes that could possibly coexist with the target molecule in the same sample and could cause real errors in analysis [81, 82]. The washing step of the developed MIPs to remove the template molecules after polymerization is a very significant and important step that can affect the successful application of the MIPs. The second most important parameter is the washing solutions. This can be a concentrated salt solution, mixtures of weak acid and sodium dodecyl sulfate, some enzymes, buffers, etc. When choosing a solvent, it should not be based only on published studies or research papers for similar molecules, it should depend on the specific study and at specific optimization to verify complete template removal. At the time of washing care should be taken that binding sites are not compromised by any solvent or conditions. This step is not just routine work; complete washing prevents the problem of template bleeding and increases the availability of binding sites. The selection of the monomer is the backbone of MIP synthesis. The selection of the monomer mainly depends on the structure elucidation of the target and its possible interactions with the monomer. Monomer selection is a somewhat time-consuming process. A computational approach to molecular imprinting is a new aid to this technology. This helps to select a specific monomer according to the target and calculate the Gibbs energy. Each specific monomer has certain desirable properties which help MIPs to minimize non-specific binding and increased hydrophilicity. It could be very beneficial if researchers explain the reasons behind their selection of functional monomers. Many researchers have addressed fluorescent MIP-based imprinting on fluorescent substrates in their studies. One should consider its the toxicity profile for the human body before seeing its application. A few studies based on quantum dot MIPs have been reported [74, 83]. These quantum dots are generally produced using heavy metals such as cadmium, tellurium, etc. Heavy metals can cause poisoning and cytotoxicity to living systems. Instead of quantum dots one can use carbon dots as an alternative. Carbon dots are less toxic, ecofriendly, and they have higher photostability and chemical inertness compare to quantum dots. The electropolymerization technique is generally used in the field of electrochemistry, and when you develop an MIP-based sensor using this technique one should aware about certain key things. The monomers on the surface of an electrode is a very widely applied and easy method of polymerization. The concentration of monomer solution should not be overloaded. The polymerization film thickness depends very much on the electrochemical scan rate and number of scan cycles. Proper polymerization films enhance sensitivity as they can bind to a large part of the target molecule. A thick or thin film can reduce the sensitivity of the sensor. The developed sensor stability should be tested to determine whether the developed sensor performance remains the same or deteriorates over repeated assays or time intervals. Based on such a study, researchers can establish a time frame in which the sensor results are considered reliable and reproducible [84].

References

[1] Haupt K, Linares A, Bompart M and Bui B T S 2012 Molecularly imprinted polymers *Top. Curr. Chem.* **325** 1–28

[2] Nestora S, Merlier F, Beyazit S, Prost E, Duma L, Baril B, Greaves A, Haupt K and Bui B T S 2016 Plastic antibodies for cosmetics: molecularly imprinted polymers scavenge precursors of malodors *Angew. Chem. Int. Ed.* **55** 6252–6

[3] Gao M, Gao Y, Chen G, Huang X, Xu X, Lv J, Wang J, Xu D and Liu G 2020 Recent advances and future trends in the detection of contaminants by molecularly imprinted polymers in food samples *Front. Chem.* **8** 616326

[4] Becskereki G, Horval H and Tóth B 2021 The selectivity of molecularly imprinted polymers *Polymers* **13** 1781

[5] BelBruno J J 2019 Molecularly imprinted polymers *Chem. Rev.* **119** 94–119

[6] Hoshino Y and Shea K J 2011 The evolution of plastic antibodies *J. Mater. Chem.* **21** 3517–21

[7] Peeters M, Eersels K, Junkers T and Wagner P 2016 *Molecularly Imprinted Catalysts* (Amsterdam: Elsevier) 253–71

[8] Esther T and Antonio M E 2020 *Solid Phase Extraction* (Amsterdam: Elsevier) 215–33

[9] Unger C and Lieberzeit P A 2021 Molecularly imprinted thin film surfaces in sensing: chances and challenges *React. Funct. Polym.* **161** 104855

[10] Zaidi S A 2015 Recent developments in molecularly imprinted polymer nanofibers and their applications *Anal. Methods* **7** 7406–15

[11] Farooq S, Nie J, Cheng Y, Yan Z, Li J, Bacha S A S, Mushtag A and Zhang H 2018 Molecularly imprinted polymers' application in pesticide residue detection *Analyst* **143** 3971–89

[12] Gui R, Guo H and HuiJin H 2019 Preparation and applications of electrochemical chemosensors based on carbon-nanomaterial-modified molecularly imprinted polymers *Nanoscale Adv.* **1** 3325–63

[13] Soufi G J, Iravani S and Varma R S 2021 Molecularly imprinted polymers for the detection of viruses: challenges and opportunities *Analyst* **146** 3087–100

[14] Wen W-C, Chung C-H, Hsin H-H, Ting T-H and Ching C-C 2008 Chromatographic characterization of molecularly imprinted polymers *Anal. Bioanal. Chem.* **390** 1101–9

[15] Mayes A G and Mosbach K 1997 Molecularly imprinted polymers: useful materials for analytical chemistry? *TrAC Trends Anal. Chem.* **16** 321–32

[16] Regal P, Díaz-Bao M, Barreiro R, Cepeda A and Fente C 2012 Application of molecularly imprinted polymers in food analysis: clean-up and chromatographic improvements *Cent. Eur. J. Chem.* **10** 766–84

[17] Maria G and Grzegorz S 2019 Application of Molecularly Imprinted Polymers (MIP) and Magnetic Molecularly Imprinted Polymers (mag-MIP) to Selective Analysis of Quercetin in Flowing Atmospheric-Pressure Afterglow Mass Spectrometry (FAPA-MS) and in Electrospray Ionization Mass Spectrometry (ESI-MS) *Molecules* **24** 1–15

[18] Vaneckova T, Vanickova L, Tvrdonova M, Pomorski A, Krężel A, Vaculovic T, Kanicky V, Vaculovicova M and Adam V 2019 Molecularly imprinted polymers coupled to mass spectrometric detection for metallothionein sensing *Talanta* **198** 224–9

[19] Obiles R, Premadasa U I, Cudia P, Erasquin U J, Berger J M, Martinez I S and Cimatu K L A 2020 Insights on the molecular characteristics of molecularly imprinted polymers as monitored by sum frequency generation spectroscopy *Langmuir* **36** 180–93

[20] Guo M, Hu Y, Wang L, Brodelius P E and Sun L 2018 A facile synthesis of molecularly imprinted polymers and their properties as electrochemical sensors for ethyl carbamate analysis *RSC Adv.* **8** 39721–30

[21] Park R, Jeon S, Jeong J, Park S-Y, Han D-W and Hong S W 2022 Recent advances of point-of-care devices integrated with molecularly imprinted polymers-based biosensors: from biomolecule sensing design to intraoral fluid testing *Biosensors* **12** 136

[22] Chen L, Xu S and Li J 2011 Recent advances in molecular imprinting technology: current status, challenges and highlighted applications *Chem. Soc. Rev.* **40** 2922–42

[23] Jia M, Zhang Z, Li J, Ma X, Chen L and Yang X 2018 Molecular imprinting technology for microorganism analysis *TrAC Trends Anal. Chem.* **106** 190–201

[24] Chen L, Wang X, Lu W, Wu X and Li J 2016 Molecular imprinting: perspectives and applications *Chem. Soc. Rev.* **45** 2137–211

[25] Mosbach K and Haupt K 1998 Some new developments and challenges in non-covalent molecular imprinting technology *J. Mol. Recognit.* **11** 62–8

[26] Whitcombe M J, Kirsch N and Nicholls I A 2014 Molecular imprinting science and technology: a survey of the literature for the years 2004–2011 *J. Mol. Recognit.* **27** 297–401

[27] Alexander C, Andersson H S, Lars L I, Ansell R J, Kirsch N, Nicholls I A, O'Mahony J and Whitcombe M J 2006 Molecular imprinting science and technology: a survey of the literature for the years up to and including 2003 *J. Mol. Recognit.* **19** 106–80

[28] Singhal A, Ranjan P, Sadique M A, Kumar N, Yadav S, Parihar A and Khan R 2022 Molecularly imprinted polymers-based nanobiosensors for environmental monitoring and analysis *Nanobiosensors for Environmental Monitoring: Fundamentals and Application* (Cham: Springer International) pp 263–78

[29] Fauzi D and Saputri F A 2019 Molecularly imprinted polymer nanoparticles (MIP-NPs) applications in electrochemical sensors *Int. J. Appl. Pharm.* **11** 35088

[30] Singh M, Shiv S, Singh S P and Patel S S 2020 Recent advancement of carbon nanomaterials engrained molecular imprinted polymer for environmental matrix *Trends Environ. Anal. Chem.* **27** e00092

[31] Wackelig J and Schirhagl R 2016 Applications of molecularly imprinted polymer nano-particles and their advances toward industrial use: a review *Anal. Chem.* **88** 250–61

[32] Dong X, Zhang C, Du X and Zhang Z 2022 Recent advances of nanomaterials-based molecularly imprinted electrochemical sensors *Nanomaterials* **12** 1913

[33] Zhong C, Yang B, Jiang X and Li J 2018 Current progress of nanomaterials in molecularly imprinted electrochemical sensing *Crit. Rev. Anal. Chem.* **48** 15–32

[34] Zarejousheghani M, Rahimi P, Borsdorf H, Zimmermann S and Joseph Y 2021 Molecularly imprinted polymer-based sensors for priority pollutants *Sensors* **21** 2406

[35] Parisi O I, Francomano F, Dattilo M, Patitucci F, Prete S, Amone F and Puoci F 2022 The evolution of molecular recognition: from antibodies to molecularly imprinted polymers (MIPs) as artificial counterpart *J. Funct. Biomater.* **13** 12

[36] Khan M A R, Moreira F T C, Riu J and Sales M G F 2016 Plastic antibody for the electrochemical detection of bacterial surface proteins *Sensors Actuators* B **233** 697–704

[37] Haupt K, Rangel P X M and Bui B T S 2020 Molecularly imprinted polymers: antibody mimics for bioimaging and therapy *Chem. Rev.* **120** 9554–82

[38] Cieplak M and Kutner W 2016 Artificial biosensors: how can molecular imprinting mimic biorecognition? *Trends Biotechnol.* **34** 922–41

[39] Nasrullah S, Mazhar U, Muhammad H and Joong K P 2012 A Brief Overview of Molecularly Imprinted Polymers: From Basics to Applications *J. Pharm. Res.* **5** 3309–17

[40] Nikesh B S, Vinayak K, Prakash A M, Ajay V R and Krishnan K A 2015 A historical perspective and the development of molecular imprinting polymer-a review *Chem. Int.* **1** 202–10

[41] Haupt K 2010 Plastic antibodies *Nat. Mater.* **9** 612–4

[42] Hoshino Y, Koide H, Urakami T, Kanazawa H, Kodama T, Oku N and Shea K J 2010 Recognition, neutralization, and clearance of target peptides in the bloodstream of living mice by molecularly imprinted polymer nanoparticles: a plastic antibody *J. Am. Chem. Soc.* **132** 6644–5

[43] Takeuchi T and Sunayama H 2018 Beyond natural antibodies—a new generation of synthetic antibodies created by post-imprinting modification of molecularly imprinted polymers *Chem. Commun.* **54** 6243–51

[44] Ma X-H, Li J-P, Wang C and Xu G-B 2016 A review on bio-macromolecular imprinted sensors and their applications *Chin. J. Anal. Chem.* **44** 152–9

[45] Rimmer S 1998 Synthesis of molecular imprinted polymer networks *Chromatographia* **46** 470–4

[46] Simon P G, Hazim F E S, Sabha H, Rieke F, Rebecca K M, Philippa H, Mark S and Subrayal M R 2019 Evaluation of Molecularly Imprinted Polymers as Synthetic Virus Neutralizing Antibody Mimics *Front. Bioeng. Biotechnol.* **7** 1–7

[47] Ranjan P, Singhal A, Yadav S, Kumar N, Murali S, Sanghi S K and Khan R 2021 Rapid diagnosis of SARS-CoV-2 using potential point-of-care electrochemical immunosensor: toward the future prospects *Int. Rev. Immunol.* **40** 126–42

[48] Scheller F W, Zhang X, Yarman A, Wollenberger U and Gyurcsányi R E 2019 Molecularly imprinted polymer-based electrochemical sensors for biopolymers *Curr. Opin. Electrochem.* **14** 53–9

[49] Singhal A, Yadav S, Sadique M A, Khan R, Kaushik A, Sathish N and Srivastava A K 2022 MXene-modified molecularly imprinted polymer as an artificial bio-recognition platform for efficient electrochemical sensing: progress and perspectives *Phys. Chem. Chem. Phys.* **24** 19164–76

[50] Singhal A, Singh A, Shrivastava A and Khan R 2023 Epitope imprinted polymeric materials: application in electrochemical detection of disease biomarkers *J. Mater. Chem.* B **11** 936–54

[51] Singhal A, Parihar A, Kumar N and Khan R 2022 High throughput molecularly imprinted polymers based electrochemical nanosensors for point-of-care diagnostics of COVID-19 *Mater. Lett.* **306** 130898

[52] Singhal A, Sadique M A, Kumar N, Yadav S, Ranjan P, Parihar A, Khan R and Kaushik A K 2022 Multifunctional carbon nanomaterials decorated molecularly imprinted hybrid polymers for efficient electrochemical antibiotics sensing *J. Environ. Chem. Eng.* **10** 107703

[53] El-Schich Z, Zhang Y, Feith M, Beyer S, Sternbæk L, Ohlsson L, Stollenwerk M and Wingren A G 2020 Molecularly imprinted polymers in biological applications *Biotechniques* **69** 406–19

[54] Ansell R J 2015 Characterization of the binding properties of molecularly imprinted polymers *Adv. Biochem Eng Biotechnol.* **150** 51–93

[55] Zaidi S A 2016 Molecular imprinted polymers as drug delivery vehicles *Drug Deliv.* **23** 2262–71

[56] Xu L, Huang Y-A, Zhu Q-J and Ye C 2015 Chitosan in molecularly-imprinted polymers: current and future prospects *Int. J. Mol. Sci.* **16** 18328–47

[57] Xie L, Xiao N, Li L, Xie X and Li Y 2019 An investigation of the intermolecular interactions and recognition properties of molecular imprinted polymers for deltamethrin through computational strategies *Polymers* **11** 1872–86

[58] Merkoçi A and Alegret S 2002 New materials for electrochemical sensing IV. Molecular imprinted polymers *TrAC Trends Anal. Chem.* **21** 717–25

[59] Zaidi S A 2016 Latest trends in molecular imprinted polymer based drug delivery systems *RSC Adv.* **6** 88807–19

[60] Sarpong K, Xu W, Huang W and Yang W 2019 The development of molecularly imprinted polymers in the clean-up of water pollutants: a review *Am. J. Anal. Chem.* **10** 202–26

[61] Meng Z, Chen W and Mulchandani A 2005 Removal of estrogenic pollutants from contaminated water using molecularly imprinted polymers *Environ. Sci. Technol.* **39** 8958–62

[62] Pichon V and Chapuis-Hugon F 2008 Role of molecularly imprinted polymers for selective determination of environmental pollutants—a review *Anal. Chim. Acta.* **622** 48–61

[63] Azizi A and Bottaro C S 2020 A critical review of molecularly imprinted polymers for the analysis of organic pollutants in environmental water samples *J. Chromatogr.* A **1614** 460603

[64] Farooq S, Wu H, Nie J, Ahmad S, Muhammad I, Zeeshan M, Khan R and Muhammad M 2022 Application, advancement and green aspects of magnetic molecularly imprinted polymers in pesticide residue detection *Sci. Total Environ.* **804** 150293

[65] Murray A and Örmeci B 2012 Application of molecularly imprinted and non-imprinted polymers for removal of emerging contaminants in water and wastewater treatment: a review *Environ. Sci. Pollut. Res.* **19** 3820–30

[66] Singh J and Mehta A 2020 Rapid and sensitive detection of mycotoxins by advanced and emerging analytical methods: a review *Food Sci. Nutr.* **8** 2183–204

[67] Ahmad R B, Nahal A, Abdul A K, Ijaz G, Suresh G and Muhammad B 2021 Molecularly imprinted polymers-based adsorption and photocatalytic approaches for mitigation of environmentally-hazardous pollutants —a review *Environ. Chem. Eng.* **9** 1–75

[68] Ramström O, Skudar K, Haines J, Patel P and Brüggemann O 2001 Food analyses using molecularly imprinted polymers *J. Agric. Food Chem.* **49** 2105–14

[69] Liu G, Huang X, Li L, Xu X, Zhang Y, Lv J and Xu D 2019 Recent advances and perspectives of molecularly imprinted polymer-based fluorescent sensors in food and environment analysis *Nanomaterials* **9** 1030

[70] Cao Y, Feng T, Xu J and Xue C 2019 Recent advances of molecularly imprinted polymer-based sensors in the detection of food safety hazard factors *Biosens. Bioelectron.* **141** 111447

[71] Tarannum N, Khatoon S and Dzantiev B B 2020 Perspective and application of molecular imprinting approach for antibiotic detection in food and environmental samples: a critical review *Food Control* **118** 107381

[72] Choi J R, Yong K W, Choi J Y and Cowie A C 2019 Progress in molecularly imprinted polymers for biomedical applications *Comb. Chem. High Throughput Screening* **22** 78–88

[73] Adumitrăchioaie A, Tertiş M, Cernat A, Săndulescu R and Cristea C 2018 Electrochemical methods based on molecularly imprinted polymers for drug detection. A review *Int. J. Electrochem. Sci.* **13** 2556–76

[74] Mostafa A M, Barton S J, Wren S P and Barker J 2021 Review on molecularly imprinted polymers with a focus on their application to the analysis of protein biomarkers *TrAC Trends Anal. Chem.* **144** 116431

[75] Hillberg A L, Brain K R and Allender C J 2005 Molecular imprinted polymer sensors: implications for therapeutics *Adv. Drug Deliv. Rev.* **57** 1875–89

[76] Liu R and Poma A 2021 Advances in molecularly imprinted polymers as drug delivery systems *Molecules* **26** 3589

[77] Zaidi S A Z 2020 Molecular imprinting: a useful approach for drug delivery *Mater. Sci. Energy Technol.* **3** 72–7

[78] Alaa F N, Thomas J W N and Webster J 2021 The promising use of nano-molecular imprinted templates for improved SARS-CoV-2 detection, drug delivery and research *J. Nanobiotechnol* **19** 1–14

[79] Piotr L 2013 Molecularly imprinted polymers as the future drug delivery devices *Acta Pol. Pharm.* **70** 601–9

[80] Piletsky S A and Turner A P F 2002 Electrochemical sensors based on molecularly imprinted polymers *Electroanalysis* **14** 317–23

[81] Haupt K 2001 Molecularly imprinted polymers in analytical chemistry *Analyst* **126** 747–56

[82] Liu Y, Wang F, Tan T and Lei M 2007 Study of the properties of molecularly imprinted polymers by computational and conformational analysis *Anal. Chim. Acta.* **581** 137–46

[83] Tóth B, Pap T, Horvath V and Horvai G 2007 Which molecularly imprinted polymer is better? *Anal. Chim. Acta.* **591** 17–21

[84] Szatkowska P, Koba K, Koslinski P and Szablewski M 2013 Molecularly imprinted polymers' applications: a short review *Mini-Rev. Org. Chem.* **10** 400–8

IOP Publishing

Molecularly Imprinted Polymers for Environmental Monitoring
Fundamentals and applications
Raju Khan and Ayushi Singhal

Chapter 2

Synthesis and characterization of molecularly imprinted polymers

K Uma, Suman Pawar and L R Hemanth

A molecularly imprinted polymer (MIP) is a polymer that has been processed using the molecular interactions and imprinting the same by optimized parameters, the action leave openings in the polymer matrix with an affinity for a selected 'template' molecule and these act as artificial receptors for a targeted molecule. They are analogues of the natural antibody–antigen systems. MIPs are specifically used as sensors. In these applications, the polymers are paired with a reporting system, which may be electrical, electrochemical, optical, or gravimetric. The presence of the targeted molecule effects a change in the reporting agent, and a calibrated amount of the target is recorded.

2.1 Synthesis of imprinted polymers

Imprinted polymers are synthesized using a wide range of methods. These methods are generally grouped on the state of the raw materials, namely the solid, liquid, and gaseous states. Emulsion, sol–gel, and precipitation, respectively, are common examples synthesis methods. The reaction state of the processing material is also a basis for grouping synthesis methods. Solid phase processing, co-precipitate method, and irradiation are a few examples. In this chapter the synthesis of imprinted polymers based on raw materials namely monomers, polymers, co-blocks, and initiators are explained. Figure 2.1 shows the classification of MIP synthesis methods. The features of them are individually discussed in the chapter.

2.1.1 Imprinted polymers from monomers

Monomers and polymers are the key constituents of imprinted structures. The polymerization of these gives the structures of the essential molecules for individual applications. The methods can involve monomers as the constituents and then polymerizing them to give the structures. Monomers such as methyl methacrylate,

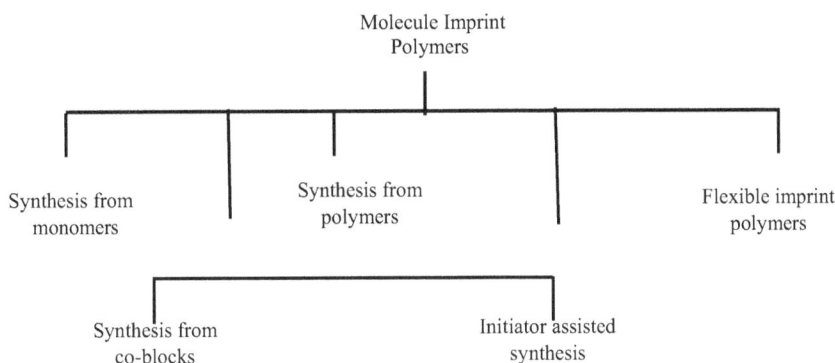

Figure 2.1. Classification of MIP synthesis methods.

ethyl ether ketone, and ethylene which are polymerized under constraints have been used as the basic raw materials. Figure 2.2 shows some of the functional monomers used in the synthesis of MIPs.

A simple process involves methacrylic acid (MAA) as a functional monomer and ethylene glycol dimethacrylate (EGDMA) as the raw materials contributing as crosslinking agents to form template molecules [1, 2].

The steps involve the essential chemical interactions of easy binding, template removal from the matrix, and controlled reactions to retain essential template molecules in the polymers. Monomer–template adduct methods can be considered for monomer-based synthesis groups. Template removal is used as the essential step to form weak interactions involving van der Waals forces, H_2 bonding, π–π interactions, dipole–dipole interactions, and ion–dipole exchanges [3]. The methods require an excess amount of monomers to complete the interactions in the chain. The processing conditions are optimized for different raw materials. The nature of material such as acidic, basic, neutral and the functional groups in the structures are key factors in synthesis methods. Figure 2.3. shows monomers with different functional groups. The temperatures, setting time are optimized based on these structures to get the MIPs.

2.1.2 Imprinted polymers from polymers

Imprinted polymers from polymers as the main raw material are also used in the synthesis methods. Polymers in their direct form are used for processing. Monomers and other constituents in minor quantities are also used in the synthesis, however, the quantity of polymers dominates the others. These constituents are mixed and processed by various methods.

The processing time for the synthesis is lower compared to the use of raw monomers, as the structures are already polymerized. The reaction is essential to bind the polymers with other constituents to produce the imprinted structures for applications. Melt mixing is one of the examples for consideration. The mechanical properties of carbon nanotube (CNT)-based polymer composites are necessary for lightweight applications and impregnation is carried out on the surface of CNTs. The initial step involved subjecting the carbon nanotubes to plasma enhanced

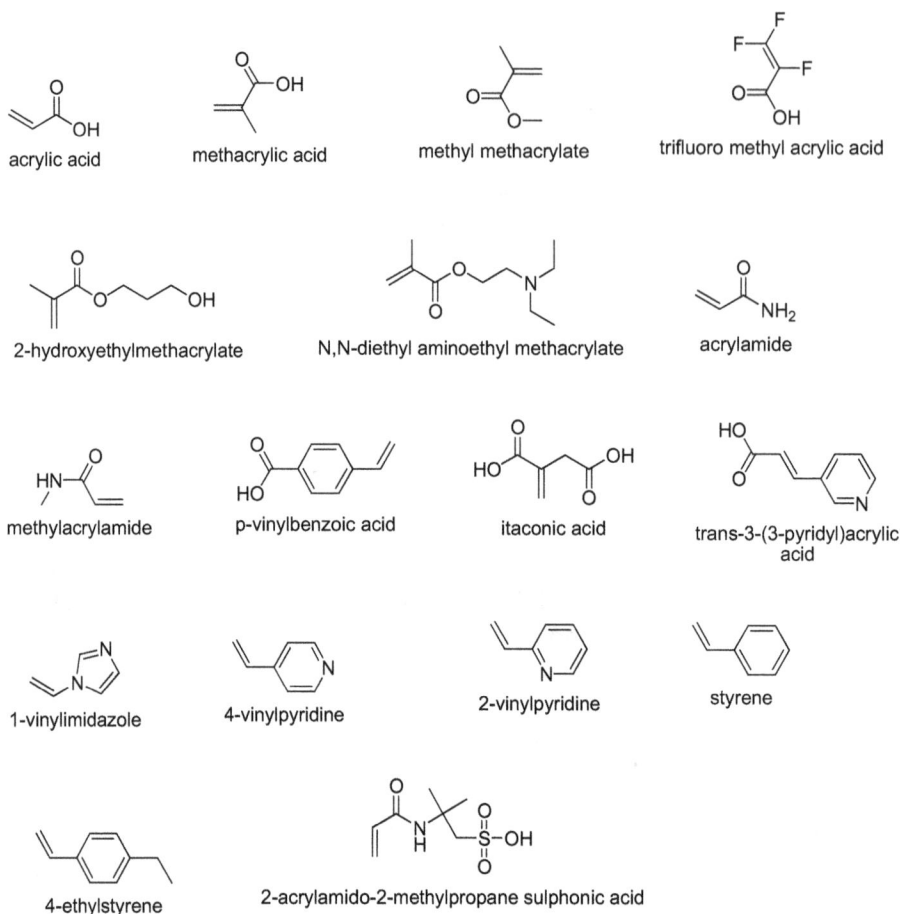

Figure 2.2. Functional monomers used in the synthesis of MIPs.

chemical vapor deposition to obtain an intermediate product. The intermediates were further utilized for nanocomposites by melt-mixing and extruding at high temperatures and pressures. Polymers with an improved modulus and tensile toughness were obtained [4]. The batch foaming process can be considered as another example. CNTs mixed with polymethyl methacrylate (PMMA) were subjected to melting by hot pressing for considerable times and temperatures to obtain the product. Supercritical carbon dioxide was used during the foaming step for a long duration to prepare the foams, which led to improved electromagnetic absorbing properties of the materials [4].

2.1.3 Imprinted polymers from co-blocks

Co-blocks are essential in polymerization for effective reactions. They are also often called 'additives'. These are limiting reactants essential for the complete trans-formations to obtain the polymers. The reaction rate, temperature, and nature of the

Functional monomers in covalent imprinting

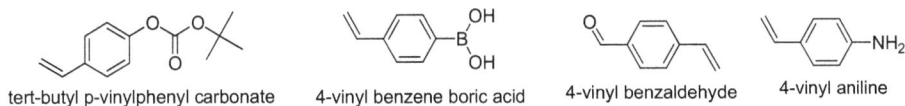

tert-butyl p-vinylphenyl carbonate   4-vinyl benzene boric acid   4-vinyl benzaldehyde   4-vinyl aniline

Functional monomers in non-covalent imprinting

Acidic

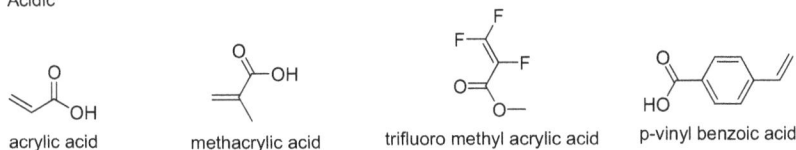

acrylic acid   methacrylic acid   trifluoro methyl acrylic acid   p-vinyl benzoic acid

Basic

2-vinyl pyridine   1-vinylimidazole   4-vinyl pyridine   allylamine

Neutral

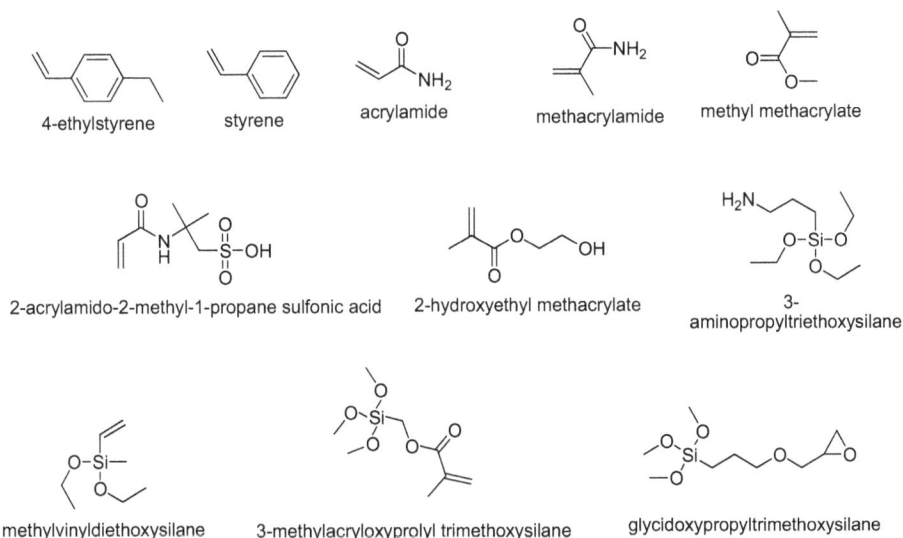

4-ethylstyrene   styrene   acrylamide   methacrylamide   methyl methacrylate

2-acrylamido-2-methyl-1-propane sulfonic acid   2-hydroxyethyl methacrylate   3-aminopropyltriethoxysilane

methylvinyldiethoxysilane   3-methylacryloxyprolyl trimethoxysilane   glycidoxypropyltrimethoxysilane

Figure 2.3. Chemical structures of some typical functional monomers employed for the fabrication of MIPs.

co-blocks affect the synthesis conditions. Cross-linked polymers are more stable and are used on various core structures.

The covalent imprinting technique has been experimented for synthesis (figure 2.3). The tuning of the properties is carried out prior to the polymerization.

The template and monomer raw materials are tuned to have covalent bonding to provide molecular recognition. Raw materials with cross-linkers are processed to yield an insoluble rigid network in the presence of alcohols, carboxylic acids, and other functional group materials. These are the templates extracted to achieve the interactions to obtain the materials. The semi-covalent imprinting method is an example of this type of approach [5]. The steps involved the use of a linker group between the template and the monomer which is expected to vanish during the bond rebinding of template molecule. The use of cholesterol and 4-vinylphenol carbonic acid as the template and sacrificial molecules substantiate the role of co-blocks during the synthesis [6, 7]. The advantages of these methods involve broad polymerization conditions and molecular polymerization.

2.1.4 Initiator assisted synthesis

The monomers essentially undergo polymerization to provide the products. The monomerization depends on their state and reaction parameters. The monomers sometimes fail to open their essential rings during the steps. Thus, initiators are added with the monomers. The nature of the initiators and their concentrations are crucial in initiator-based synthesis. Up to 2%–15% initiators are added to initiate the polymerization and the temperature also has an effect on the reaction. Temperatures from −5 °C to 60 °C have been used during the synthesis. Crosslinking polymerization has been employed as synthesis method. This approach involved templates with their monomers and cross-linked them using an initiator. An electrospinning process has been used as one of the methods. The process involved methyl methacrylate (MMA), multi-walled CNTs, and an initiator. Initiators were also employed for polymerization to initiate organic conversion with the retention of functional and dependent groups [6]. Common initiators employed for polymerization are tert-butyl peroxide and tetra-hydrafuran [7]. The use of monomers and initiators for radical polymerization has been a common approach. A functionalized radical initiator was conceptualized and synthesized using templates. The raw materials were mixed and the pH was used as a parameter. The solutions were N_2 purged and polymerization was carried out under specified temperatures for times above 20 h. Washing and drying steps were carried out to obtain the desired materials.

2.1.5 Flexible imprinted polymers

Imprinted polymers generally exhibit rigid shape and this is useful for some applications. However, some applications demand flexible properties of MIP. In order to meet the requirements flexible imprinted polymers are synthesized. The transformation into flexible imprinted polymers requires additional processing steps and reagents. These steps shift the properties of the material. The properties such as stress, conductivity, elasticity and others are determined by varying the processing conditions. Bulk polymerization can be studied as an example. Monolithic polymers are primarily synthesized by bulk polymerization. The drying conditions yield uneven shapes and structure. These are further subjected to powders to facilitate as raw materials for flexible polymers.

2.1.6 Resist modified imprinted polymers

These are electrical and application-oriented polymer synthesis methods. They require the addition of silica or other constituents for special applications. The resist plays a key role in patterning and electrical applications. Synthesis by precipitation polymerization is one example. Fang *et al* prepared MIP microspheres with photoresponsive (PR) template binding via precipitation polymerization using an azo-functional monomer with a 4-{(4-methacryloyloxy)-phenylazo} pyridine. These MIPs had a diameter of 1.33 mm and a polydispersity index of 1.15 [8]. Researchers have already established a sol–gel method for the preparation of MIPs. Typically, sol–gel-based materials are prepared by acid- or base-catalyzed hydrolysis of SiH_4, followed by polycondensation of the $\equiv Si–OH$ into a $(R_2SiO)_n$ network. An innovative photo-responsive monomer bearing a siloxane polymerizable group and $C_{12}H_{10}N_2$ moieties were used by Jiang *et al* [9]. This $C_{12}H_{10}N_2$ monomer was used to prepare responsive MIPs with exact binding sites for 2,4-dichlorophenoxyacetic acid, through H_2 bonding interaction. Similarly, Tang *et al* designed a fluorine-substituted PR functional monomer [10].

For the PR molecular recognition approach, metal ions can be incorporated in the imprinting process as the template. Various approaches have been employed for the synthesis of metal-ion imprinted polymers: (i) crosslinking of linear chain polymers containing metal-binding ligands with bifunctional reagents, (ii) copolymerization of metal complexes containing polymer disabling ligands with crosslinkers, and (iii) surface imprinting at the interface of water-in-oil emulsions through assembly with amphiphilic monomers [4, 11]. These are usually complexed by a transition metal, polymerizable ligands, and a target molecule [12]. The oxidation state of the metal ion and the characteristics of the ligand influence the strength of the interaction. Due to the difficulties related to the synthesis and separation of tertiary or higher metal complexes, metal-ion mediated imprinting has been applied to a lesser extent than metal ion imprinting. The polymers obtained by such a process have been used for the development of ion selective sensors in remediation studies [13–15], for the building of catalytic spots [16], for imprinting histidine-containing short peptides [17], and for the specific imprinting of sugars [18, 19].

These photonic-magnetic responsive molecularly imprinted microspheres (PM-MIMs) have noteworthy advantages, such as high adsorption capacity and fast binding kinetics, and good separation and processing efficiency. A group of researchers studied an MIP based on 4-methacrylyloxy azobenzene as a functional monomer in favor of branched cyclodextrins, 6-O-α-D-maltosyl-cy-cyclodextrin (G2-β-CD).

The phototoxicity of UV rays limits the use of conventional azobenzene-based P-MIPs in the biomedical field. The visible-light-responsive surface molecularly imprinted polymer (VSMIP) showed greater drug dosing, improved specificity, and a quicker drug release rate and photoisomerization rate compared to an acyclovir (ACV)-permanent, light-sensitive molecular surface non-imprinted polymer (VSNIP). The Baimani group presented a polymer dendrimer (PD) method to improve the ability to identify polymer matrix recognition sites. They synthesized P-MIP on modified magnetic nanoparticles in the presence of PD using the reactivity between $C_6H_{10}O_3$ allies and $C_3H_4O_2$ for more yield by surface polymerization [22].

The molecularly imprint (MI) yield identification sites with the steric and chemical information of target molecules in a cross-linked polymeric matrix. The efficacy of MI is reliant on the nature of bonding in the template–monomer pre-polymerized complex [23–26], the form of imprinted materials [27], and the stiffness of the polymeric network [28, 29], and avoids the major drawbacks associated with traditional imprinting protocols, such as (i) most printed polymers are cross-linked polymers, (ii) it is difficult to separate the template molecules on the inner surface of large polymer materials [29], (iii) this slows down the kinetics [30, 31], (iv) rigid polymer systems reduce the selectivity of conformational recognition [32], (v) uncontrolled polymerization suffers from the heterogeneity of imprinted cavities [33], and (vi) bonding with analytical MIP materials is usually difficult due to poor assembly ability on the surface of the carrier, which limits its use in bioassays [34]. In most analytical applications, this limitation and the uniform binding site of non-covalent MIPs can be tolerated; numerous synthesis approaches have been presented to improve the uniformity of MIP binding sites.

The introduction of unique nanotechnology into the MI procedure has involved much consideration due to the extensive use of these nanomaterials in chemo-sensors and bio-assays [35, 36]. Nanomaterials have unique electrical, mechanical, catalytic, optical, and magnetic properties [34] that distinguish them from many materials synthesized using the same chemical composition. The good dispersion capacity and surface-to-density ratio of nanomaterials provide large adsorption sites for the enrichment of target species. In particular, chemical immobilization of biological receptors, such as enzymes and antibodies or organic functional groups in nanostructures, increases the sensitivity and selectivity of identification [34]. Chemical/biosensors based on functionalized inorganic or organic nanomaterials have been discovered for sensing countless analytes, such as quantum dots for fluorescence detection of nitro-aromatic explosives, enzyme activity [37], and functionalized carbon nanotubes for DNA and electrochemical recognition [38] (figure 2.4).

2.2 Surface characterization techniques for MIPs

Analytical instruments such as such as Fourier transform infrared (FTIR), nuclear magnetic resonance (NMR), and chromatography analysis are used for the confirmation of MIPs. The opening sizes and specific surface zones of the MIP can be found using Brunauer–Emmett–Teller (BET) analysis. Thermogravimetric analysis (TGA) is used to find the thermal stability. The surface analysis of MIPs is performed using scanning electron microscopy (SEM) and transmission electron microscopy (TEM) [40].

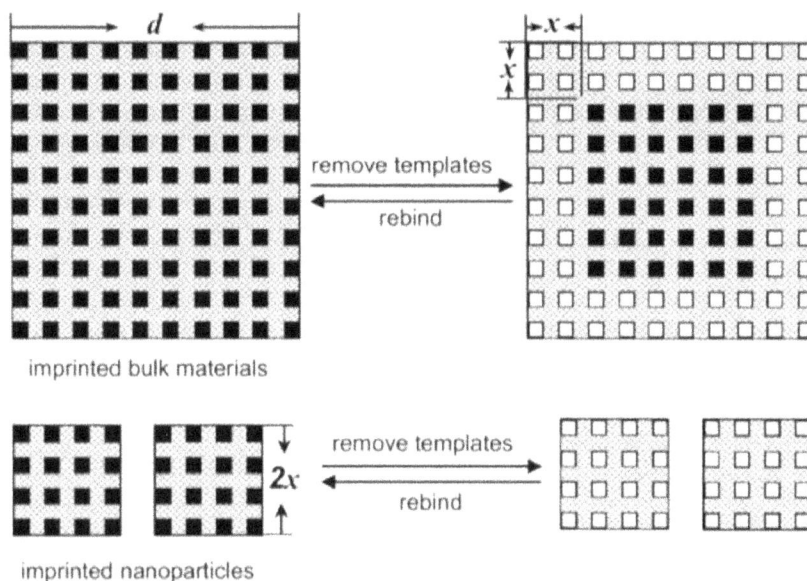

Figure 2.4. Schematic illustration of the distribution of binding cavities in the imprinted massive materials and the nanosized, imprinted particles after the elimination of templates. (Reproduced with permission from [39]. Copyright 2021 Elsevier.)

2.2.1 FTIR spectroscopy

FTIR is an important tool for analyzing the functional groups in the structures. The instrument scans the sample and results are interpreted for standard wavelengths. The tool employs the standard light, which is passed through a blank for calibration. The same conditions are used for the sample analysis. The tool can be used in the normal environment and also in vacuum. A few examples of MIPs analysed by FTIR are described. In one of the reported articles, to confirm the interaction of $C_{12}H_{10}OC_8H_8$, $C_6H_4(CH=CH_2)_2$, that is, DVB (divinyl benzene) in the MIPS were successfully analyzed by FTIR. The FTIR spectra for non-imprinted polymers and MIPs were obtained. A broad peak at 3500–3200 cm^{-1} was observed, which was attributed to the vibration modes of the O-H bond of MIPs (figures 2.5 and 2.6). This was useful in understanding the interactions between the monomers and the template. The FTIR spectra is also useful in analyzing the loss of functional groups, modifications or new bonds arising, indicating the formation of desired structures. The loss of templates can be confirmed by the loss of essential peaks in the spectra.

The advantages of using FTIR for analysis of MIPs are as follows:

1. Analysis of formation of new structures and loss of functional groups in the structures.
2. Clear observations of all functional groups.

However, the knowledge of all the wavelengths of functional group and their associated wavelengths are essential.

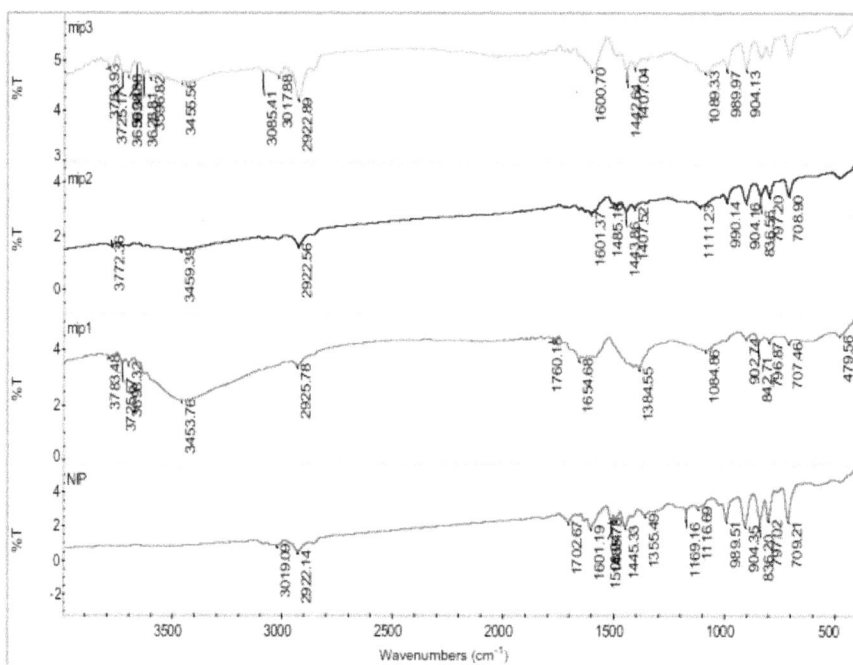

Figure 2.5. FTIR of MIPs and NIPs. (Reproduced with permission from [41]. Copyright 2019 Springer. CC BY 4.0.)

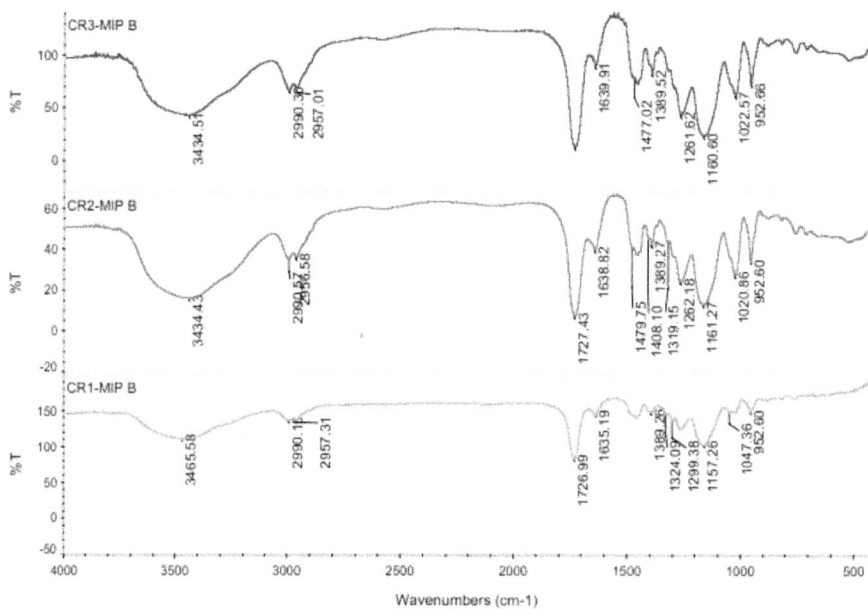

Figure 2.6. FTIR spectra of CR-MIPs. (Reproduced with permission from [42]. Copyright 2020 Springer. CC BY 4.0.)

2.2.2 NMR spectroscopy

The structural analysis can also be carried out by NMR spectroscopy. In this technique the chemical shifts in the samples are compared and interpreted. The spectra gives the information about interactions inside the structures at the atomic level. The interactions in the MIP between choloramphenol (CAP) and monomeric methacrylic acid (MAA) are considered as an example [44]. There was a change in a hydrogen atom of the monomer from 5.71 to 5.86 ppm and also the change in the amino hydrogen atom of the phenolic group from 5.67 to 5.65 ppm (Figure 2.7). The changes observed as chemical shifts indicate the formation of H2 bonding leading to a stable complex. The similar shifts observed in different structures indicate the successful formation of MIPs by the employed processes.

2.2.3 Chromatography

The chromatographic technique is useful for analysis of selective sites in the structures. The chromatograph indicates the volume and presence of considered species in the polymers. One example highlighted is the uptake of ethinylestradiol [EE] in MIPs and NIPs. The adsorptive volume of EE was higher in the NIPs while the volume was low in the MIPs. The same standards were used for both samples. The lower values indicate more absorption by formed MIPs, thus indicating the formation of structures in the employed process [45].

2.2.4 BET analysis

The Brunauer-Emmett-Teller (BET) surface area analysis is a useful tool for predicting the surface area, porosity, and pore size distribution of samples. In this technique, an inert gas is adsorbed on the sample surface and the properties are evaluated. The samples (a few milligrams) are considered, degassed and tested. Different isotherms are obtained and features such as microporosity, mesoporous and macroporous, are evaluated. For MIPs it is common to have stepwise multilayer adsorption on the surface. The MIPs were synthesized and the effect of solvents were studied. The congo red MIPS (CR-MIPS) indicated better surface area, pore radius and pore volume than NIP [46] (see Table 2.1).

2.2.5 TGA

TGA is a useful tool for analysis of samples to get the physio-chemical behavior of samples. A known amount of sample is placed in the TGA system and the effect of

Table 2.1. BET of CR1-MIP and CR1-NIP.

Properties	CR1-MIP (magnitude)	CR1-NIP (magnitude)
Surface area ($m^2 g^{-1}$)	390	92.6
Average pore radius (Å)	8.881	1.374
Total pore volume (c.c. g^{-1})	1.735	0.344

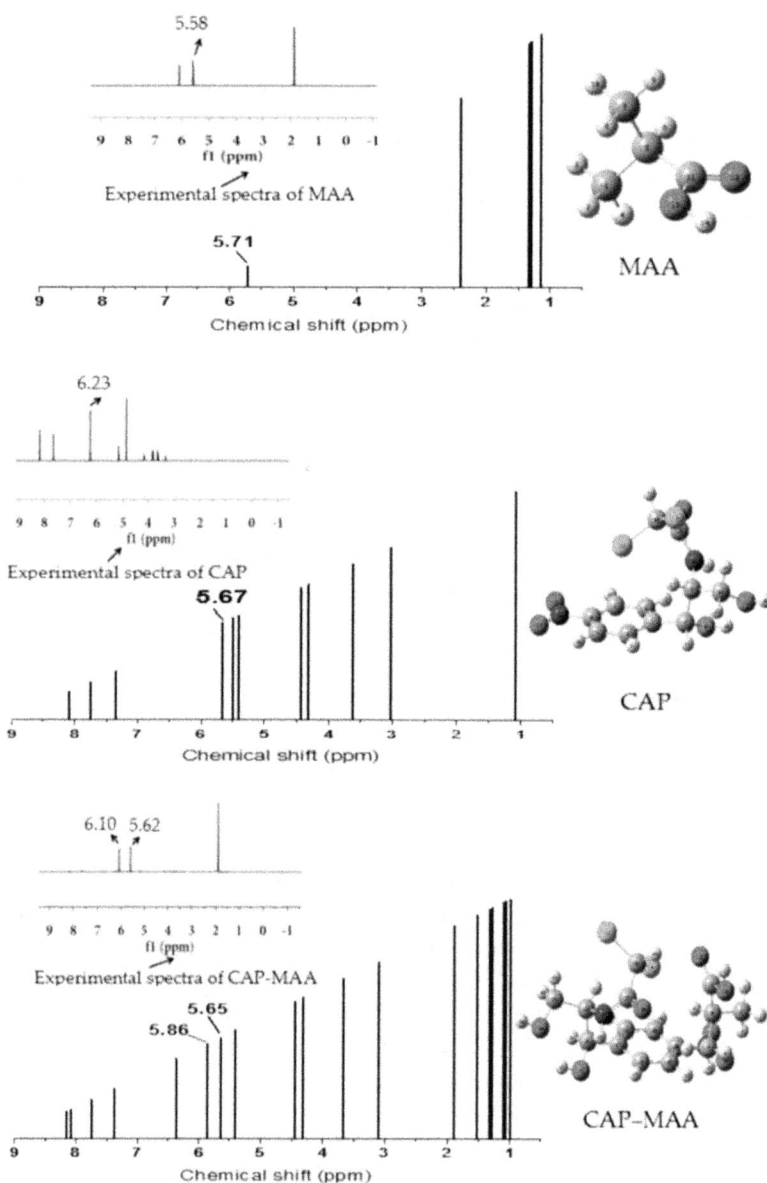

Figure 2.7. ^{1}H-NMR spectra of MAA, CAP, and CAP–MAA. (Reproduced from [44]. Copyright 2020 the authors. CC BY 4.0.)

heat with time is evaluated. The decomposition of MIP was evaluated by TGA [47]. Figure 2.8 shows the decomposition of standard substance at 280 °C and polymer decomposition at 440 °C for MIP for furaltadone (MIP-F3) and MIP for nitro-furantoin (MIP-N3). These temperatures vary for individual MIPs [47].

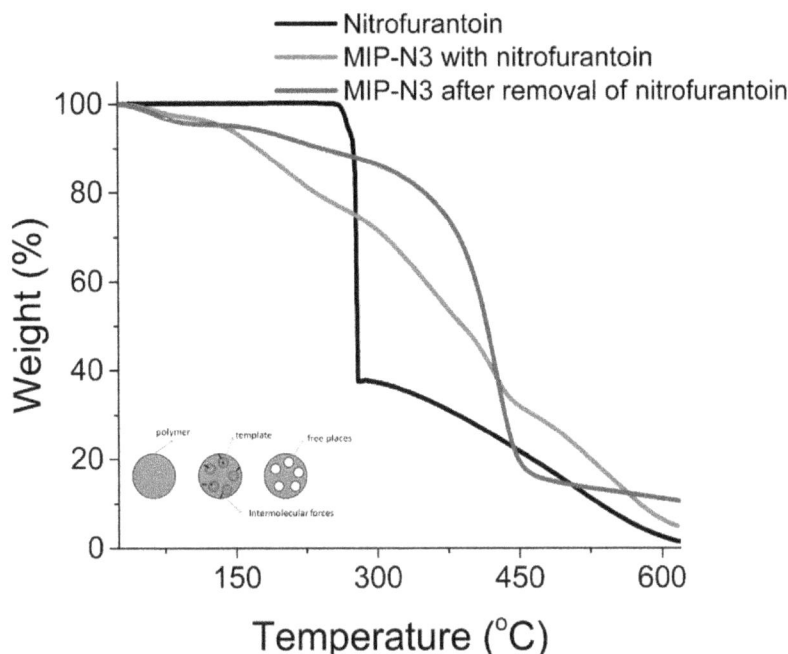

Figure 2.8. TGA for the synthetized MIPs. (Reproduced from [47] with permission from The Royal Society of Chemistry. CC BY 3.0.)

2.2.6 Isotherms governing MIPs

MIPs are homogeneous or heterogeneous materials containing binding sites with a wide array of binding affinities and selectivities. The binding behavior of MIPs is modeled by the heterogeneous Langmuir-Freundlich (LF) isotherm. In contrast, the LF model enabled direct comparisons of the binding characteristics of MIPs that have very different underlying distributions and were measured under different conditions. The binding parameters can be calculated directly using the LF fitting coefficients that yield a measure of the total number of binding sites, mean binding affinity, and heterogeneity. Alternatively, solution of the Langmuir adsorption integral for the LF model enabled direct calculation of the corresponding affinity spectrum from the LF fitting coefficients from a simple algebraic expression, yielding a measure of the number of binding sites. The ability of the LF isotherm to model MIPs suggests that a unimodal heterogeneous distribution is an accurate approximation of the distribution found in homogeneous and heterogeneous MIPs.

2.2.7 SEM and TEM analysis

The surface information and morphology of MIPs are evaluated by SEM and TEM. The sample preparations and analytical skills are essential for the analysis. MIPs based on Si nanospheres were observed by SEM [49]. The preparation method uses monomers, a cross-linker and an initiator. SEM of mesoporous Si shows spheres of

Figure 2.9. SEM images of (a) mesoporous SiO_2 and (b) MSN@MIPs. (Reproduced from [50]. Copyright 2021 Yuxuan Ma *et al*, published by De Gruyter. CC BY 4.0.)

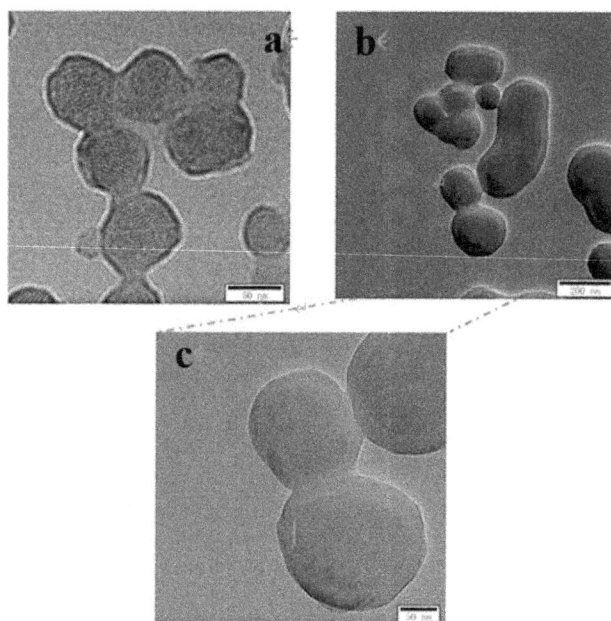

Figure 2.10. TEM images of (a) MSNs and (b and c) MSN@MIPs. (Reproduced from [50]. Copyright 2021 Yuxuan Ma *et al*, published by De Gruyter. CC BY 4.0.)

90 nm; the mesoporous material is shown in figure 2.9 (a). The same material with MIP shows an increase in size with spheres of 150 nm. The MIPs are seen in figure 2.9 (b). These images are evident for concluding the formation of MIPs by the employed processes.

TEM is also used for the evaluation of MIPs. The same material, mesoporous silica, shows a globular-shaped morphology with particles from 50–100 nm Figure 2.10. The structures with MIPs were bigger in size indicating the polymer attachment to the surface of the silica. An average particle size of 150 nm was obtained. These morphological informations are crucial for a good MIP method.

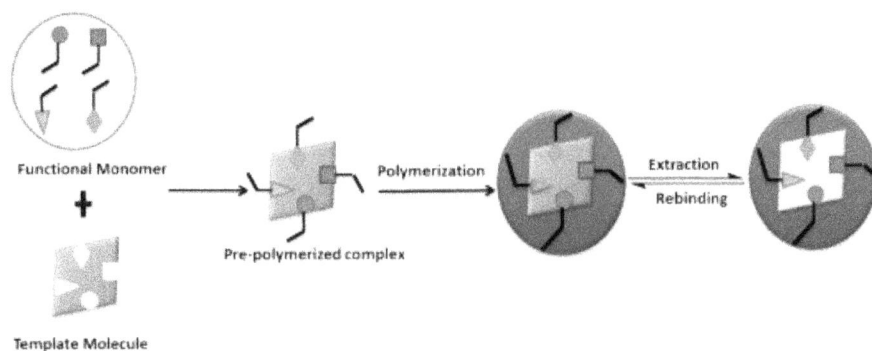

Figure 2.11. Schematic the representation of molecular imprinting process. (Reproduced with permission from [39]. Copyright 2021 Elsevier.)

2.3 Applications of imprinted polymers

MIPs are being applied in an increasing number of applications, as 'tailor-made' separation materials, antibody–receptor binding site mimics in recognition and assay systems, enzyme mimics in catalytic uses, recognition elements in sensors, and in facilitated chemical synthesis [51] (figure 2.11).

To date, they have been studied most broadly as separation components for the analysis of several compounds, including drugs [52, 53], pesticides [54], and amino acids [55]. As an extremely precise detection technology, MIPs are also used for the separation of isomers and enantiomers [56], solid extraction [57], biochemical sensors [58], and chemosensors [35], in simulating enzyme-catalysed pharmaceutical analysis [59], in sorbents, and in membrane separation technologies [60]. The applications of MIP are widely exploited in every technology today.

2.4 Challenges in synthesizing imprinted polymers

The development of imprints became available in 1931 [61], but research on MI was scarce until the 1980s. From various reviews that describe the improvements made over the years, it becomes clear that MI is a very promising and quickly growing technology, with numerous applications such as analytical separation, enzyme-like catalysis, chemical sensors, and drug delivery [62]. The MI technique has proven to be predominantly used for molecules with lower molecular weight (<1500 Da), and during the last five years the number of articles on the imprinting of larger (bio) templates has been increasing greatly [63]. However, intensifying the procedure toward imprinted materials for selectivity towards of proteins, DNA, viruses, and bacteria looks to be extremely challenging [64]. MI is a very challenging area of research. The ability to replace costly and complex affinity materials using MIPs has great potential and very fascinating materials have appeared in the last few years. Despite the relevant works in the field, there is still lack of industrial applications of MIPs.

The number of research papers in this field is still relatively small. Until 2003, fewer than ten research papers were published per year, which shows the problems faced when trying to imprint huge and sensitive biomolecules [65]. Initially, for low molecular weight compounds (LMWC), highly cross-linked gels are used to protect the imprint cavity after removal of the template. Nevertheless, for large template molecules, high crosslink densities encumber the weight transfer of the template extremely, leading to sluggish template elimination and rebinding kinetics or, in the worst case, permanent entrapment of the template in the polymer network due to physical immobilization [66]. Furthermore, due to the solubility properties and sensitive structural nature of bio-macromolecules (BMMs), imprinting can generally only be performed in an aqueous atmosphere, which limits the choice of monomers. Moreover, H_2 bonding interactions strongly contribute to the affinity of MIPs for LMWCs and, aprotic solvents, but are hindered in H_2O. In addition, BMMs are extremely complex. Physicochemical properties such as charge or hydrophobicity can vary greatly in different regions of, e.g. the protein template, whereas similar regions may be present in other templates. This could lead to high a specific binding and cross-reactivity of the imprinted polymer.

Nano-MIPs possess superior properties to bulk polymers. The soluble nature of these materials allows applications such as imaging and drug delivery. The challenges associated with this technology include: (i) deficiency of an indication of commercial success for MIP-based assays and sensors; (ii) prerequisites for complete analysis of toxicity, biodistribution, and clearance; and (iii) demonstration of similar presentation of MIP nanoparticles to antibodies in key therapeutic applications.

References

[1] Cheong W J, Yang S H and Ali F 2013 Molecular imprinted polymers for separation science: a review of reviews *J. Sep. Sci.* **36** 609–28

[2] Vasapollo G, Sole R D, Mergola L, Lazzoi M R, Scardino A, Scorrano S and Mele G 2011 Molecularly imprinted polymers: present and future prospective *Int. J. Mol. Sci.* **12** 5908–45

[3] Arshady R and Mosbach K 1981 Synthesis of substrate-selective polymers by host–guest polymerization *Macromol. Chem. Phys.* **182** 687–92

[4] Mayes A G and Whitcombe M J 2005 Synthetic strategies for the generation of molecularly imprinted organic polymers *Adv. Drug Delivery Rev.* **57** 1742–78

[5] Whitcombe M J, Rodriguez M E, Villar P and Vulfson E N 1995 A new method for the introduction of recognition site functionality into polymers prepared by molecular imprinting: synthesis and characterization of polymeric receptors for cholesterol *J. Am. Chem. Soc.* **117** 7105–11

[6] Klein J U, Whitcombe M J, Mulholland F and Vulfson E N 1999 Template-mediated synthesis of a polymeric receptor specific to amino acid sequences *Angew. Chem. Int. Ed.* **38** 2057–60

[7] Kirsch N, Alexander C, Davies S and Whitcombe M J 2004 Sacrificial spacer and non-covalent routes toward the molecular imprinting of 'poorly-functionalized' N-heterocycles *Anal. Chim. Acta.* **504** 63–71

[8] Fang L, Chen S, Guo X, Zhang Y and Zhang H 2012 Azobenzene-containing molecularly imprinted polymer microspheres with photo- and thermoresponsive template binding properties in pure aqueous media by atom transfer radical polymerization *Langmuir* **28** 9767–77

[9] Jiang G S, Zhong S A, Chen L, Blakey I and Whitaker A 2011 Synthesis of molecularly imprinted organic–inorganic hybrid azobenzene materials by sol–gel for radiation induced selective recognition of 2,4-dichlorophenoxyacetic acid *Radiat. Phys. Chem.* **80** 130–5

[10] Tang Q, Gong C, Lam M H and Fu X 2011 Preparation of a photoresponsive molecularly imprinted polymer containing fluorine-substituted azobenzene chromophores *Sensors Actuators* B **156** 100–7

[11] Rao T P, Kala R and Daniel S 2006 Metal ion-imprinted polymers—novel materials for selective recognition of inorganics *Anal. Chim. Acta.* **578** 105–16

[12] Hall A J, Emgenbroich M and Sellergren B 2005 Imprinted polymers *Templates in Chemistry II* (Berlin: Springer) pp 317–49

[13] Moreno-Bondi M C, Navarro-Villoslada F, Benito-Pena E and Urraca J L 2008 Molecularly imprinted polymers as selective recognition elements in optical sensing *Curr. Anal. Chem.* **4** 316–40

[14] Turker A R 2012 Separation, preconcentration and speciation of metal ions by solid phase extraction *Sep. Purif. Rev.* **41** 169–206

[15] Takeuchi T, Mukawa T and Shinmori H 2005 Signaling molecularly imprinted polymers: molecular recognition-based sensing materials *Chem. Rec.* **5** 263–75

[16] Resmini M 2012 Molecularly imprinted polymers as biomimetic catalysts *Anal. Bioanal. Chem.* **402** 3021–6

[17] Hart B R and Shea K J 2001 Synthetic peptide receptors: molecularly imprinted polymers for the recognition of peptides using peptide–metal interactions *J. Am. Chem. Soc.* **123** 2072–3

[18] Ottman N, Ruokolainen L, Suomalainen A, Sinkko H, Karisola P, Lehtimäki J, Lehto M, Hanski I, Alenius H and Fyhrquist N 2019 Soil exposure modifies the gut microbiota and supports immune tolerance in a mouse model *J. Allergy Clin. Immunol.* **143** 1198–206

[19] Striegler S 2001 Selective discrimination of closely related monosaccharides at physiological pH by a polymeric receptor *Tetrahedron* **57** 2349–54

[20] Wei Y, Zeng Q, Bai S, Wang M and Wang L 2017 Nanosized difunctional photo responsive magnetic imprinting polymer for electrochemically monitored light-driven paracetamol extraction *ACS Appl. Mater. Interfaces* **9** 44114–23

[21] Alaei H S, Tehrani M S, Husain S W, Panahi H A and Mehramizi A 2018 Photo-regulated ultraselective extraction of Azatioprine using a novel photoresponsive molecularly imprinted polymer conjugated hyperbranched polymers based magnetic nano-particles *Polymer* **148** 191–201

[22] Baimani N, Aberoomand Azar P, Waqif Husain S, Ahmad Panahi H and Mehramizi A 2019 Ultrasensitive separation of methylprednisolone acetate using a photoresponsive molecularly imprinted polymer incorporated polyester dendrimer based on magnetic nanoparticles *J. Sep. Sci.* **42** 1468–76

[23] Matsui J, Fujiwara K, Ugata S and Takeuchi T 2000 Solid-phase extraction with a dibutylmelamine-imprinted polymer as triazine herbicide-selective sorbent *J. Chromatogr.* A **889** 25–31

[24] Andersson L I, Miyabayashi A, O'Shannessy D J and Mosbach K 1990 Enantiomeric resolution of amino acid derivatives on molecularly imprinted polymers as monitored by potentiometric measurements *J. Chromatogr.* A **516** 323–31

[25] Barnes P J 2006 New therapies for asthma *Trends Mol. Med.* **12** 515–20

[26] Spivak D A 2005 Optimization, evaluation, and characterization of molecularly imprinted polymers *Adv. Drug Delivery Rev.* **57** 1779–94

[27] Gao D, Zhang Z, Wu M, Xie C, Guan G and Wang D 2007 A surface functional monomer-directing strategy for highly dense imprinting of TNT at surface of silica nanoparticles *J. Am. Chem. Soc.* **129** 7859–66

[28] Ki C D, Oh C, Oh S G and Chang J Y 2002 The use of a thermally reversible bond for molecular imprinting of silica spheres *J. Am. Chem. Soc.* **124** 14838–9

[29] Markowitz M A, Kust P R, Deng G, Schoen P E, Dordick J S, Clark D S and Gaber B P 2000 Catalytic silica particles via template-directed molecular imprinting *Langmuir* **16** 1759–65

[30] Rao M S and Dave B C 1998 Selective intake and release of proteins by organically-modified silica sol–gels *J. Am. Chem. Soc.* **120** 13270–1

[31] Carter S R and Rimmer S 2002 Molecular recognition of caffeine by shell molecular imprinted core–shell polymer particles in aqueous media *Adv. Mater.* **14** 667–70

[32] Espinosa-García B M, Argüelles-Monal W M, Hernández J, Félix-Valenzuela L, Acosta N and Goycoolea F M 2007 Molecularly imprinted chitosan–genipin hydrogels with recognition capacity toward o-xylene *Biomacromolecules* **8** 3355–64

[33] Katz A and Davis M E 2000 Molecular imprinting of bulk, microporous silica *Nature* **403** 286–9

[34] Vickers N J 2017 Animal communication: when I'm calling you, will you answer too? *Curr. Biol.* **27** R713–5

[35] Boonpangrak S, Whitcombe M J, Prachayasittikul V, Mosbach K and Ye L 2006 Preparation of molecularly imprinted polymers using nitroxide-mediated living radical polymerization *Biosens. Bioelectron.* **22** 349–54

[36] Lu C H, Zhou W H, Han B, Yang H H, Chen X and Wang X R 2007 Surface-imprinted core–shell nanoparticles for sorbent assays *Anal. Chem.* **79** 5457–61

[37] Jiang H and Ju H 2007 Enzyme–quantum dots architecture for highly sensitive electrochemiluminescence biosensing of oxidase substrates *Chem. Commun.* **4** 404–6

[38] Huang F, Peng Y, Jin G, Zhang S and Kong J 2008 Sensitive detection of haloperidol and hydroxyzine at multi-walled carbon nanotubes-modified glassy carbon electrodes *Sensors* **8** 1879–89

[39] Sajini T and Mathew B 2021 A brief overview of molecularly imprinted polymers: highlighting computational design, nano and photo-responsive imprinting *Talanta Open* **4** 100072

[40] Wang L, Zhi K, Zhang Y, Liu Y, Zhang L, Yasin A and Lin Q 2019 Molecularly imprinted polymers for gossypol via sol–gel, bulk, and surface layer imprinting—a comparative study *Polymers* **11** 602

[41] Bakhtiar S, Bhawani S A and Shafqat S 2019 Synthesis and characterization of molecular imprinting polymer for the removal of 2-phenylphenol from spiked blood serum and river water *Chem. Biol. Technol. Agric.* **6** 15

[42] Shafqat S R, Bhawani S A, Bakhtiar S and Ibrahim M N 2020 Synthesis of molecularly imprinted polymer for removal of Congo red *BMC Chem.* **14** 27

[43] Dugo G, Rotondo A, Mallamace D, Cicero N, Salvo A, Rotondo E and Corsaro C 2015 Enhanced detection of aldehydes in extra-virgin olive oil by means of band selective NMR spectroscopy *Physica* A **420** 258–64

[44] Xie L, Xiao N, Li L, Xie X and Li Y 2020 Theoretical insight into the interaction between chloramphenicol and functional monomer (methacrylic acid) in molecularly imprinted polymers *Int. J. Mol. Sci.* **21** 4139

[45] Pereira A C, Braga G B, Oliveira A E, Silva R C and Borges K B 2019 Synthesis and characterization of molecularly imprinted polymer for ethinylestradiol *Chem. Pap.* **73** 141–9

[46] Esfandyari-Manesh M, Javanbakht M, Atyabi F, Badiei A and Dinarvand R 2011 Effect of porogenic solvent on the morphology, recognition and release properties of carbamazepine-molecularly imprinted polymer nanospheres *J. Appl. Polym. Sci.* **121** 1118–26

[47] Rusen E, Diacon A, Mocanu A, Rizea F, Bucur B, Bucur M P, Radu G L, Bacalum E, Cheregi M and David V 2017 Synthesis and retention properties of molecularly imprinted polymers for antibiotics containing a 5-nitrofuran ring *RSC Adv.* **7** 50844–52

[48] Umpleby R J, Baxter S C, Chen Y, Shah R N and Shimizu K D 2001 Characterization of molecularly imprinted polymers with the Langmuir– Freundlich isotherm *Anal. Chem.* **73** 4584–91

[49] Popa A, Sasca V, Kiss E E, Marinkovic-Neducin R and Holclajtner-Antunovic I 2011 Mesoporous silica directly modified by incorporation or impregnation of some heteropolyacids: synthesis and structural characterization *Mater. Res. Bull.* **46** 19–25

[50] Ma Y, Xu Y, Chen H, Guo J, Wei X and Huang L 2021 Supported on mesoporous silica nanospheres, molecularly imprinted polymer for selective adsorption of dichlorophen *Green Process. Synth.* **10** 336–48

[51] Crapnell R D, Dempsey-Hibbert N C, Peeters M, Tridente A and Banks C E 2020 Molecularly imprinted polymer based electrochemical biosensors: overcoming the challenges of detecting vital biomarkers and speeding up diagnosis *Talanta Open* **2** 100018

[52] Kouhpaei A, Shahtaheri S, Ganjali M R, Rahimi F A and Golbabaei F 2008 Molecular imprinted solid phase extraction for determination of atrazine in environmental samples *Org. Biomol. Chem.* **6** 2459e67

[53] Didaskalou C, Buyuktiryaki S, Kecili R, Fonte C P and Szekely G 2017 Valorisation of agricultural waste with an adsorption/nanofiltration hybrid process: from materials to sustainable process design *Green Chem.* **19** 3116–25

[54] Masque N, Marce R M, Borrull F, Cormack P A and Sherrington D C 2000 Synthesis and evaluation of a molecularly imprinted polymer for selective on-line solid-phase extraction of 4-nitrophenol from environmental water *Anal. Chem.* **72** 4122–6

[55] Sikiti P, Msagati T A, Mamba B B and Mishra A K 2014 Synthesis and characterization of molecularly imprinted polymers for the remediation of PCBs and dioxins in aqueous environments *J. Environ. Health Sci. Eng.* **12** 82

[56] Perez-Moral N and Mayes A G 2004 Comparative study of imprinted polymer particles prepared by different polymerisation methods *Anal. Chim. Acta.* **504** 15–21

[57] Andac M, Mirel S, Şenel S, Say R, Ersoz A and Denizli A 2007 Ion-imprinted beads for molecular recognition based mercury removal from human serum *Int. J. Biol. Macromol.* **40** 159–66

[58] Zhou T, Jorgensen L, Mattebjerg M A, Chronakis I S and Ye L 2014 Molecularly imprinted polymer beads for nicotine recognition prepared by RAFT precipitation polymerization: a step forward towards multi-functionalities *RSC Adv.* **4** 30292–9

[59] Pan G, Zhang Y, Guo X, Li C and Zhang H 2010 An efficient approach to obtaining water-compatible and stimuli-responsive molecularly imprinted polymers by the facile surface-grafting of functional polymer brushes via RAFT polymerization *Biosens. Bioelectron.* **26** 976–82

[60] Gonzato C, Courty M, Pasetto P and Haupt K 2011 Magnetic molecularly imprinted polymer nanocomposites via surface-initiated RAFT polymerization *Adv. Funct. Mater.* **21** 3947–53

[61] Polyakov M V 1931 Adsorption properties and structure of silica gel *Zh. Fiz. Khim* **2** 799–805

[62] Byrne M E, Park K and Peppas N A 2002 Molecular imprinting within hydrogels *Adv. Drug Delivery Rev.* **54** 149–61

[63] Maier N M and Lindner W 2007 Chiral recognition applications of molecularly imprinted polymers: a critical review *Anal. Bioanal. Chem.* **389** 377–97

[64] Spivak D A and Shea K J 2001 Investigation into the scope and limitations of molecular imprinting with DNA molecules *Anal. Chim. Acta.* **435** 65–74

[65] Takeuchi T and Hishiya T 2008 Molecular imprinting of proteins emerging as a tool for protein recognition *Org. Biomol. Chem.* **6** 2459–67

[66] Valdebenito A, Espinoza P, Lissi E A and Encinas M V 2010 Bovine serum albumin as chain transfer agent in the acrylamide polymerization. Protein–polymer conjugates *Polymer* **51** 2503–7

IOP Publishing

Molecularly Imprinted Polymers for Environmental Monitoring
Fundamentals and applications
Raju Khan and Ayushi Singhal

Chapter 3

Theoretical and computational approaches and strategies in molecularly imprinted polymer development

Manish S Sengar, Priya Kumari and Neha Sengar

This chapter provides a brief introduction to molecular imprinting technology, and its structure, function, and preparation in the context of theoretical and computational approaches. It involves the co-polymerization of functional and cross-linking monomers in the presence of a molecular template that acts as the imprint molecule. It is increasingly being adopted for synthetic polymers involving biomimetic receptor systems that bind analytes with high affinity and specificity. The theoretical and computational approaches are explained in terms of multivariate analysis, electronic structure calculations, and molecular dynamics. The integration of these approaches in diagnostic and prognostic capabilities contributes to the advancement of molecularly imprinted materials.

3.1 Introduction

There has been growing interest in advanced analytical chemistry, in particular developing chemical sensors and biosensors. This is due to the need to meet new demands and opportunities in areas such as environmental analysis, clinical diagnostics, illicit drug detection, food analysis and production monitoring, chemical warfare agents, and geno-toxicity. A developing area in recent years is drug screening.

The recognition element is the central part of a chemical or biosensor, which remains in close proximity with an interrogative transducer. The role of the recognition element is to recognize specifically and bind the analyte of interest in a complex matrix or sample. The transducer in turn converts the chemical signal produced upon analyte binding to an easily detectable output signal. Biosensors depend on biological entities such as enzymes, receptors, antibodies, or whole cells which act as recognition elements. With the advancement of phage display antibody libraries and recombinant antibodies, a suitable recognition element can be invented for analytes where a natural receptor does not exist. Different attempts have been made in this direction to replace natural receptors with smaller and highly stable counterparts. This has led to the design of bioengineered antibody fragments, for example, single-chain variable fragments (ScFv) which are already in use for biosensors [1]. From phage display libraries, small protein domains such as the R-helical domain have been used for specific binding to target proteins [2]. For biosensors, new types of semisynthetic receptors are nucleic acids and peptide nucleic acids [3]. However, the poor physical and chemical stability of biomolecules sometimes limits their use in harsh conditions, although there is great interest in them for designing biosensors for environmental applications [1].

Artificial receptors can be obtained for small target analytes via rational design and chemical synthesis [4], however, if the analyte is large and more complex in nature this approach is unfavorable. In such cases, other techniques may be preferable, for example combinatorial chemistry [5] involving the design of biomimetic ligands for proteins or the preparation of tailor-made receptors by templating with a target analyte (figure 3.1). In this regard one technique, i.e. molecular imprinting, is increasingly being adopted for synthetic polymers. It is a modified approach involving the use of biomimetic receptor systems which are capable of binding analytes of interest with high affinity and specificity. During the imprinting process, the resulting binding sites possess affinity and selectivity approaching that of antigen–antibody systems. Therefore, molecularly imprinted polymers are dubbed 'antibody mimics' [6]. These mimics exhibit clear advantages over real antibodies for applications in sensor technology owing to their highly cross-linked polymeric nature and intrinsic stability, facilitating applications in harsh environments such as in acid and basic conditions, in organic solvents, or at high temperatures and pressures. Additionally, they are cheaper to produce and possess long durability if stored in a dry state at room temperature.

Recently, the great potential of this technology has been observed specifically after the development of synthetic organic polymers as imprinting matrices. Now,

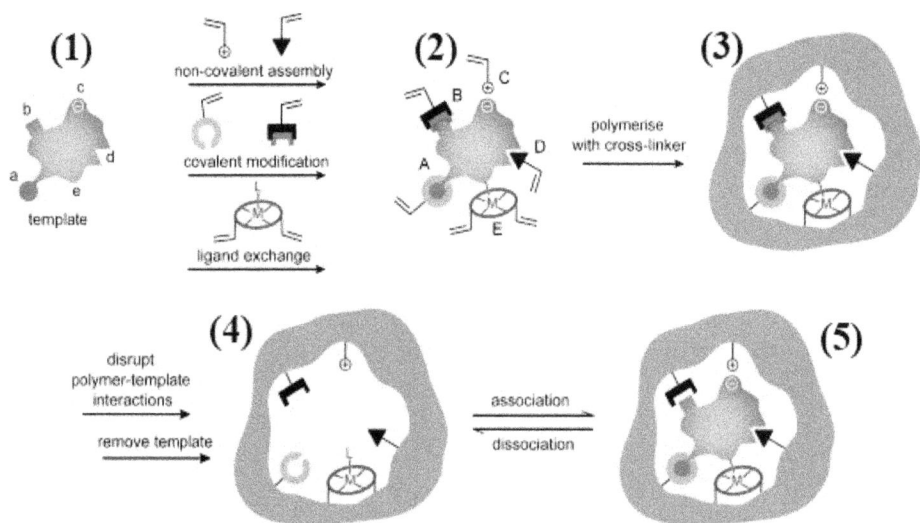

Figure 3.1. Schematic of MIP development using a template.

there has been a rapid advancement towards molecularly imprinted polymers (MIPs) as recognition elements for sensors.

3.2 Molecular imprinting technology

Molecular imprinting technology is used for synthetic polymers and involves the co-polymerization of functional and cross-linking monomers (figures 3.2 and 3.3) in the presence of a molecular template that acts as the imprint molecule. Initially, complex formation between the functional monomers and imprint molecule takes place followed by polymerization, resulting in the adherence of functional groups by the highly cross-linked polymeric structure. Imprint molecule removal reveals the binding sites which are complementary to the size and shape of the analyte. Thus a molecular memory is generated into the polymer which in turn has the capacity of rebinding the analyte with high specificity.

There are two different approaches for molecular imprinting. First, non-covalent interactions can facilitate the pre-polymerization complex formation between the imprint molecule and functional monomers, hence involving self-assembly. The second approach is covalent coupling of monomers to the imprint molecule, thereby synthesizing a polymerizable derivative of the imprint molecule. Covalent imprinting protocols have received a great deal of attention due to the high stability of the covalent bonds [7]. Additionally, the relative yield of binding sites is higher in comparison to the non-covalent protocols. The covalent approach was provided by Wulff and co-workers [8] and the non-covalent imprinting approach was initiated by Mosbach and co-workers [9] and is more flexible in terms of the choice of functional monomers, the use

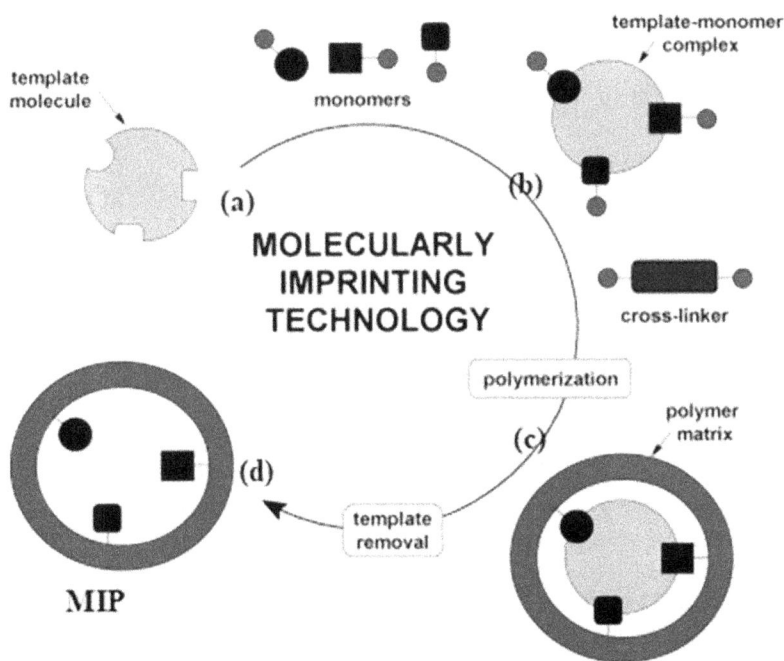

Figure 3.2. Diagram describing the different stages of molecular imprinting technology.

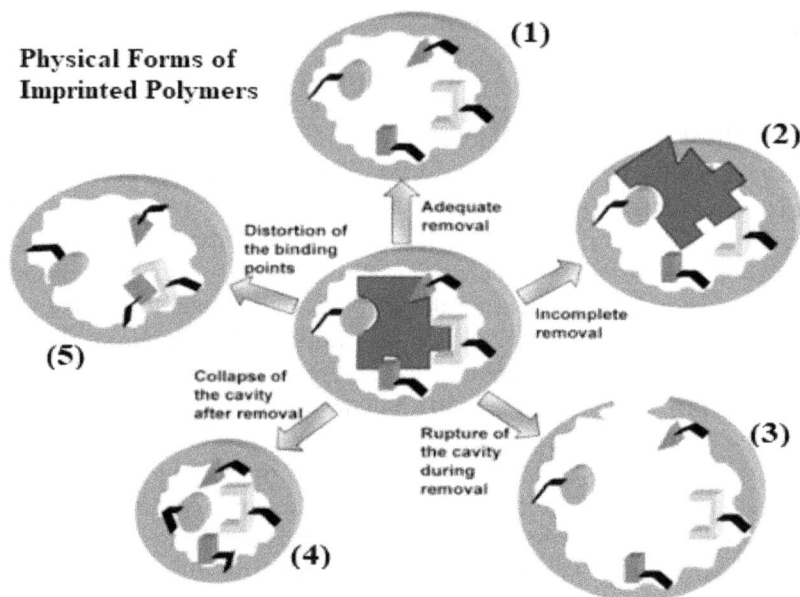

Figure 3.3. Schematic diagram of the physical forms of molecular imprinted polymers. (Reproduced from [94]. Copyright 2011 the authors. CC BY 4.0.)

of imprinted materials, and possible target molecules. However, they are helpful in natural processes because biomolecular interactions are largely non-covalent in nature. Recently, hybrid protocols have been proposed that combine the advantages of both covalent and no-covalent imprinting. For instance, by the use of covalent and non-covalent interactions, a tripeptide of Lys-Trp-Asp has been imprinted [1]. The interaction of peptides with polymers takes place during rebinding [10].

3.2.1 The physical forms of imprinted polymers

Different physical forms of MIPs (figure 3.4) are available so that different final applications can be achieved. Generally, these are based on organic polymers prepared via radical polymerization using monomers with acrylic or vinyl groups. The vast abundance of functional monomers with vinyl and acrylic polymers makes them popular for use in imprinting. They can make hydrogen bonds, hydrophobic contacts, and can be either positively or negatively charged. Recently, more highly developed sensors have been put forward with specialized functions for analyte detection. Generally, an inert solvent in a polymerization mixture dissolves all ingredients, including imprint molecules and also results in the formation of a highly porous structure. This formed structure allows the elution of the imprint molecule and also allows the analyte to access the imprinted sites [1]. However, it has been shown that a solvent is not required in cases where immobilization of the imprint molecules on a solid support is performed before polymerization [11].

The most common method of preparing MIPs is through solution polymerization involving mechanical grinding of the resulting monolithic block to form small

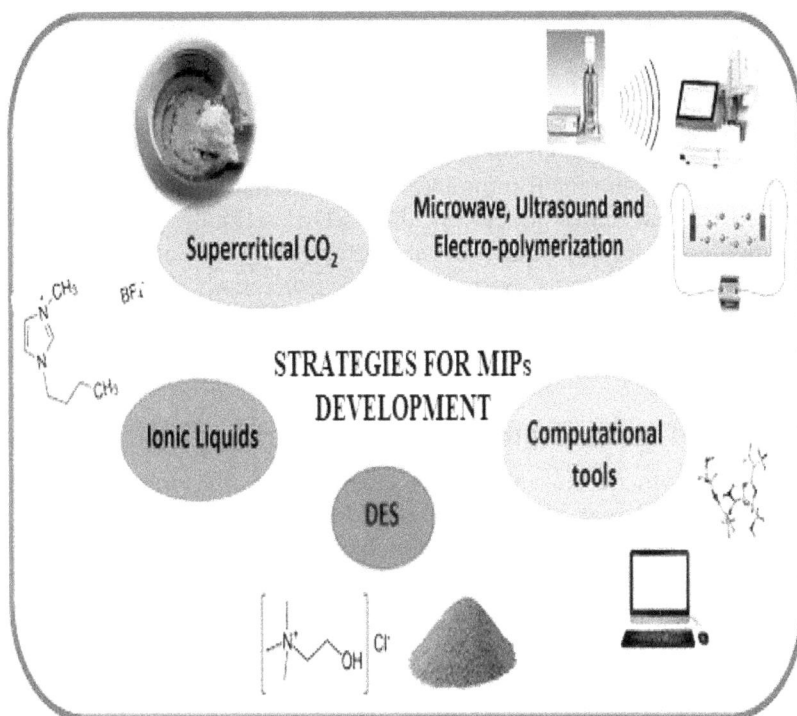

Figure 3.4. Diagram of the theoretical and computational strategies for MIP development.

particles with diameters in the micrometer range. Alternatively, particles can also be synthesized directly in the form of spherical beads with controlled diameters. In case of biphasic systems, using liquid perfluorocarbons is preferred in place of water as a continuous phase since water can cause disadvantageous effects on non-covalent interactions. By including iron oxide nanoparticles, magnetic beads can be prepared in this way. Another simple method where mechanical grinding is not required for preparing imprinted supports is dispersion polymerization, which yields aggregated spherical particles. The polymerization reaction can also be performed *in situ*, for instance directly inside a chromatography column or in capillary [12]. Imprinted membranes for sensor applications can be prepared in the form of thin cross-linked imprinted polymer films using the precipitation of linear polymers in the pores of an inert support membrane [13].

3.2.2 Target molecules and applications

The molecular imprinting method is applicable to diverse analytes and this is one of its most attractive features. It is now well established that the imprinting of small and organic molecules (for example, amino acids, steroids, sugars, pharmaceuticals, etc) is a routinely used process. Similarly, larger organic compounds can also be imprinted using the above mentioned approaches, however, it still remains a challenge [14]. It has been found that porogenic solvent is helpful in introducing molecular memory into the polymer and that it can be regarded as a molecular imprinting process [15]. The same can be applied for metal and other ions that can be used as templates for inducing unique arrangements of functional groups in organic polymers and silica [16]. Combinatorial approaches for MIP synthesis have been adapted for obtaining an optimized polymer for a particular target analyte. The molar ratios of functional monomers and ingredients of imprinting materials can be varied via automated processes. MIPs were originally applied as stationary phases in high-performance liquid chromatography (HPLC) and chiral separation [17]. Subsequently, their use has been extended to other analytical techniques such as capillary electro-chromatography [1, 12], thin-layer chromatography [18], solid-phase extraction [19], chemical sensors, and immunoassay type binding assays [6].

3.3 Theoretical and computational strategies for MIP development

Due to rapid growth in the applications of molecular imprinting, different tools regarding MIPs synthesis and design in terms of *in silico* studies have been undertaken. These tools provide insights into the pre-polymerization mixture events, polymer–ligand interactions, and polymer morphology from an atomic point of view. The field of applying computational strategies to molecular imprinting has developed steadily since the beginning of the twenty-first century. This growth has been facilitated through experimental validation. By using the available empirical data of polymer–ligand interactions, knowledge regarding the molecular events involved in MIP behavior has been extracted. For explaining and understanding the pre-polymerization events and polymer recognition features, thermodynamic models are applied. Recently, a stochastic algorithm was applied for studying the pre-polymerization monomer–template equilibria and it was a probability based approach (figure 3.5).

In recent years, probability based stochastic simulations (PSS) for the pre-polymerization of monomer–template equilibria [20] have been utilized in this research area and an example of this is the stochastic algorithm [21] for simulating the heterogeneity of pre-polymerization solution and setting down monomer units in lattice array. In experiments the distribution of simulated affinity was close to that of

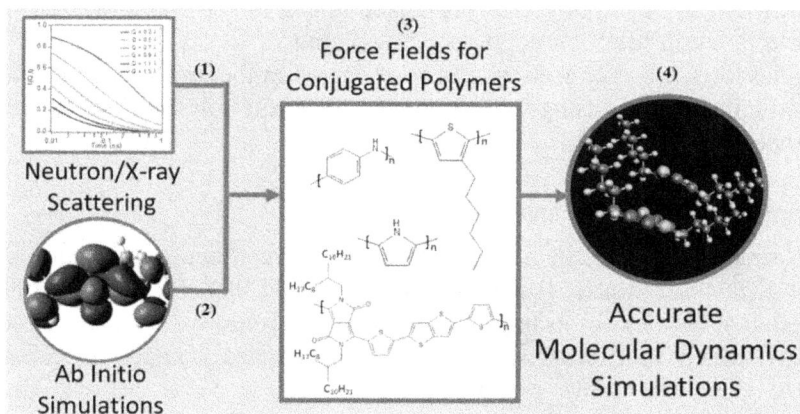

Figure 3.5. A schematic diagram of an accurate MD simulation with its four primary stages.

the measured MIPs. A theoretical and computational model describes the pre-polymerization template–monomer matrix and its subsequent development, which is also defined by mathematical modeling [22]. The main reasons for the recent increasing use of computational strategies for MIPs are affordability, computational power, and the presence of appropriate softwares [23]. This facilitates the use of multi-correlational analysis for the calculation of electronic structure, full systems with all the design from molecular dynamics (MD) simulation aspects, and the evaluation and synthesis of MIPs. These computational tools depend on three multi-correlational analyses which are defined below with their application in MIP technology.

3.3.1 Electronic structure calculations

For designing and evaluating MIPs, the use of computational methods based on electronic structure calculations, for example density functional theory (DFT), *ab initio*, and semi-empirical strategies, is popular. The class of computational methods which are used is quantum chemistry, which notably solves the electronic Schrödinger equation involving atomic coordinates and the number of electrons studied. However, this study is impossible for systems containing few electrons and, hence, approximations are undertaken. The electronic wave-function is approximated by *ab initio* methods while the Hamiltonian operator is approximated by semi-empirical and DFT methods. The computational demand and accuracy increases from semi-empirical (including only valence electrons and certain parameters obtained from experiment) to DFT (electron density calculation) and finally to *ab initio* methods. To provide acceptable approximations for the studied system, different approaches, parameters, and basis sets are selected for a reasonable timeframe. Generally, on comparing quantum mechanical calculations of isolated molecules with molecular complexes, information regarding the type and strength of interaction can be obtained. Hence, this approach is useful in evaluating template–monomer combinations [24].

3.3.2 Molecular dynamics

Molecular dynamics is considered to have been introduced by Alder and Wainwright during a seminal study on the hard sphere simulation of gaseous argon in 1957, although its development is closely related to that of Monte Carlo simulations [25]. The force fields in MD simulations (figure 3.6) are described by a set of parameters and equations for the forces present on and between interacting atoms and molecules [26]. The solving of Newton's equations of motion allows for motion simulation or system dynamics. MD simulations are helpful in developing both software and force fields and have been applied increasingly to different research areas, for example, surface studies, biomolecular interactions, solvents, protein folding, phospholipid bilayers, DNA conformation, and membrane transport of drugs. The commonly used force fields are GAFF, OPLS, AMBER, CHARMM, and GROMOS. MD simulations need fewer computational resources in comparison to electronic structure methods for systems of a comparable sizes, hence enabling the study of multimolecular systems such as MIP pre-polymerization mixtures involving thousands of molecules. Electrons are explicitly not considered, therefore MD simulations do not account for processes involving the movement of

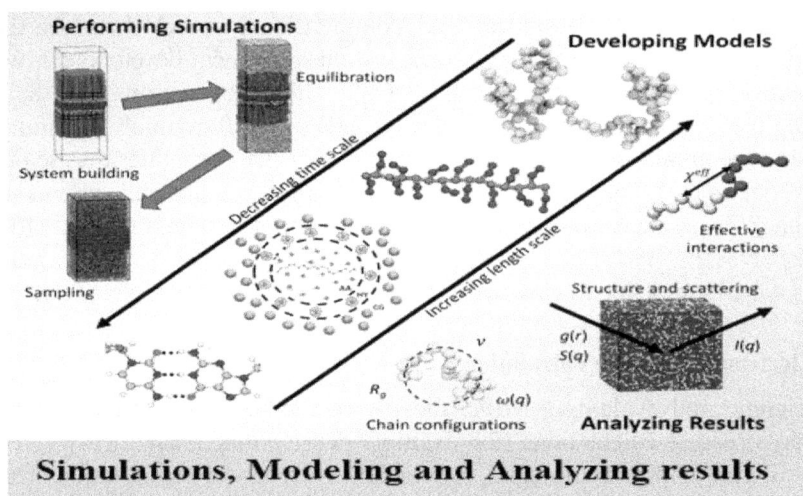

Figure 3.6. A schematic for the multivariate analysis of MIPs. (Reproduced with permission from [95]. Copyright 2022 Elsevier.)

electrons, such as bond breaking or formation. However, prominent information regarding the non-covalent interactions involved in pre-polymerization mixtures as well as in MIP binding site models are obtained [24].

3.3.3 Multivariate analysis

The application of molecular imprinting with infinite combinations of pre-polymerization components, conditions for polymerization, evaluation parameters, and analytical responses, has led towards multivariate analysis (figure 3.7) [27]. This includes an advanced level of modeling in terms of molecular interactions and energies. The aim here is to develop mathematical models that can simultaneously correlate different experimental variables to one or more properties of the MIPs. The resulting models can be used for optimization of analytical parameters, polymer recipes, or for finding correlations and patterns hidden in large datasets. The application of multivariate methods lies in determining the parameters used for the study followed by selection of an experimental design which allows for simultaneous evaluation of the parameters. Generally, a training or pilot experimental set is performed to determine the variables that have the greatest effect on the outcome. This is followed by a more focused and second experimental design to develop models for optimization or prediction. The commonly used experimental designs are Box–Behnken, full or fractional factorial, Doehlert, Plackett–Burman, and central composite design [24]. The calibration or fitting of experimental data to mathematical models can be done using different methods. For MIP studies, the mathematical models used are partial least squares regression (PLSR), multiple linear regression (MLR), principal component analysis (PCA), and artificial neural networks (ANN).

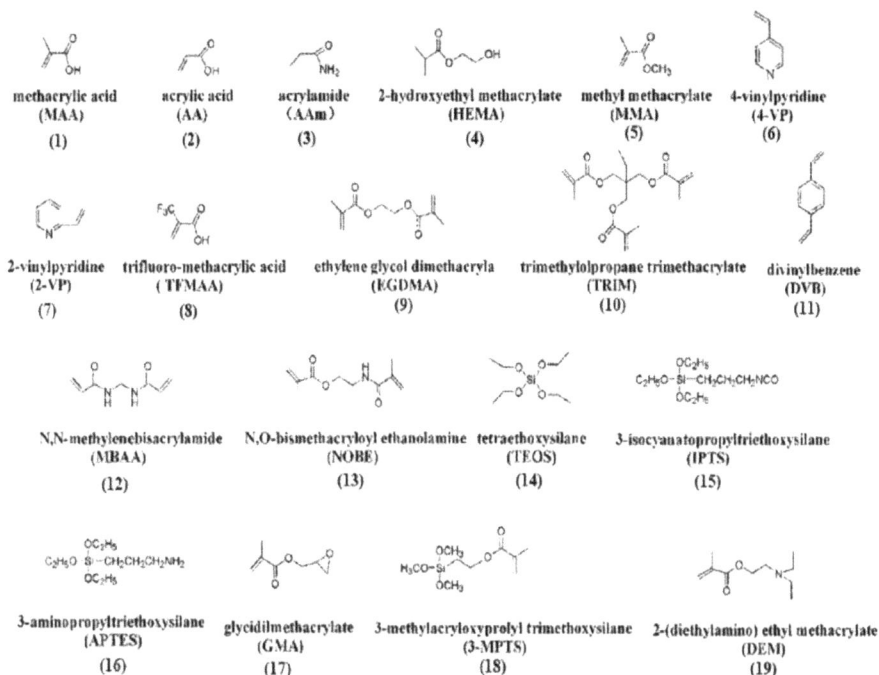

Figure 3.7. Structures of commonly used functional monomers and cross-linkers of MIPs. (Reproduced with permission from [96]. Copyright 2010 The Royal Society of Chemistry.)

3.4 The pre-polymerization stage

The computational approaches used for the pre-polymerization stages are predominantly based on electronic structure methods or MD simulations.

3.4.1 Electronic structure calculations

With increasing system size under investigation, the demands for hardware resources and simulation are increasing. The general use of electronic structure-determining methods in MIP studies is for characterizing the template–monomer complexes and to find the most suitable functional monomer and optimized stoichiometry. Among the electronic structure based methods, fewer computational resources are required in semi-empirical approaches. The two most commonly used semi-empirical methods for MIP development are PM3 [24, 28] and AM1 [29]. The *ab initio* and DFT methods are more demanding in terms of computational approaches and also provide higher accuracy. Various methods have been utilized in investigations concerning template-monomer complexes, depending on the basis sets and theoretical levels. The DFT methods are commonly applied [30] while *ab initio* methods are used in fewer cases [31].

These studies have been focused on the interaction of single functional monomers (SFMs) with a single template (ST). A study focused on DFT with semi-empirical based calculations that were compared to the characterizations of monomer–template (M–T) interaction [32]. The increasing accessiblity of computational power has been accompanied by different reports in the literature, such as quantum chemical calculations in the design and characterization of MIPs. Owing to the high requirement for resources of these calculations, most studies are focused primarily on pre-polymerization mixtures subsets, restricted interaction sets, or *in vacuo* isolated non-solvated molecular complexes. It is expected that the combination of electronic structure and MD simulations with increasing availability of computational power and the discovery of novel mixed strategies will enhance the use of electronic structure methods in the design and study of molecular imprinting systems [24].

3.4.2 Molecular dynamics

MD is highly suitable for studying liquid systems by assuming that the recognition properties of the MIP originate from pre-polymerization interactions [33–35]. The information about the strengths and types of all pre-polymerization interactions can be obtained and correlated with MIP recognition performance from the molecular trajectories or resulting data. The computational cost is lower for force field methods in comparison to quantum chemical calculations. The MD simulations are also applicable to very large systems with the inclusion of solvent molecules [24].

In an approach introduced by Piletsky *et al*, for assessing the interaction energy with template ephedrine in both charged and neutral states, 20 functional monomers were used initially [36]. The chosen monomers were used for polymer synthesis and also subjected to further MD simulations together with the cross-linker, template, and solvent, resulting in correlation between observed interactions and experimental binding data. This approach has been adapted by various researchers [37]. In several reports, the same approach has been used for evaluating and characterizing M–T interactions via MD and docking simulations [38]. Despite the rapid development of computer software and hardware, the multimolecular simulations involving templates, multiple copies of monomers, and an explicit solvent are still not suitable for electronic structure methods alone. Several studies have reported the combined use of MD simulations and quantum chemical calculations for studying different aspects of pre-polymerization mixtures [39–41]. The other aspects of the pre-polymerization stage in MD-based approaches include studies on the structural stability of protein epitopes for template screening [42], the mapping of potential monomer interaction sites of a protein target., acrylamide-derived monomer docking and interaction energy post-docking calculations [43], and studies of Dengue virus as a support matrix for interaction with templates for creating larger binding sites [44]. Different reports have been attempted for correlating the structural and physical properties of dummy ligands and templates with rebinding properties [45], and studying the effects of composition on material properties and template interaction via coarse-grained simulations [46]. Furthermore, for mimicking chromatography in a virtual capillary, large-scale MD simulations have been performed [47].

3.4.3 Multivariate analysis

Generally, the optimization and analysis of MIPs are univariate in nature. This involves the optimization and evaluation of one parameter and the results are carried forward for the optimization of the next parameter and so forth. It may turn out to be false or local and may not always be ideal as an identified optima [27]. The interdependency of variables and inherent flexibility in MIP synthesis makes this stage a better candidate for multivariate optimization [24, 48]. Different studies have been reported on the application of different multivariate methods and experimental designs for optimizing synthesis methods and polymer composition [49, 50].

3.5 The polymerization stage

It is the least studied stage of molecular imprinting in general as well as in terms of computational treatment. The literature available on imprinting is abundant with experimental correlations between pre-polymerization mixture composition and MIP recognition properties. This provides support for the assumption that the states of the template–functional monomer complexes are preserved in the polymer matrix. However, less evidence exists about the fate of these complexes once polymerization has been initiated, although NMR studies have depicted their retention during polymerization [51]. These are examples of polymerization studies via computational methods, in particular using molecular dynamics. In force field-based simulations, the development of reactive force fields and other solutions enable bond-formation and breaking, which further help in filling this knowledge gap [52].

3.5.1 Electronic structure calculations

Despite the fact that these calculations can accurately describe the electron's movement and chemical bond breaking and formation, unreasonable computer demands arise from the number of molecules required for a meaningful representation of the polymerization process of MIPs. In the near future, technical development may allow such calculations, However, to the best of our knowledge, no examples have been published so far [24].

3.5.2 Molecular dynamics

Few attempts have been made in the direction of applying MD for studying MIP polymerization. In a work by Yungerman and Srebnik studying binding site imperfection formation in MIPs, coarse-grained Monte Carlo procedures were used [53]. For the study, monomers and templates were modeled as Lennard-Jones spheres and rigid dumbbells, respectively. The imprinting using simulation is based on shape and size. Some similar studies have been reported that combine Monte Carlo simulation of hard spheres with statistical mechanics [54] or mean field theory [55]. For replicating MIP recognition on the molecular level, it is necessary to perform atomic simulations. All atom based MD simulations are reported by Henthorn and Peppas for the formation of glucose-imprinted polymers [56]. In this, 160 functional monomers (2-hydroxyethyl methacrylate), 160 template molecules, 300 cross-linkers (ethylene glycol dimethacrylate, EGDMA), 20 initiator molecules, and 800 water molecules were diffused and relaxed using MD simulation followed by certain reaction steps such as initiation, propagation, and termination. It was repeated till the quenching of all radicals. The binding of ligands to the resulting polymer models was compared with experimental data. In a series of protein-imprinted polymer studies, Srebnik and co-workers combined a similar reaction scheme with lattice Monte Carlo simulations [57]. Schauperl and Lewis attempted to simulate the polymerization reaction for xanthine MIPs [58]. Initiating with one or more template molecules, cross-linkers and monomers were added in a stepwise manner to the system and allowed the formation of new bonds with the growing polymer chain. The energy minimization and MD allowed for optimal host–guest interaction. The simulations were continued until the threshold energy was reached. This resulting polymer model was used to explain the binding site heterogeneity. Hyunh *et al* reported the simulation of electropolymerized MIPs selective for 6-thioguanine [59]. For the MD simulation, a system containing one template molecule, six cross-linking monomers, and two functional monomers was used. In the monomers, predetermined 'radical positions' form bonds if they lie within a 3 Å distance until no additional bonds are formed. The MD simulation was continued until saturation and the equilibrated system was replicated eight times. Except for density, which was very similar to that of the polymers prepared by electropolymerization, no other analysis of the resultant model was reported. A similar algorithm was developed for polymerization reaction simulation by Cowen and co-workers during MD simulations of the pre-polymerization mixture [60]. In brief, after attainment of system equilibrium, within a suitable distance new bonds are formed between 'reactive' atoms followed by another round for minimization of energy. This process was repeated until there was no possibility of more reactions [24].

3.5.3 Multivariate analysis

For optimizing polymerization conditions, the application of a multivariate approach is rare. An example is the comparison of bulk polymerization with surface molecular imprinting for optimized polymerization temperature studies and the role of insulin-imprinted magnetic nanoparticles [61], investigating the effect of polymerization temperature and time on the diameter of 5-fluorouracil-imprinted MIP nanoparticles synthesized through precipitation polymerization [62], and the electro-polymerization of ketorolac tromethamine MIPs on paper graphite electrodes involving the optimization of the number of cycles and scan rate [63]. It has been found in different reports that to optimize polymer synthesis, the polymerization parameters need to be included in the initial experimental design. However, these parameters were omitted from further optimization when they exhibited no significant effects on the outcomes [24].

3.6 MIP structure and function

Multivariate analysis is used in post-polymerization computational studies of MIP properties, although reports on quantum chemical calculations or MD simulations have also been provided.

3.6.1 Electronic structure calculations

For studying the aspects of MIP–template recognition, electronic structure methods have been used in some cases. Examples are the PM3 calculations for the binding site model of nicotinamide [64], AM1 calculations for explaining the recognition differences in different buffers [65], DFT studies for explaining the selectivity of MIPs based on phenylurea herbicide [66], confirming the binding site structure in a catalytic silica MIP [67], for the adsorption mechanism of 5-fluorouracil MIP [68], the interaction of poly-pyrrole models with tryptophan or glyphosate [69], both DFT and *ab initio* studies for studying the binding site models in cetirizine and hydroxyzine MIPs [70], and *ab initio* studies for the binding site model of phenolic compounds [71].

3.6.2 Molecular dynamics

Methods based on force fields have been reported for simulating the aspects of ligand or template rebinding to MIPs. In various studies, the approximation of polymer models has been achieved by equilibrating templates with functional monomers of linear chains, followed by binding energy analysis and other recognition aspects [72, 73] or docking [74]. Docking procedures for studying MIP models selectivity have been optimized semi-empirically by Terracina *et al* [75]. Pyridine MIP models were first equilibrated by a system containing EGDMA, chloroform, pyridine, and methacrylic acid [76]. The removal of the solvent and template and the positions of monomers were fixed. For investigating the adsorption of toluene, benzene, and pyridine, Monte Carlo simulations were applied further. The MIP binding site models were created via MD simulations by Sobiech *et al* from pre-formed functional/template/cross-linking monomer clusters [77]. After attaining equilibration, removal of the template, and polymerization of the system via the replacement of double-bonds by new single bonds were accomplished. A similar approach was used by Gajda *et al* for mimicking the binding site of aripiprazole [78]. Finally, a computational approach was developed by Curk *et al* for deriving binding site models and evaluating the rebinding of templates. These strategies used a range of parameters such as the number of templates in interaction points, monomer concentrations, and combinations of material properties with grand canonical Monte Carlo simulations describing multiple interaction sites in templates [22].

3.6.3 Multivariate analysis

The polymer performance is influenced by the experimental conditions employed while evaluating or applying MIPs. The possibilities of combinations of experimental parameters, for example, pH, flow rate, solvent, incubation time, analyte concentration, etc, are nearly infinite, thereby making this area highly suitable for multivariate optimization. Therefore, various combinations of response surface modeling and experimental designs were used for optimizing parameters when using MIPs for separation, adsorption, or sensing applications [79–85]. The improvements of optimized parameters have been done in some studies by using them as the input in artificial neural network (ANN) models [86]. In some other cases, when evaluating MIPs, a multivariate approach has been used to reveal hidden correlations in the obtained data. Various combinations of PLSR and PCA methods have been employed for the interpretation of MIP binding data resulting from surface-enhanced Raman spectroscopy (SERS), for correlating the MIP-quartz crystal shape frequency curves to different analytes, for recognition of patterns in responses from different MIP-sensor arrays, for binary mixtures via cyclic voltammetry measurements [79, 87], for the processing of fluorescence data [88], and to correlate the results with HPLC data [87]. These combinations have also been used in polymer morphology and pre-polymer-ization interactions from MD simulations and for correlating bupivacaine–MIP binding with rebinding solvent properties. Similarly, PCA was used for examining the relation between non-specific water and specific analyte sorption in an iprodione-imprinted MIP for its application in aqueous media [24, 89].

Chemometrics is a separate branch of multivariate analysis of chemical data, where large numbers of molecular descriptors and structure-derived properties are generated for molecule sets followed by correlation with other targeted properties. Therefore, different studies attempt to correlate the molecular descriptors and MIP binding data via various multivariate tools. Rossetti *et al* investigated PLS models for correlating molecular descriptors and solid-phase extraction retention data for a biomarker series involving pro-gastrin-releasing peptides in elucidation of recog-nition mechanisms. For the coupling of structural and molecular parameters of a quercetin MIP to its adsorption selectivity, Liu *et al* employed PCA and MLR models [90]. PCA was employed for correlating the chromatographic selectivity of MIP based on pentachlorophenol as a template and 52 related phenols with 16 AM1-derived molecular descriptors by Baggiani *et al* [91]. The chromatography data obtained from this study was further subjected to PLSR modeling using 25 descriptors that improved the selectivity and prediction capability of the model. Nantasemat *et al* created ANN models for MIP selectivity prediction and bisphenol A MIPs by using molecular descriptors for templates, HPLC mobile phases, functional monomers, and cross-linkers (figure 3.3) via using the reported literature [92, 93]. Likewise, for correlating analyte recognition with molecular descriptors, some models have been reported for MIPs selective for erythromycin, penicillin G, and milk lactose [24].

3.7 Conclusions and perspectives

The prominent increase in the number of reports in the literature describing computational studies of MIPs followed the availability, development, and afford-ability of both software and hardware. This has enabled the use of such tools for both their diagnostic and prognostic abilities for the development of molecularly imprinted materials. The growing awareness and development for use of these tools is reinforced via validation through experimental studies. In the near future, the combination of easier access to computational tools and the valuable insights achieved from their use is expected to enhance their prevalent use in the field of molecular imprinting.

References

[1] Balaji R, Boileau S, Guérin P and Grande D 2004 Feature article: design of porous polymeric materials from miscellaneous macromolecular architectures: an overview *Polym. News* **29** 205–12

[2] Nord K, Gunneriusson E, Ringdahl J, Stahl S, Uhlen M and Nygren P A 1997 Binding proteins selected from combinatorial libraries of an α-helical bacterial receptor domain *Nat. Biotechnol.* **15** 772–7

[3] Wang J 1998 DNA biosensors based on peptide nucleic acid (PNA) recognition layers. A review *Biosens. Bioelectron.* **13** 757–62

[4] Lehn J M 1995 *Supramolecular Chemistry* (Weinheim: Wiley)

[5] Li R X, Dowd V, Steward D J, Burton S J and Lowe C R 1998 Design, synthesis, and application of a protein A mimetic *Nat. Biotechnol.* **16** 190–5

[6] Vlatakis G, Andersson L I, Muller R and Mosbach K 1993 Drug assay using antibody mimics made by molecular imprinting *Nature* **361** 645–7

[7] Whitcombe M J, Rodriguez M E, Villar P and Vulfson E N J 1995 A new method for the introduction of recognition site functionality into polymers prepared by molecular imprinting: synthesis and characterization of polymeric receptors for cholesterol *Am. Chem. Soc.* **117** 7105–11

[8] Wulff G 1995 Molecular imprinting in cross-linked materials with the aid of molecular templates—a way towards artificial antibodies *Angew. Chem. Int. Ed. Engl.* **34** 1812–32

[9] Mosbach K and Ramström O 1996 The emerging technique of molecular imprinting and its future impact on biotechnology *Biotechnology* **14** 163–70

[10] Klein J U, Whitcombe M J, Mulholland F and Vulfson E N 1999 Template-mediated synthesis of a polymeric receptor specific to amino acid sequences *Angew. Chem. Int. Ed. Engl.* **38** 2057–60

[11] Yilmaz E, Haupt K and Mosbach K 2000 The use of immobilized templates—a new approach in molecular imprinting *Angew. Chem. Int. Ed. Engl.* **39** 2115–8

[12] Schweitz L, Andersson L I and Nilsson S 1997 Capillary electrochromatography with predetermined selectivity obtained through molecular imprinting *Anal. Chem.* **69** 1179–83

[13] Dzgoev A and Haupt K 1999 Enantioselective molecularly imprinted polymer membranes *Chirality* **11** 465–9

[14] D'Souza S M, Alexander C, Carr S W, Waller A M, Whitcombe M J and Vulfson E N 1999 Directed nucleation of calcite at a crystal-imprinted polymer surface *Nature* **398** 312–6

[15] Dickert F L, Forth P, Lieberzeit P and Tortschanoff M 1998 Molecular imprinting in chemical sensing—detection of aromatic and halogenated hydrocarbons as well as polar solvent vapors *Fresenius' J. Anal. Chem.* **360** 759–62

[16] Sasaki D Y, Rush D J, Daitch C E, Alam T M, Assink R A, Ashley C S, Brinker C J and Shea K J 1998 Molecular imprinted receptors in sol–gel materials for aqueous phase recognition of phosphates and phosphonates *ACS Symp. Ser.* **703** 314–22

[17] Sellergren B and Shea K J J 1993 Influence of polymer morphology on the ability of imprinted network polymers to resolve enantiomers *Chromatographia* **635** 31–49

[18] Kriz D, Berggren-Kriz C, Andersson L I and Mosbach K 1994 Thin-layer chromatography based on the molecular imprinting technique *Anal. Chem.* **66** 2636–9

[19] Sellergren B 1994 Direct drug determination by selective sample enrichment on an imprinted polymer *Anal. Chem.* **66** 1578–82

[20] Wu X, Carroll W R and Shimizu K D 2008 Stochastic lattice model simulations of molecularly imprinted polymers *Chem. Mater.* **20** 4335–46

[21] Veitl M, Schweiger U and Berger M L 1997 Stochastic simulation of ligand–receptor interaction *Comput. Biomed. Res.* **30** 427–50

[22] Curk T, Dobnikar J and Frenkel D 2016 Rational design of molecularly imprinted polymers *Soft Matter* **12** 35–44

[23] Nicholls I A, Andersson H S, Golker K, Henschel H, Karlsson B C G, Olsson G D, Rosengren A M, Shoravi S, Wiklander J G and Wikman S 2013 *Molecular Imprinting—Principles and Applications of Micro- and Nanostructured Polymers* ed Y Lei (Singapore: Pan Stanford) pp 71–104

[24] Nicholls I A, Golker K, Olsson G D, Suriyanarayanan S and Wiklander J G 2021 The use of computational methods for the development of molecularly imprinted polymers *Polymers* **13** 2841

[25] Fermi E, Pasta P, Ulam S and Tsingou M 1955 *Studies of Nonlinear Problems* (Berkeley, CA: University of California Press)

[26] Leach A R 2001 *Molecular Modelling: Principles and Applications* 2nd edn (Harlow: Pearson) pp 353–409

[27] Esbensen K H (ed) 2002 *Multivariate Data Analysis: in Practice* 5th edn (Oslo: Camo Process) p 16

[28] Li P, Rong F, Xie Y, Hu V and Yuan C 2004 Study on the binding characteristic of s-naproxen imprinted polymer and the interactions between templates and monomers *J. Anal. Chem.* **59** 939–44

[29] Fu Q, Sanbe H, Kagawa C, Kunimoto K K and Haginaka J 2003 Uniformly sized molecularly imprinted polymer for (s)-nilvadipine. Comparison of chiral recognition ability with HPLC chiral stationary phases based on a protein *Anal. Chem.* **75** 191–8

[30] Peng M, Li H, Long R, Shi S, Zhou H and Yang S 2018 Magnetic porous molecularly imprinted polymers based on surface precipitation polymerization and mesoporous SiO_2 layer as sacrificial support for efficient and selective extraction and determination of chlorogenic acid in Duzhong brick tea *Molecules* **23** 1554

[31] He H, Gu X, Shi L, Hong J, Zhang H, Gao Y, Du S and Chen L 2014 Molecularly imprinted polymers based on SBA-15 for selective solid-phase extraction of baicalein from plasma samples *Anal. Bioanal. Chem.* **407** 509–19

[32] Chen L, Lee Y K, Manmana Y, Tay K S, Lee V S and Rahman N A 2015 Synthesis, characterization, and theoretical study of an acrylamide-based magnetic molecularly imprinted polymer for the recognition of sulfonamide drugs *Polymers* **15** 141–50

[33] Alexander C, Andersson H S, Andersson L I, Ansell R J, Kirsch N, Nicholls I A, O'Mahony J and Whitcombe M J 2006 Molecular imprinting science and technology: a survey of the literature for the years up to and including 2003 *J. Mol. Recognit.* **19** 106–80

[34] Piletsky S A, Panasyuk T L, Piletskaya E V, Nicholls I A and Ulbricht M 1999 Receptor and transport properties of imprinted polymer membranes—a review *J. Membr. Sci.* **157** 263–78

[35] Mayes A and Whitcombe M 2005 Synthetic strategies for the generation of molecularly imprinted organic polymers *Adv. Drug. Del. Rev.* **57** 1742–78

[36] Piletsky S A, Karim K, Piletska E V, Turner A P F, Day C J, Freebairn K W and Legge C 2001 Recognition of ephedrine enantiomers by molecularly imprinted polymers designed using a computational approach *Analyst* **126** 1826–30

[37] Piletska E V, Turner N W, Turner A P and Piletsky S A 2005 Controlled release of the herbicide simazine from computationally designed molecularly imprinted polymers *J. Controlled Release* **108** 132–9

[38] Toro M J U, Marestoni L D and Sotomayor M D P T 2015 A new biomimetic sensor based on molecularly imprinted polymers for highly sensitive and selective determination of hexazinone herbicide *Sensors Actuators* B **208** 299–306

[39] Terracina J J, Sharfstein S T and Bergkvist M 2018 *In silico* characterization of enantio-selective molecularly imprinted binding sites *J. Mol. Recognit.* **31** 2612

[40] Dong C, Li X, Guo Z and Qi J 2009 Development of a model for the rational design of molecular imprinted polymer: computational approach for combined molecular dynamics/quantum mechanics calculations *Anal. Chim. Acta* **647** 117–24

[41] Mamo S K, Elie M, Baron M G and Gonzalez-Rodriguez J 2020 Computationally designed perrhenate ion imprinted polymers for selective trapping of rhenium ions *ACS Appl. Polym. Mater.* **2** 3135–47

[42] Altintas Z, Takiden A, Utesch T, Mroginski M A, Schmid B, Scheller F W and Sussmuth R D 2019 Integrated approaches toward high-affinity artificial protein binders obtained via computationally simulated epitopes for protein recognition *Adv. Funct. Mater.* **29** 1807332

[43] Sullivan M V, Dennison S R, Archontis G, Reddy S M and Hayes J M 2019 Toward rational design of selective molecularly imprinted polymers (MIPs) for proteins: computational and experimental studies of acrylamide based polymers for myoglobin *J. Phys. Chem.* B **123** 5432–43

[44] Sukjee W, Tancharoen C, Yenchitsomanus P T, Gleeson M P and Sangma C 2017 Small-molecule dengue virus co-imprinting and its application as an electrochemical sensor *Chem. Open* **6** 340–4

[45] Cai Y, He X, Cui P L, Liu J, Li Z B, Jia B J, Zhang T, Wang J P and Yuan W Z 2019 Preparation of a chemiluminescence sensor for multi-detection of benzimidazoles in meat based on molecularly imprinted polymer *Food Chem.* **280** 103–9

[46] Zadok I and Srebnik S 2018 Coarse-grained simulation of protein-imprinted hydrogels *J. Phys. Chem.* B **122** 7091–101

[47] Zink S, Moura F A, Autreto P, Galvao D S and Mizaikoff B 2018 Virtually imprinted polymers (VIPs): understanding molecularly templated materials via molecular dynamics simulations *Phys. Chem. Chem. Phys.* **20** 13145–52

[48] Kempe H and Kempe M 2004 Novel method for the synthesis of molecularly imprinted polymer bead libraries *Macromol. Rapid Commun.* **25** 315–20

[49] Nezhadali A and Mojarrab M 2014 Computational study and multivariate optimization of hydrochlorothiazide analysis using molecularly imprinted polymer electrochemical sensor based on carbon nanotube/polypyrrole film *Sensors Actuators* B **190** 829–37

[50] Davies M P, De Biasi V and Perrett D 2004 Approaches to the rational design of molecularly imprinted polymers *Anal. Chim. Acta* **504** 7–14

[51] Svenson J, Karlsson J G and Nicholls I A 2004 ^{1}H Nuclear magnetic resonance study of the molecular imprinting of (−)-nicotine: template self-association, a molecular basis for cooperative ligand binding *J. Chromatogr.* A **1024** 39–44

[52] Senftle T P *et al* 2016 The ReaxFF reactive force-field: development, applications and future directions *Comput. Mater.* **2** 15011

[53] Yungerman I and Srebnik S 2006 Factors contributing to binding-site imperfections in imprinted polymers *Chem. Mater.* **18** 657–63

[54] Cheng S and Van Tassel P R 2001 Theory and simulation of the available volume for adsorption in a chain molecule templated porous material *J. Chem. Phys.* **114** 4974–81

[55] Srebnik S (ed) 2002 *Characterization of Porous Solids* vol 144 6th edn (Amsterdam: Elsevier) pp 43–50

[56] Henthorn D B and Peppas N A 2003 Molecular simulations of recognitive polymer networks prepared by biomimetic configurational imprinting as responsive biomaterials *MRS Online Proc. Libr.* **787** 7–15

[57] Yankelov R, Yungerman I and Srebnik S 2017 The selectivity of protein-imprinted gels and its relation to protein properties: a computer simulation study *J. Mol. Recognit.* **30** 2607

[58] Schauperl M and Lewis D W 2015 Probing the structural and binding mechanism heterogeneity of molecularly imprinted polymers *J. Phys. Chem.* B **119** 563–71

[59] Huynh T P, Wojnarowicz A, Sosnowska M, Srebnik S, Benincori T, Sannicolo F, D'Souza F and Kutner W 2015 Cytosine derivatized bis(2,2'-bithienyl)methane molecularly imprinted polymer for selective recognition of 6-thioguanine, an antitumor drug *Biosens. Bioelectron.* **70** 153–60

[60] Piletska E V, Guerreiro A, Mersiyanova M, Cowen T, Canfarotta F, Piletsky S, Karim K and Piletsky S 2020 Probing peptide sequences on their ability to generate affinity sites in molecularly imprinted polymers *Langmuir* **36** 279–83

[61] Goudarzi F and Hejazi P 2020 Comprehensive study on the effects of total monomers' content and polymerization temperature control on the formation of the polymer-layer in preparation of insulin-imprinted magnetic nanoparticles *Eur. Polym. J.* **126** 109541

[62] Madadian-Bozorg N, Zahedi P, Shamsi M and Safarian S 2018 Poly (methacrylic acid)-based molecularly imprinted polymer nanoparticles containing 5-fluourouracil used in colon cancer therapy potentially *Polym. Adv. Technol.* **29** 2401–9

[63] Nezhadali A and Biabani M 2020 Melamine recognition: molecularly imprinted polymer for selective and sensitive determination of melamine in food samples *Anal. Bioanal. Electrochem.* **12** 48–62

[64] Wu L and Li Y 2004 Study on the recognition of templates and their analogues on molecularly imprinted polymer using computational and conformational analysis approaches *J. Mol. Recognit.* **17** 567–74

[65] Lai E P C and Feng S Y 2003 Molecularly imprinted solid phase extraction for rapid screening of metformin *Microchem. J.* **75** 159–68

[66] Wang J, Guo R, Chen J, Zhang Q and Liang X 2005 Phenylurea herbicides-selective polymer prepared by molecular imprinting using *N*-(4-isopropylphenyl)-*N'*-butyleneurea as dummy template *Anal. Chim. Acta* **540** 307–15

[67] Tada M, Sasaki T and Iwasawa Y 2004 Design of a novel molecular-imprinted Rh–Amine complex on SiO$_2$ and its shape-selective catalysis for α-methylstyrene hydrogenation *J. Phys. Chem.* B **108** 2918–30

[68] Li L F, Chen L, Zhang H, Yang Y Z, Liu X G and Chen Y K 2016 Temperature and magnetism bi-responsive molecularly imprinted polymers: preparation, adsorption mechanism and properties as drug delivery system for sustained release of 5-fluorouracil *Mater. Sci. Eng.* C **61** 158–68

[69] Mazouz Z, Rahali S, Fourati N, Zerrouki C, Aloui N, Seydou M, Yaakoubi N, Chehimi M M, Othmane A and Kalfat R 2017 Highly selective polypyrrole MIP-based gravimetric and electrochemical sensors for picomolar detection of glyphosate *Sensors* **17** 2586

[70] Azimi A and Javanbakht M 2014 Computational prediction and experimental selectivity coefficients for hydroxyzine and cetirizine molecularly imprinted polymer based potentiometric sensors *Anal. Chim. Acta* **812** 184–90

[71] Mukawa T, Goto T, Nariai H, Aoki Y, Imamura A and Takeuchi T 2003 Novel strategy for molecular imprinting of phenolic compounds utilizing disulfide templates *J. Pharm. Biomed. Anal.* **30** 1943–7

[72] Pavel D and Lagowski J 2005 Computationally designed monomers and polymers for molecular imprinting of theophylline and its derivatives. Part I *Polymer* **46** 7528–42

[73] Lv Y, Lin Z, Feng W, Zhou X and Tan T 2007 Selective recognition and large enrichment of dimethoate from tea leaves by molecularly imprinted polymers *Biochem. Eng. J.* **36** 221–9

[74] Monti S, Cappelli C, Bronco S, Giusti P and Ciardelli G 2006 Towards the design of highly selective recognition sites into molecular imprinting polymers: a computational approach *Biosens. Bioelectron.* **22** 153–63

[75] Terracina J J, Bergkvist M and Sharfstein S T 2016 Computational investigation of stoichiometric effects, binding site heterogeneities, and selectivities of molecularly imprinted polymers *J. Mol. Model.* **22** 139

[76] Herdes C and Sarkisov L 2009 Computer simulation of volatile organic compound adsorption in atomistic models of molecularly imprinted polymers *Langmuir* **25** 5352–9

[77] Sobiech M, Zolek T, Lulinski P and Maciejewska D 2016 Separation of octopamine racemate on (R,S)-2-amino-1-phenylethanol imprinted polymer—experimental and computational studies *Talanta* **146** 556–67

[78] Khajeh M and Sanchooli E 2011 A pre-concentration procedure employing a new imprinted polymer for the determination of copper in water *Int. J. Environ. Anal. Chem.* **91** 1310–9

[79] Gholivand M B, Shariati-Rad M, Karimian N and Torkashvand M 2012 A chemometrics approach for simultaneous determination of cyanazine and propazine based on a carbon paste electrode modified by a molecularly imprinted polymer *Analyst* **137** 1190–8

[80] Tarley C R T, Segatelli M G and Kubota L T 2006 Amperometric determination of chloroguaiacol at submicromolar levels after on-line preconcentration with molecularly imprinted polymers *Talanta* **69** 259–66

[81] Gore P M, Khurana L, Siddique S, Panicker A and Kandasubramanian B 2018 Ion-imprinted electrospun nanofibers of chitosan/1-butyl-3-methylimidazolium tetrafluoroborate for the dynamic expulsion of thorium (IV) ions from mimicked effluents *Environ. Sci. Pollut. Res.* **25** 3320–34

[82] de Oliveira G F, Hudari F F, Pereira F M V, Zanoni M V B and da Silva J L 2020 Carbon nanotube-based molecularly imprinted voltammetric sensor for selective diuretic analysis of dialysate and hemodialysis wastewater *Chem. Electro. Chem.* **7** 3006–16

[83] Dorraji P S, Noori M and Fotouhi L 2019 Voltammetric determination of adefovir dipivoxil by using a nanocomposite prepared from molecularly imprinted poly(o-phenylenediamine), multi-walled carbon nanotubes and carbon nitride *Microchim. Acta* **186** 427

[84] Hatamluyi B, Hashemzadeh A and Darroudi M 2020 A novel molecularly imprinted polymer decorated by CQDs@HBNNS nanocomposite and UiO-66-NH$_2$ for ultra-selective electrochemical sensing of Oxaliplatin in biological samples *Sensors Actuators* B **307** 127614

[85] Tarley C R T and Kubota L T 2005 Molecularly-imprinted solid phase extraction of catechol from aqueous effluents for its selective determination by differential pulse voltammetry *Anal. Chim. Acta* **548** 11–9

[86] Nezhadali A, Motlagh M O and Sadeghzadeh S 2018 Spectrophotometric determination of fluoxetine by molecularly imprinted polypyrrole and optimization by experimental design, artificial neural network and genetic algorithm *Spectrochim. Acta, Part* A **190** 181–7

[87] Chatterjee T N, Das D, Roy R B, Tudu B, Sabhapondit S, Tamuly P, Pramanik P and Bandyopadhyay R 2018 Molecular imprinted polymer based electrode for sensing catechin (+C) in green tea *IEEE Sens. J.* **18** 2236–44

[88] Valero-Navarro A, Damiani P C, Fernández-Sánchez J F, Segura-Carretero A and Fernández-Gutiérrez A 2009 Chemometric-assisted MIP-optosensing system for the simultaneous determination of monoamine naphthalenes in drinking waters *Talanta* **78** 57–65

[89] Bitar M, Roudaut G, Maalouly J, Brandes S, Gougeon R D, Cayot P and Bou-Maroun E 2017 Water sorption isotherms of molecularly imprinted polymers. Relation between water binding and iprodione binding capacity *React. Funct. Polym.* **114** 1–7

[90] Liu D L, Chen Z B, Du X Y and Liu Z 2017 Study of structural parameters on the adsorption selectivity of a molecularly imprinted polymer *J. Macromol. Sci.* A **54** 622–8

[91] Baggiani C, Anfossi L, Giovannoli C and Tozzi C 2004 Multivariate analysis of the selectivity for a pentachlorophenol-imprinted polymer *J. Chromatogr.* B **804** 31–41

[92] Nantasenamat C, Isarankura-Na-Ayudhya C, Naenna T and Prachayasittikul V 2007 Quantitative structure-imprinting factor relationship of molecularly imprinted polymers *Biosens. Bioelectron.* **22** 3309–17

[93] Nantasenamat C, Naenna T, Ayudhya C I N and Prachayasittikul V 2005 Quantitative prediction of imprinting factor of molecularly imprinted polymers by artificial neural network *J. Comput. Aided Mol. Des.* **19** 509–24

[94] Lorenzo R A, Carro A M, Alvarez-Lorenzo C and Concheiro A 2011 To remove or not to remove? The challenge of extracting the template to make the cavities available in molecularly imprinted polymers (MIPs) *Int. J. Mol. Sci.* **12** 4327–47

[95] Jayan J S, Deeraj B D S, Saritha A and Joseph K 2022 Theoretical modeling and simulation of elastomer blends and nanocomposites *Elastomer Blends and Composites: Principles, Characterization, Advances, and Applications* ed M Rangappa *et al* (Amsterdam: Elsevier) pp 243–67

[96] Chen L 2010 Recent advances in molecular imprinting technology: current status, challenges and highlighted applications *Chem. Soc. Rev.* **40** 2922–42

IOP Publishing

Molecularly Imprinted Polymers for Environmental Monitoring
Fundamentals and applications
Raju Khan and Ayushi Singhal

Chapter 4

The role of nanomaterials in molecularly imprinted polymers for disease diagnosis

Cem Erkmen, Rafia Nimal, Muhammad Siddiq, Afzal Shah and Bengi Uslu

When it comes to human health, early disease identification is crucial for both selecting the best course of action and determining required safeguards to take. Therefore the most cutting-edge analytical methods are needed to produce high-throughput analysis findings. Molecularly imprinted polymers (MIPs) have been combined with various nanomaterials, including carbon nanomaterials, metal oxide nanomaterials, metal nanoparticles, biomaterials, and polymers, to improve their analytical performance. In this chapter MIP sensors based on nanomaterials for disease-specific biomarkers are summarized. The chapter also mentions the advantages that these sensors will offer in the future and the steps that can be taken to improve current capabilities.

4.1 Fundamentals of nanomaterials in molecularly imprinted polymers (MIPs)

Today's chemical and materials production industry has access to methods that enable material manipulation on an atomic or molecular scale to create objects with a diameter of no more than a few nanometers. A nanometer (nm) is 1×10^{-9} meters (a billionth of a meter). Nanotechnology is the name given to the production and processing techniques used to create these materials, and nanomaterials are the finished products. The term nanomaterial is mostly used for materials with a length between 1 and 100 nm [1].

The sciences of chemistry, physics, biology, and engineering all use nanotechnology today. The particles' small size endows them with certain very advantageous properties. One of these characteristics include a significant increase in surface area when a material is ground into particles with a diameter of a few nanometers or less. The rate of any reaction taking place on the material's surface increases as a result of this increase in surface area. Nanomaterials' small size also makes it possible to combine them closely with other materials to improve the material's qualities [1, 2].

The basic properties of nanomaterials such as high electrocatalytic activity, large surface area, high conductivity, and chemical stability increase the use of nanomaterials in sensor design. In addition, adding nanomaterials to sensor designs provides many advantages such as low cost, easy preparation, high sensitivity, selectivity for analytes, short detection time, and long stability [2].

Depending on their characteristics, nanoparticles (NPs) can be categorized in a variety of ways. For instance, spherical, tube, dendrimer, rod, and prism type NPs are categorized according to their shape. Due to their excellent electrical conductivity, high mechanical strength, high chemical stability, optical properties, and low toxicity, carbon nanomaterials or organic NPs containing carbon nanotubes, graphene, carbon nanofibers, and graphene quantum dots and fullerene are trusted nanomaterials for integration into sensor systems. Due to their size, shape, content, structure, and other characteristics as well as their chemical, physical, optical, and optoelectronic properties, inorganic NPs made of metal or metal oxide nanomaterials are frequently utilized in the design of sensors. Noble metal nanomaterials such as gold, silver, and platinum are among the most preferred nanomaterials due to their optoelectronic properties. Additionally, they have great chemical stability, simple production, and good biocompatibility with biomolecules (such as antibodies, aptamers, and enzymes). Magnetic NPs are useful nanomaterials used to isolate the target analyte from different matrices. Because of their distinctive qualities, including high surface area, homogeneous size, low cost, and biocompatibility, these NPs are employed as an alternative in sensor applications. The most often utilized magnetic NPs are made of iron and iron oxide. They can be coupled with different recognition elements to boost their sensitivity for sensor applications. Additionally, magnetic NPs can minimize the need for organic solvents and sample preparation steps [3–5].

Currently, in biological imaging, diagnostics, and sensor design, quantum dots (QDs) are of interest as next-generation fluorescent colloidal NPs. The main benefit of QDs over conventional fluorescent materials in the development of sensors is their size-tuned optical properties, which allow the detection of distinct analytes using varying sizes of QDs. Additionally, in fluorescence studies, while boosting the method's brightness by 100 times compared to organic dyes, they also improved the method's repeatability and sensitivity. Moreover, a successful combination of two nanomaterials (hybrid nanomaterials) not only offers the benefits of each nanomaterial but also boosts sensitivity through an exceptional synergistic effect [6, 7].

4.2 MIP sensors for disease diagnosis purposes

In the development of sensors, high sensitivity, selectivity, and usability are crucial. MIPs serve as the foundation for a significant class of novel sensing devices. MIPs are created as synthetic, molecule-specific receptors for chemical recognition. Due to their high affinity and selectivity to a predefined target molecule, resistance to extreme temperature, pressure, and pH changes, and long-term stability in many applications compared to structurally similar compounds, MIPs have grown in popularity in recent years in the design of new sensors. Additionally, MIPs offer benefits including simplicity, low preparation costs, the potential for reuse, microbial resistance in comparison to biological identification systems, and adaptability to different transducers. Today, a key method for creating synthetic receptors with recognition sites is molecular imprinting technology. The idea behind MIPs is to create artificial substances that can replicate the selective binding of target molecules, mimicking of antibodies and enzymes. In this framework, one attempts to produce artificial voids in synthetic polymers for the recognition of the analyte. The goal of molecular imprinting is to construct synthetic receptors that can recognize the analyte (template/target molecule) virtually as well as biological systems [8, 9].

The thickness of the shielding in the exhibiting polymer, the electron transfer ability between the recognition sites and the electrode surface, and the electro-catalytic corrosion resistance on the electrodes of the templates are factors that affect the protective elements of the sensitivity of MIPs-based sensors. To make more effective MIP sensors, nanomaterials, particularly nanocomposites, are employed as support materials. This method keeps the preserved cavities while increasing the capacity for electron transfer among the nearby electrodes. It has been in particular suggested that combining carbon materials such as graphene and carbon nanotubes with MIPs can result in a considerable increase of the acquired signal for the intended purpose. Therefore, when identifying target biological molecules from biological fluids including serum, plasma, urine, and saliva that contain a wide variety of biological molecules, nanomaterial-based MIPs sensors offer superiority in terms of sensitivity and selectivity [10, 11].

4.3 Assembled nanomaterials on MIPs for improved disease diagnosis purposes

The introduction of nanomaterials in MIP-based materials leads to improved stability of the resulting transducer. Moreover, combining MIPs with various nanocarriers or functional materials, etc, improves its functionality. In diagnostics, this has been largely employed to fabricate novel nanoprobes with wide-ranging therapeutic properties that have huge prospects for early detection and drug treatment of subclinical illnesses [12].

4.3.1 Electrochemical sensors

Various MIP-based electrochemical sensor types are used for disease diagnosis purposes. Among them amperometric and potentiometric sensors are commonly used for these purposes. Amperometric sensors, owing to their simplicity and easy fabrication, have been the most employed sensors. Amperometric sensors based on MIPs measure the rate at which electrochemically active analytes move mass to an electrode coated with MIPs [13]. Potentiometric sensors measure the difference in potential between the working electrode and a reference electrode under conditions of near-zero current flow [14]. The selectivity of these sensors can be improved by

MIP-based ion exchangers or neutral carriers. Since transistors are semiconductor devices that are sensitive to the surface electric gradient or charge at the gate electrode, chemically sensitive field-effect transistors can also be added to these devices [15]. The silicon chip's surface experiences a binding event between the analyte template and the MIPs that shifts the surface potential and changes the current, enabling the monitoring of the reaction's rate. When an MIP is recognized for the related analyte, conductometric sensors evaluate the time-dependent change in conductivity that results. Impedance is the overall electrical resistance to the passage of an alternating current through a certain medium while observing the sensor response brought on by an MIP. It usually results in a decrease in impedance and an increase in conductivity and capacitance.

The incorporation of NPs or nanostructured coatings improves the sensitivity and selectivity of electrochemical sensors to a great extent [16]. Nanocarbon based assemblies such as graphene demonstrate unique mechanical and electrical properties. In comparison to other carbon-based nanomaterials such as carbon nanotubes and carbon dots (CDs), graphene has a number of advantages thanks to its large surface area, biocompatibility, and ease of manufacture. Lately, the use of graphene in the field of sensor devices has risen steeply [17]. The adsorption of molecules causes an increase in the carrier concentration, which enhances the conduction through a nanochannel used in the detecting method. In study performed by Bai *et al* to create an MIP sensor for artemisinin detection, *in situ* polymerization of a monomer mixture was carried out on the surface of a graphene-modified glassy carbon electrode (GCE). Under optimal conditions, the developed sensor demonstrated good selectivity, sensitivity, and cross-reactivity against its analogs. The developed sensor, with superior reusability and stability, determined that the limit of detection (LOD) for artemisinin was 2.0 nM [18]. Sol–gel MIPs with multiwalled carbon nanotubes (MWCNTs) can also be used to improve sensor signals. MWCNTs are excellent nanomaterials because to their extraordinarily large surface area, strong adsorptive ability, and unique electron transfer capabilities [18].

Among carbon nanomaterials, graphene oxide (GO) has a high density of functional groups that can be functionalized. Reduced GO (rGO), which provides good conductivity and keeps the surface for functionalization, can be created by electrochemically reducing GO. Additionally, MIPs have been employed in conjunction with these characteristics of GO and rGO to improve the sensors' analytical performance [19]. Hexacyanoferrate (HCF) and β-cyclodextrin (β-CD)-functionalized rGO linked MIP sensor design was produced by Goyal *et al* (figure 4.1). The prepared sensor was used to determine the cortisol level, which is utilized for the early detection and prevention of chronic diseases. In particular, rGO, which increases the electroactive surface area, makes CDs immobilized on surfaces more effectively, and the prepared sensors exhibit more sensitive results. The prepared PPy-CD-rGO/GCE MIPs was effective for detecting cortisol with a LOD of 19.3 pM in biological samples, with a linear response to logarithmic cortisol concentration throughout a large range from 5 pg ml^{-1} to 5000 ng ml^{-1} [20].

Gold nanoparticles (AuNPs) can be a great option to increase the sensitivity of electrochemical sensors. They can amplify the electrode surface by increasing the

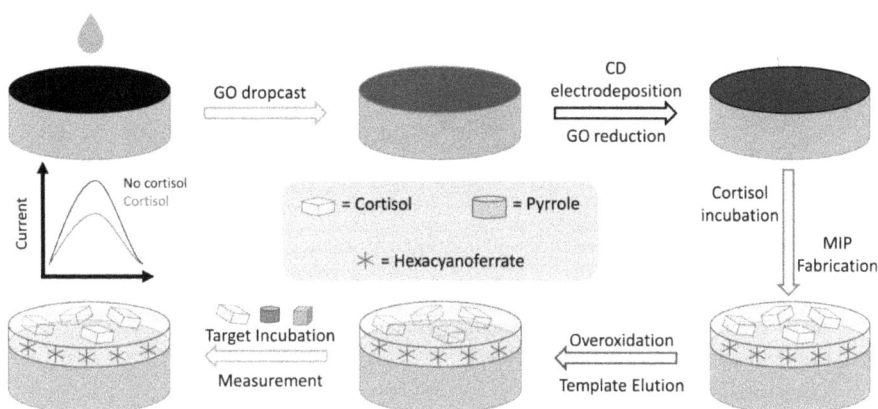

Figure 4.1. Schematic of cortisol MIP sensor fabrication based on electropolymerization of pyrrole on CD rGO/GCE doped with cortisol and HCF. (Reproduced with permission from [20]. Copyright 2022 American Chemical Society.)

electron transfer process between electroactive species and accelerating electrochemical processes [21]. It was reported that a conductive 3D network composite film was produced after polymerization of a nanoengineered voltammetric MIP sensor based on AuNPs covered with 3-thiophene acetic acid for the detection of adenine. This sensor fabrication exhibited high selectivity and sensitivity for adenine detection with an LOD value of 0.99 nM [22].

In addition to having excellent conductivity, AuNPs also provide a large specific surface area. In sensors design, by increasing the loading amount of protein and ionic liquid monomer, this can aid to improve the imprinting process' effectiveness and increase the sensitivity of the imprinted sensor. Wu *et al* developed an MIP-based sensor for the ultra-sensitive detection of the liver cancer biomarker alpha-fetoprotein (AFP) on an electrode surface modified with AuNPs and using an ionic liquid as a functional monomer. Electrochemical deposition of AuNPs on the electrode surface was achieved by cyclic voltammetry technique using a chloroauric acid precursor. Then, the sensor design was completed by following the steps of polymerization of the ionic liquid on the surface with the self-assembling process technique, removal of the AFP molecules from the surface, and rebinding of the AFP molecules (figure 4.2). The MIP sensor exhibited a good linear response to AFP in the concentration range from 0.03 ng ml^{-1} to 5 ng ml^{-1} while operating under optimal conditions. Moreover, the LOD level was found to be 2 pg ml^{-1} [23].

Redox indicators have traditionally been used in conjunction with MIPs for indirect detection of analyte in solution. A decrease in the absorptivity of redox indicators such as ferricyanide or ferrocene caused by the analyte binding to an MIP-coated electrode correlates with analyte concentration. For instance, the concentration of aspartic acid has been determined by using a template to displace ferricyanide that has been adsorbed in MIP binding sites [24]. A specific interaction that modifies the redox probe's diffusion rate, which can be measured as a function of the Faradaic current, may produce morphological changes in the polymer structure.

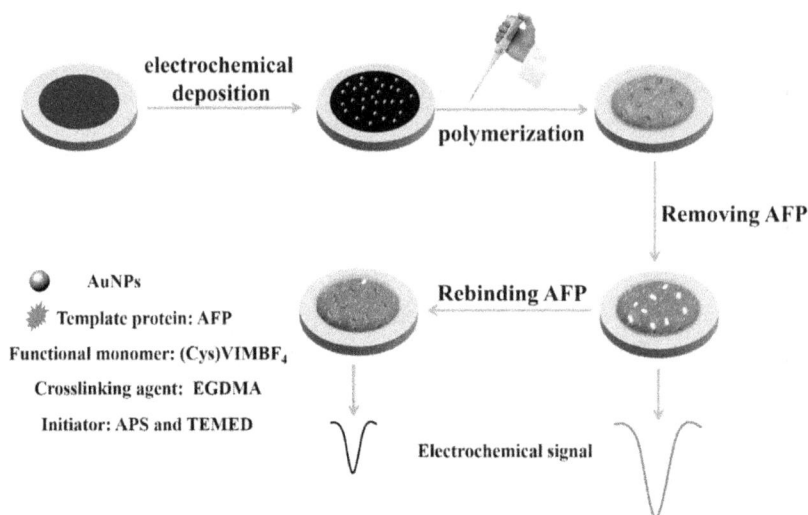

Figure 4.2. Schematic illustration of the alpha-fetoprotein (AFP) imprinted sensor fabrication and electrochemical responses. (Reproduced with permission from [23]. Copyright 2019 the authors. CC BY 4.0.)

In a different work, scopoletin electropolymerization on gold electrodes was used to create an MIP nanofilm sensor for transferrin detection [25]. To monitor how the sensor response varied based on the permeability changes of the MIP film and the redox couple ferri/ferrocyanide, cyclic voltammetry (CV), square-wave voltammetry, and surface plasmon resonance (SPR) were used. The ferrocene functionalized vinyl group was used by Udomsap and colleagues as a monomer capable of establishing aromatic stacking interactions with a benzo[*a*]pyrene (BaP) template [25]. Due to the environment-responsive reversible oxidation to a ferricenium ion, these interactions caused noticeable changes in the ferrocene's redox characteristics after BaP detection. With this technique, BaP concentrations as low as 90 nM could be found [25].

MIP sensors made of nanomaterials are very helpful for biological samples. How to properly quantify the linear range of 1 nM to 1 mM of cocaine in blood serum using a cocaine potentiometric sensor based on MIP NPs has been described. A temperature-sensitive MIP electrochemical sensor for the detection of bovine serum albumin (BSA) was also proposed by Wei and colleagues. The sensor operated on a GCE-mounted thermoresponsive memory hydrogel. The MIP underwent a reversible morphological change in response to an external temperature stimulation, and this change was observed by CV and electrochemical impedance spectrometry (EIS) in an aqueous solution [26]. The proposed BSA sensor also performed admirably with milk samples and had high recovery rates [26].

MIP magnetic composites (Fe_3O_4–MIPs) are applicable for a wide range of molecules. Applications of MIP magnetic composites for 17-β-estradiol have recently been published [27]. The 17-β-estradiol oxidation current was amplified using square-wave voltammetry in the Fe_3O_4–MIP sensor's operating mode to an LOD of 20 nM [27]. Another study claimed the development of a disposable graphite paste electrode-based MIP sensor to track blood heparin levels. The voltammetric sensor demonstrated great selectivity in a range of 0–8 unit ml^{-1} of heparin and negligible cross-reactivity when tested in physiological saline and bovine whole blood with 5 mM ferrocyanide [28].

Screen-printed electrodes (SPEs) and disposable plastic fiber optics have been discovered to work with MIP sensors. Sensors created using modified MIP on a gold SPE surface were used to detect the cancer biomarkers vascular endothelial growth factor (VEGF) and epidermal growth factor receptor (EGFR). For the EGFR and VEGF detection with picomolar LOD level, a potentiometric stripping analysis method was used. The SPE-based sensor could be included into microfluidic analysis and lab-on-a-chip platforms [29]. Considering the roles of nanomaterials in the design of MIP-based electrochemical sensors, selected studies for biomarkers of various diseases from different biological samples are presented in table 4.1. Sensor design, analysis conditions, and some important validation characteristics are presented in this table.

4.3.2 Optical sensors

Optical sensors work on the principle of measuring optical properties (photons, etc) and then converting them into measurable electronic signal (electrons). MIP-based sensors yield an optical shift during the binding event, which is then transduced into an electronic signal and measured. Based on their function, MIP optical sensors can be divided into two categories. The first category is MIP-affinity sensors while the second category is optoelectronic MIP sensors. MIP-affinity sensors are largely employed for the analytes possessing endogenous optical properties in the form of refractive index, optical absorbance, or fluorescence, etc. A change in these optical properties is monitored to obtain an insight of a binding event, for example fluorescence quenching. The working mechanism of optoelectronic MIP sensors, on the other hand, depends on the optically active reporter monomers that are able to sense changes in their surroundings and react to the presence of analyte. Optoelectronic MIP molecular reporters must possess considerably large molar absorption coefficients and greater quantum yields. Moreover, molecular reporters ought to be photochemically and thermally stable, producing an optical response dependent on the concentration of the target analyte and not on the broad-based side reactions [60, 61].

Incorporation of various nanomaterials improve the properties of MIP-based detectors. Fluorescence monitoring, owing to its simple set-up and low LOD, has become a popular technique in recent years. The incorporation of fluorescent CDs in the microgel matrix enhances its LOD. CDs not only exhibit tremendous photo-stability but are also chemically inert and possess exceptional electron–hole transfer and thus are suitable fluorophores for biosensing. In a work where single-pot free

Table 4.1. Key recent examples of nanomaterials assembled on MIPs for improved disease diagnosis purposes.

Target analyte	Related disease	Sensing platform	Method	Application	Linear range	LOD	Reference
SARS-CoV-2	SARS	Au-TEF/ncovNP-MIP	DPV	Nasopharyngeal swab	2.22–111 fM	15 fM	[30]
HIV-1	Herpes simplex virus infection	GCE/MIP/CNF-Bi	DPV	Plasma	0.002–0.05 ng ml^{-1}	0.0003 ng ml^{-1}	[31]
Dengue	Dengue fever	SPCE/MIP	EIS	Serum	50–200 μg l^{-1}	29.3 μg l^{-1}	[32]
FMDV-SAT-2	Foot-and-mouth disease virus serotype South-Africa territories-2	SAT-2 VIP	LSV	Clinical samples	0.04–0.96 ng ml^{-1}	0.1 ng ml^{-1}	[33]
C-reactive protein (CRP)	Cardiovascular disease	WS$_2$-doped peptide-imprinted polymer-coated electrodes	CV	Serum	1.0 pg ml^{-1}–1.0 ng ml^{-1}	NS	[34]
Insulin	Diabetes	G-quadruplex/redox label-modified electrode	Amperometry	NS	NS	0.02 μg ml^{-1}	[35]
Cardiac troponin I	Myocardial damage	MIP–MWCNT–GS–GCE	CV–DPV	Human serum	0.005–60 ng cm^{-3}	0.8 pg cm^{-3}	[36]
HIV-p24	AIDS	MIP–MWCNT–GCE	CV–DPV	Human serum	0.0001–2 ng cm^{-3}	0.083 pg cm^{-3}	[37]
Insulin	Diabetes	EMMIP–GCE	CV–DPV	Human serum	0.01–1 nM	3–17 pM	[38]
Ganciclovir	Cytomegalovirus (CMV) infection	Fe$_3$O$_4$/MWCNT/GCE	DPAS voltammetry	Serum, urine	0.05–500 μM	0.0015 μM	[39]
Metronidazole	Skin infections	MIP–MWCNT–GCE	CV–DPV	Tablet, fish tissue	0.171–205 μg l^{-1}	49.2 ng l^{-1}	[40]

Analyte	Disease/application	Electrode/material	Method	Sample	Linear range	LOD	Ref.
Naproxen	Stroke	MIP–MWCNTs–EME	CV–DPV	Urine, plasma, wastewater	1–1000 μg l^{-1}	0.3 μg l^{-1}	[41]
Salbutamol	COPD	MIP/Ag–N–RGO	CV–DPV	Urine, pork	0.03–20 μM	7 nM	[42]
Human blood biomarker	—	Amino-modified truncated 2′ fluoro-RNA aptamer	Dielectric sensor	Human plasma	NS	40 fM	[43]
Paracetamol	—	(P(NVC-EHA-AA), PNEA) containing carbazole group	CV–DPV	Tablet, human urine	1 μM–0.1 mM	0.3 μM	[44]
Indoxyl sulfate	Chronic kidney diseases	PANI-G-NiS$_2$ MIP/ITO electrode	DPV	Human serum, urine	1.0 pM–6.0 mM	0.286 pg ml^{-1}	[45]
Dopamine	Neurological disorders	S-MoSe$_2$/NSG/Au/MIPs/GCE	DPV	Human serum	0.05–1000 μM	0.02 μM	[46]
L-tryptophan	Alzheimer's, Parkinson's, and some cancers	AuNPs@PVP@SiO$_2$MIP/Gr	LSV	Solution for infusion	1–350 μM	0.3 μM	[47]
Glutathione	Kwashiorkor, seizure, Alzheimer's and Parkinson's diseases	m-GEC/MIP	DPV	Rat liver samples	1–10 μM and 10–1400 μM	7 nM	[48]
Neuron specific enolase	Cancer	AuNP/MIP	SWV	Serum	25–4000 pg ml^{-1}	25 pg ml^{-1}	[49]
Cortisol	Post-traumatic stress disorders, Cushing's disease, and chronic fatigue	AuNP/MIP	SWV	Artificial saliva	1 pM–500 nM	200 fM	[50]

(Continued)

Table 4.1. (*Continued*)

Target analyte	Related disease	Sensing platform	Method	Application	Linear range	LOD	Reference
Bilirubin	Liver function and liver toxicity	MIP/AuFe$_2$O$_3$-GrCNT/GCE	DPV	Saliva and serum	3.7–130 nM	1.54 nM	[51]
2-aminoadipic acid	Diabetes	Si-GO/MIP	DPV	Serum	0.10×10^{-11}–3.00×10^{-11} M	0.45×10^{-11} M	[52]
SARS-CoV-2-RBD	Coronavirus disease	MIP/MP-Au-SPE	EIS	Saliva	2.0–40.0 pg ml^{-1}	0.7 pg ml^{-1}	[53]
Uric acid	Gout, renal disease, and Lesch–Nyhan syndrome	AuNPs @PVP@SiO$_2$MIP/GCE	DPV	Urine	5–100 μM and 100–1000 μM	0.4 μM	[54]
Follicle-stimulating hormone	Polycystic ovary syndrome	NiCo$_2$O$_4$/rGO/MIP	EIS	Clinical samples	0.1 pM–1 μM	0.1 pM	[55]
Lactic acid	Cancer	AuE/rGO-AgNPs/MIP	CV	Serum	10–250 μM	0.726 μM	[56]
CA 15-3	Breast cancer	CNE/AuNP/MIP	Chronoamperometry	Serum	5–35 U ml^{-1}	1.16 U ml^{-1}	[57]
Tryptophan	Schizophrenia, Alzheimer's, Parkinson's diseases and depressions	MIP/CuCo$_2$O$_4$@BC/GCE	DPV	Serum, urine	0.01–1 μM and 1–40 μM	0.003 μM	[58]
Sarcosine	Prostate cancer	Fe$_3$O$_4$@ZIF-8@MIP	CV	Urine	1–100 pM	0.4 pM	[59]

Abbreviations: BC: biomass carbon; CNT: carbon nanotubes; CNE: carbon nanotube electrode; COPD: chronic obstructive pulmonary disorder; CV: cyclic voltammetry; EIS: electrochemical impedance spectroscopy; GCE: glassy carbon electrode; GO: graphene oxide; AuNPs: gold nanoparticles; LSV: linear sweep voltammetry; MP-Au-SPE: macroporous gold screen-printed electrode; m-GEC: magnetic graphite–epoxy composite; MWCNTs: multiwalled carbon nanotubes; NS: not stated; PANI-G: polyaniline-graphene; SPE: screen-printed electrode; Si-GO: silylated graphene oxide; AgNPs: silver nanoparticles; AuNPs@PVP@SiO$_2$: polysilicate on polyvinylpyrrolidone coated gold nanoparticles; S-MoSe$_2$/NSG: MoSe$_2$ with N, S co-doped graphene; SIP: surface-imprinted polymer; CNF-Bi: carbon nanofragment and bismuth oxides; PS: polysulfone; PVP: polyvinylpyrrolidone; SPCE: screen-printed carbon electrodes; WS$_2$: tungsten disulfide; SARS: severe acute respiratory syndrome; SWV: square-wave voltammetry; ZIF-8: zeolitic imidazolate framework-8.

radical polymerization was used to combine nontoxic CDs during the imprinting process, the resulting hybrid MIP gels exhibited superior sensitivity and selectivity for glucose detection compared to non-imprinted polymer (NIP) hybrid microgels over chosen range of 0–30 mM under physiological conditions [62, 63].

In recent years an increased interest has arisen in combining QDs to MIPs to develop optical sensors as QDs possess unique intrinsic optical properties such as widespread absorption spectra with constricted emission, thus making them potential candidates for the optical detection of numerous analytes. QD-MIP-based sensors are superior to QD-based immunosensors owing to their chemical stability, cost effectiveness, and ability to detect smaller sized analytes. Consequently, QD-MIPs have flourished in the field of bioanalysis and pharmaceutical analysis [64].

QD-MIPs have been utilized in one study to develop a sensitive fluorescence detection probe by radical polymerization of acrylamide and bisacrylamide for the detection of myoglobin. The sensitivity of this MIP-QD probe was up to femtomolar concentrations of myoglobin in human samples, below the cut-off values for regulating myocardial infarction. Another study reported a thermoresponsive MIP fluorescent nanoprobe NP sensor for trypsin prepared by solid-phase synthesis. The resulting fluorescent nano-MIPs showed a nanomolar LOD in phosphate buffer and urine samples with no cross-reactivity [65].

Magnetic NPs (Fe_3O_4 MNPs) have gained attention recently not only for sample pretreatment and purification but also for their capacity to improve sensor performance. Nanocomposites containing Fe_3O_4 MNPs and QDs facilitate magnetic separation and provide a fast fluorescent response due to their fast electron transfer rate. In this way, the analysis and detection process is greatly simplified and it can improve the detection performance and sensitivity. In addition, combining these nanocomposites with MIPs increases the binding kinetics, binding capacity, and specificity of biological molecules. A highly selective analysis of molecularly imprinted molecules can be achieved in a single compound system. In their study, Zhang *et al* proposed a fluorescence sensor based on MNP/QD@MIPs for the detection of lysozyme (figure 4.3). Under optimal conditions, the developed MIP-based sensor provided good linearity over a lysozyme concentration range of 0.2–2.0 μM and a detection limit of 4.53×10^{-3} μM for lysozyme [66].

Considering the roles of nanomaterials in the design of MIP-based optical sensors for biomarkers and drugs of various diseases from different biological samples, selected studies are presented in table 4.2. Sensor designs, analysis conditions, and some important validation characteristics are presented in this table.

4.4 Conclusion, challenges, and prospects

Many diseases brought on by an expanding global population threaten public health and incur significant financial expenditures. Traditional methods can be used in regular analysis for disease diagnosis, but their utility is constrained by their complexity and restricted availability. Therefore, this necessitates the development of sensitive analytical methods for the early and sensitive determination of diseases to reduce the burden of diseases caused by many adverse conditions such as viruses,

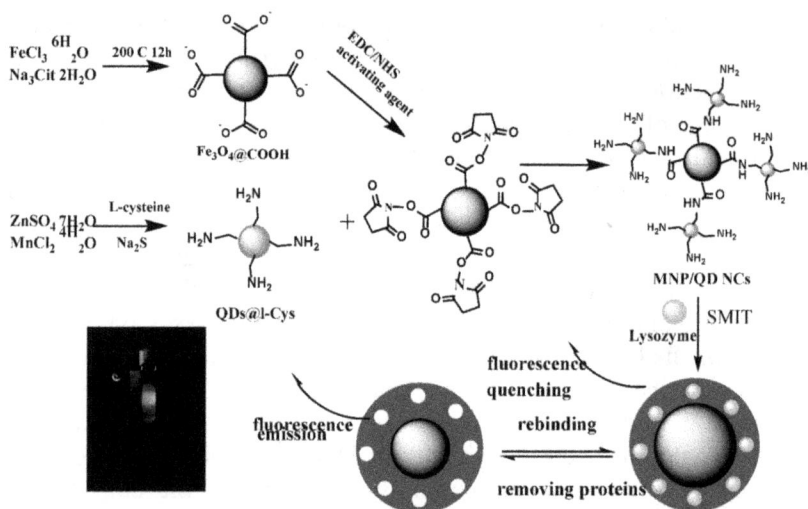

Figure 4.3. Schematic of the synthesis of an MNP/QD@MIP-based sensor. SMIT: surface molecular imprinting technology. (Reproduced with permission from [66]. Copyright 2021 the authors. CC BY 4.0.)

bacteria, and malnutrition and to protect the quality of life of people. Since MIP-based sensors have excellent selectivity, sensitivity, ease of application, portability, and low cost, they may be able to solve this issue. The nanomaterials' excellent physicochemical properties, sizes, dimensions, conductivity, and morphologies make them extremely particular, and selective materials for sensing platforms. The benefits provided by nanomaterial supported MIP sensors are also helpful for the analysis of biological substances. Due to their affordability, chemical stability, and biocompatibility, metallic nanoparticles are frequently used in electrochemical and optical approaches. The combined nanomaterials, which are created by combining several materials and giving them improved capabilities, produce more accurate analytical results and more stable biorecognition components on the electrode surfaces or in the solution environment for MIP sensors, according to literature studies. It is possible to create quick sensing devices that are portable, sensitive, accurate, cost-effective, and simple to use in terms of functionality and miniaturization by combining these MIP sensors with various methods and microfluidic devices.

Nucleic acid amplification methods or better immuno-recognition protocols that provide more selective results must be developed for analyses of biomarkers for disease diagnosis. Additionally, nanomaterial-based MIP sensors can be included in microfluidic systems. However, this might also make the systems more complex. Therefore, the way that samples are prepared today generally presents problems. In the future, this issue can be solved by creating multifunctional devices by merging sample preparation techniques with MIP sensors based on nanomaterials. Consequently, further work is required to turn sensors that offer automated, quick, or *in situ* detection analysis of biomarkers into commercially viable products for both the economy and public health.

Table 4.2. Key recent examples of analyte detection by optical MIP sensors.

Target analyte	Sensing material	Detection method	Linear range	LOD	Ref.
Paracetamol	FMIP nanoparticles	Fluorescence	4–1000 μM	1 μM	[67]
Diclofenac	SPR-2 gold sensor chips	SPR	1.24–80 ng ml^{-1}	0.8 ng ml^{-1}	[68]
Prednisolone	Molecularly imprinted nanoparticles	SPR	NS	5 ppb	[69]
Dopamine	Si-MG-MIP-CL	Optical	8.0–200.0 ng ml^{-1}	1.5 ng ml^{-1}	[70]
Galectin-3	Au/MPS/(PDDA/GO)$_n$	SPR	10.0–50.0 ng ml^{-1}	2.0 ng ml^{-1}	[71]
Rutin	BA-CdTe@MIPs QDs	Fluorescence	0.1–30 μM	0.04 μM	[72]
Diazepam and its metabolites	MIP Mn-doped ZnS QDs	Fluorescence	0.3–250 μM	8.78×10^{-2} μM	[73]
Mycotoxins type AFB1	Coupling nanocarbon and Nb	Direct non-competitive detection technique	0.01–100 ng ml^{-1}	3.3 pg ml^{-1}	[74]
Mycotoxins OTA	Gold nanorod (GNR)	LSPR	lower than 1 nM	NS	[75]
Zearalenone	MIPPy film was prepared by electropolymerization of pyrrole onto the bare Au chip in the presence of a template zearalenone molecule	SPR	0.3–3000 ng ml^{-1}	0.3 ng ml^{-1}	[76]
Low molecular weight mycotoxins	Composite films of Au–MIP	SPR	1 nM–1 mM	NS	[74]
Mycotoxins AFB1	Aflatoxin B1–acetylcholinesterase	SPR	NS	0.008 μM	[77]
Trypsin	Carbon dots anchored SiO$_2$ nanoparticles surface modified with disulfanylpropanoic acid and vinyl groups	SPR	NS	9.8 nM	[78]
Tetracycline	CdS QDs	Fluorescence detection	0.05–10 μM	5 nM	[79]
CA19–9	MIP@QDs	Fluorescence detection	2.76×10^{-2}–5.23 $\times 10^2$ U ml^{-1}	5.83×10^{-4} U ml^{-1}	[80]
Interleukin-2	MIP-QDs	Fluorescence detection	35 fg ml^{-1}–39 pg ml^{-1}	5.91 fg ml^{-1}	[81]
Dopamine	GQD/Pin-Bac@MIPs	Fluorescence detection	5×10^{-9}–1.2 $\times 10^{-6}$ M	2.5 nM	[82]
Cortisol	AuNP-MIP	SPR	0.01–100 ppb	0.0087 ppb	[83]
Lysozyme	MNP/QD@MIPs	Fluorescence detection	0.2–2.0 μM	4.53×10^{-3} μM	[66]

Abbreviations: MNP: magnetic nanoparticles; Si-MG-MIP-CL: graphene oxide-molecular imprinted polymer silanized Fe$_3$O$_4$ nanoparticles; Gal3: galectin-3; Au/MPS: thiolated Au surface modified by self-assembling; PDDA: poly(diallyldimethylammonium chloride); Pin-Bac: poly(indolylboronic acid); GO: graphene oxide; 3ABA: 3-aminephenylboronic acid; SPR: surface plasmon resonance; BA-CdTe@MIPs QDs: boronate affinity-based surface-imprinted CdTe quantum dots; NS: not stated.

References

[1] Khan I, Saeed K and Khan I 2019 Nanoparticles: properties, applications and toxicities *Arab. J. Chem.* **12** 908–31

[2] Sudha P N, Sangeetha K, Vijayalakshmi K and Barhoum A 2018 Nanomaterials history, classification, unique properties, production and market *Emerging Applications of Nanoparticles and Architectural Nanostructures: Current Prospects and Future Trends* (Amsterdam: Elsevier)

[3] Lu X, Wang C and Wei Y 2009 One-dimensional composite nanomaterials: synthesis by electrospinning and their applications *Small* **5** 2349–70

[4] Bai H, Li C and Shi G 2011 Functional composite materials based on chemically converted graphene *Adv. Mater.* **23** 1089–115

[5] Dave S, Das J and Ghosh S 2022 Advanced nanomaterials for point of care diagnosis and therapy *Advanced Nanomaterials for Point of Care Diagnosis and Therapy* (Amsterdam: Elsevier)

[6] Bozal-Palabiyik B, Kurbanoglu S, Erkmen C and Uslu B 2021 Future prospects and concluding remarks for electroanalytical applications of quantum dots *Electroanalytical Applications of Quantum Dot-Based Biosensors* (Amsterdam: Elsevier) pp 427–50

[7] Aftab S, Shah A, Erkmen C, Kurbanoglu S and Uslu B 2021 Quantum dots: synthesis and characterizations *Electroanalytical Applications of Quantum Dot-Based Biosensors* (Amsterdam: Elsevier)

[8] Bozal-Palabiyik B, Erkmen C and Uslu B 2019 Molecularly imprinted electrochemical sensors: analytical and pharmaceutical applications based on ortho-phenylenediamine polymerization *Curr. Pharm. Anal* **16** 350–66

[9] Lahcen A A and Amine A 2019 Recent advances in electrochemical sensors based on molecularly imprinted polymers and nanomaterials *Electroanalysis* **31** 188–201

[10] Rahman S *et al* 2022 Molecularly imprinted polymers (MIPs) combined with nanomaterials as electrochemical sensing applications for environmental pollutants *Trends Environ. Anal. Chem.* **36** e00176

[11] Chen L, Xu S and Li J 2011 Recent advances in molecular imprinting technology: current status, challenges and highlighted applications *Chem. Soc. Rev.* **40** 2922–42

[12] Poma A, Turner A P F and Piletsky S A 2010 Advances in the manufacture of MIP nanoparticles *Trends Biotechnol* **28** P629–37

[13] Lange U, Roznyatovskaya N V and Mirsky V M 2008 Conducting polymers in chemical sensors and arrays *Anal. Chim. Acta.* **614** 1–26

[14] Wadie M, Marzouk H M, Rezk M R, Abdel-Moety E M and Tantawy M A 2022 A sensing platform of molecular imprinted polymer-based polyaniline/carbon paste electrodes for simultaneous potentiometric determination of alfuzosin and solifenacin in binary co-formulation and spiked plasma *Anal. Chim. Acta.* **1200** 339599

[15] Iskierko Z *et al* 2015 Extended-gate field-effect transistor (EG-FET) with molecularly imprinted polymer (MIP) film for selective inosine determination *Biosens. Bioelectron.* **74** 526–33

[16] Zhu C, Yang G, Li H, Du D and Lin Y 2015 Electrochemical sensors and biosensors based on nanomaterials and nanostructures *Anal. Chem.* **87** 230–49

[17] Hao Z *et al* 2021 An intelligent graphene-based biosensing device for cytokine storm syndrome biomarkers detection in human biofluids *Small* **17** 2101508

[18] Bai H, Wang C, Chen J, Peng J and Cao Q 2015 A novel sensitive electrochemical sensor based on *in situ* polymerized molecularly imprinted membranes at graphene modified electrode for artemisinin determination *Biosens. Bioelectron.* **64** 352–8

[19] Tang Q, Zhou Z and Chen Z 2013 Graphene-related nanomaterials: tuning properties by functionalization *Nanoscale* **5** 4541–83

[20] Goyal A and Sakata T 2022 Development of a redox-label-doped molecularly imprinted polymer on β-cyclodextrin/reduced graphene oxide for electrochemical detection of a stress biomarker *ACS Omega* **7** 33491–9

[21] Pingarrón J M, Yáñez-Sedeño P and González-Cortés A 2008 Gold nanoparticle-based electrochemical biosensors *Electrochim. Acta.* **53** 5848–66

[22] Wang X, Zheng Y and Xu L 2018 An electrochemical adenine sensor employing enhanced three-dimensional conductivity and molecularly imprinted sites of Au NPs bridged poly(3-thiophene acetic acid) *Sensors Actuators* B **255** 2952–8

[23] Wu Y, Wang Y, Wang X, Wang C, Li C and Wang Z 2019 Electrochemical sensing of α-fetoprotein based on molecularly imprinted polymerized ionic liquid film on a gold nano-particle modified electrode surface *Sensors* **19** 3218

[24] Ozcelikay G *et al* 2019 Electrochemical MIP sensor for butyrylcholinesterase *Polymers* **11** 1970

[25] Udomsap D *et al* 2018 Electrochemical molecularly imprinted polymers as material for pollutant detection *Mater. Today Commun* **17** 458–65

[26] Wei Y *et al* 2022 Self-cleaning electrochemical protein-imprinting biosensor with a dual-driven switchable affinity for sensing bovine serum albumin *Talanta* **237** 122893

[27] Lahcen A A, Baleg A A, Baker P, Iwuoha E and Amine A 2017 Synthesis and electro-chemical characterization of nanostructured magnetic molecularly imprinted polymers for 17-β-estradiol determination *Sensors Actuators* **241** 698–705

[28] Yoshimi Y, Yagisawa Y, Yamaguchi R and Seki M 2018 Blood heparin sensor made from a paste electrode of graphite particles grafted with molecularly imprinted polymer *Sensors Actuators* B **259** 455–62

[29] Johari-Ahar M, Karami P, Ghanei M, Afkhami A and Bagheri H 2018 Development of a molecularly imprinted polymer tailored on disposable screen-printed electrodes for dual detection of EGFR and VEGF using nano-liposomal amplification strategy *Biosens. Bioelectron.* **107** 26–33

[30] Raziq A, Kidakova A, Boroznjak R, Reut J, Öpik A and Syritski V 2021 Development of a portable MIP-based electrochemical sensor for detection of SARS-CoV-2 antigen *Biosens. Bioelectron.* **178** 113029

[31] Ma Y, Liu C, Wang M and Wang L S 2019 Sensitive electrochemical detection of gp120 based on the combination of NBD-556 and gp120 *Talanta* **196** 486–92

[32] Zhang C *et al* 2019 A disposable electrochemical sensor based on electrospinning of molecularly imprinted nanohybrid films for highly sensitive determination of the organotin acaricide cyhexatin *Microchim. Acta.* **186** 504

[33] Hussein H A, El Nashar R M, El-Sherbiny I M and Hassan R Y A 2021 High selectivity detection of FMDV-SAT-2 using a newly-developed electrochemical nanosensors *Biosens. Bioelectron.* **191** 113435

[34] Liu K H *et al* 2022 Sensing of C-reactive protein using an extended-gate field-effect transistor with a tungsten disulfide-doped peptide-imprinted conductive polymer coating *Biosensors* **12** 31

[35] Lian K *et al* 2022 Insulin quantification towards early diagnosis of prediabetes/diabetes *Biosens. Bioelectron.* **203** 114029

[36] Ma Y *et al* 2017 MIPs-graphene nanoplatelets-MWCNTs modified glassy carbon electrode for the determination of cardiac troponin I *Anal. Biochem.* **520** 9–15

[37] Ma Y, Shen X L, Zeng Q, Wang H S and Wang L S 2017 A multi-walled carbon nanotubes based molecularly imprinted polymers electrochemical sensor for the sensitive determination of HIV-p24 *Talanta* **164** 121–7

[38] Ansari S 2017 Combination of molecularly imprinted polymers and carbon nanomaterials as a versatile biosensing tool in sample analysis: recent applications and challenges *TrAC-Trends Anal. Chem.* **93** 134–51

[39] Paimard G, Gholivand M B and Shamsipur M 2016 Determination of ganciclovir as an antiviral drug and its interaction with DNA at Fe_3O_4/carboxylated multi-walled carbon nanotubes modified glassy carbon electrode *Meas. J. Int. Meas. Confed* **77** 269–77

[40] Liu Y *et al* 2015 Fabrication of highly sensitive and selective electrochemical sensor by using optimized molecularly imprinted polymers on multi-walled carbon nanotubes for metronidazole measurement *Sensors Actuators* B **206** 647–52

[41] Tahmasebi Z, Davarani S S H and Asgharinezhad A A 2016 An efficient approach to selective electromembrane extraction of naproxen by means of molecularly imprinted polymer-coated multi-walled carbon nanotubes-reinforced hollow fibers *J. Chromatogr.* A **1470** 19-26

[42] Li J *et al* 2017 Ag/N-doped reduced graphene oxide incorporated with molecularly imprinted polymer: an advanced electrochemical sensing platform for salbutamol determination *Biosens. Bioelectron.* **90** 210–6

[43] Krishnan H, Gopinath S C B, Arshad M K M, Zulhaimi H I, Anbu P and Subramaniam S 2022 Molecularly imprinted polymer enhances affinity and stability over conventional aptasensor for blood clotting biomarker detection on regimented carbon nanohorn and gold nanourchin hybrid layers *Sensors Actuators* B **363** 131842

[44] Luo J, Ma Q, Wei W, Zhu Y, Liu R and Liu X 2016 Synthesis of water-dispersible molecularly imprinted electroactive nanoparticles for the sensitive and selective paracetamol detection *ACS Appl. Mater. Interfaces* **8** 21028–38

[45] Dalal N *et al* 2022 MIP-based sensor for the detection of gut microbiota-derived indoxyl sulphate using PANI–graphene–NiS_2 *Mater. Today Chem.* **26** 101157

[46] Zang Y *et al* 2020 Fabrication of S-$MoSe_2$/NSG/Au/MIPs imprinted composites for electrochemical detection of dopamine based on synergistic effect *Microchem. J.* **156** 104845

[47] Rezaei F, Ashraf N and Zohuri G H 2023 A smart electrochemical sensor based upon hydrophilic core–shell molecularly imprinted polymer for determination of L-tryptophan *Microchem. J.* **185** 108260

[48] Santos A C F *et al* 2021 Development of magnetic nanoparticles modified with new molecularly imprinted polymer (MIPs) for selective analysis of glutathione *Sensors Actuators* B **344** 130171

[49] Pirzada M, Sehit E and Altintas Z 2020 Cancer biomarker detection in human serum samples using nanoparticle decorated epitope-mediated hybrid MIP *Biosens. Bioelectron.* **166** 112464

[50] Yeasmin S, Wu B, Liu Y, Ullah A and Cheng L J 2022 Nano gold-doped molecularly imprinted electrochemical sensor for rapid and ultrasensitive cortisol detection *Biosens. Bioelectron.* **206** 114142

[51] Parnianchi F, Kashanian S, Nazari M, Peacock M, Omidfar K and Varmira K 2022 Ultrasensitive electrochemical sensor based on molecular imprinted polymer and ferromagnetic nanocomposite for bilirubin analysis in the saliva and serum of newborns *Microchem. J.* **179** 107474

[52] Anirudhan T S, Mani A and Athira V S 2021 Molecularly imprinted electrochemical sensing platform for 2-aminoadipic acid, a diabetes biomarker *React. Funct. Polym.* **168** 105056

[53] Amouzadeh Tabrizi M, Fernández-Blázquez J P, Medina D M and Acedo P 2022 An ultrasensitive molecularly imprinted polymer-based electrochemical sensor for the determination of SARS-CoV-2-RBD by using macroporous gold screen-printed electrode *Biosens. Bioelectron.* **196** 113729

[54] Rezaei F, Ashraf N, Zohuri G H and Arbab-Zavar M H 2022 Water-compatible synthesis of core–shell polysilicate molecularly imprinted polymer on polyvinylpyrrolidone capped gold nanoparticles for electrochemical sensing of uric acid *Microchem. J.* **177** 107312

[55] Pareek S, Jain U, Balayan S and Chauhan N 2022 Ultra-sensitive nano-molecular imprinting polymer-based electrochemical sensor for follicle-stimulating hormone (FSH) detection *Biochem. Eng. J.* **180** 108329

[56] Ben Moussa F, Achi F, Meskher H, Henni A and Belkhalfa H 2022 Green one-step reduction approach to prepare rGO@AgNPs coupled with molecularly imprinted polymer for selective electrochemical detection of lactic acid as a cancer biomarker *Mater. Chem. Phys.* **289** 126456

[57] Oliveira A E F, Pereira A C and Ferreira L F 2023 Disposable electropolymerized molecularly imprinted electrochemical sensor for determination of breast cancer biomarker CA 15-3 in human serum samples *Talanta* **252** 123819

[58] Chen B *et al* 2021 A novel electrochemical molecularly imprinted senor based on CuCo$_2$O$_4$@ biomass derived carbon for sensitive detection of tryptophan *J. Electroanal. Chem.* **901** 115680

[59] Tang P, Wang Y and He F 2020 Electrochemical sensor based on super-magnetic metal–organic framework@molecularly imprinted polymer for Sarcosine detection in urine *J. Saudi Chem. Soc* **24** 620–30

[60] Rico-Yuste A and Carrasco S 2019 Molecularly imprinted polymer-based hybrid materials for the development of optical sensors *Polymers* **11** 1173

[61] Henry O Y F, Cullen D C and Piletsky S A 2005 Optical interrogation of molecularly imprinted polymers and development of MIP sensors: a review *Anal. Bioanal. Chem.* **382** 947–56

[62] Nasrullah A *et al* 2021 Imprinted polymer and Cu$_2$O–graphene oxide nanocomposite for the detection of disease biomarkers *Meas. Sci. Technol.* **32** 105111

[63] Ahmad O S, Bedwell T S, Esen C, Garcia-Cruz A and Piletsky S A 2019 Molecularly imprinted polymers in electrochemical and optical sensors *Trends Biotechnol.* **37** P294–309

[64] Yarman A, Kurbanoglu S, Erkmen C, Uslu B and Scheller F W 2021 Quantum dot-based electrochemical molecularly imprinted polymer sensors: potentials and challenges *Electroanalytical Applications of Quantum Dot-Based Biosensors* (Amsterdam: Elsevier)

[65] Dong Z M, Cheng L, Zhang P and Zhao G C 2020 Label-free analytical performances of a peptide-based QCM biosensor for trypsin *Analyst* **145** 3329–38

[66] Zhang X, Tang B, Li Y, Liu C, Jiao P and Wei Y 2021 Molecularly imprinted magnetic fluorescent nanocomposite-based sensor for selective detection of lysozyme *Nanomaterials.* **11** 1575

[67] Huang J, Tong J, Luo J, Zhu Y, Gu Y and Liu X 2018 Green synthesis of water-compatible fluorescent molecularly imprinted polymeric nanoparticles for efficient detection of paracetamol *ACS Sustain Chem. Eng.* **6** 9760–70

[68] Altintas Z, Guerreiro A, Piletsky S A and Tothill I E 2015 NanoMIP based optical sensor for pharmaceuticals monitoring *Sensors Actuators B* **213** 305–13

[69] Sari E, Üzek R, Duman M, Alagöz H Y and Denizli A 2018 Prism coupler-based sensor system for simultaneous screening of synthetic glucocorticosteroid as doping control agent *Sensors Actuators* B **260** 432–44

[70] Duan H, Li L, Wang X, Wang Y, Li J and Luo C 2015 A sensitive and selective chemiluminescence sensor for the determination of dopamine based on silanized magnetic graphene oxide-molecularly imprinted polymer *Spectrochim. Acta, Part* A **139** 374–9

[71] Primo E N, Kogan M J, Verdejo H E, Bollo S, Rubianes M D and Rivas G A 2018 Label-free graphene oxide-based surface plasmon resonance immunosensor for the quantification of galectin-3, a novel cardiac biomarker *ACS Appl. Mater. Interfaces* **10** 23501–8

[72] Li D, Wang N, Wang F and Zhao Q 2019 Boronate affinity-based surface-imprinted quantum dots as novel fluorescent nanosensors for the rapid and efficient detection of rutin *Anal. Methods* **11** 3212–20

[73] Samadi-Maybodi A and Malekaneh M 2022 New fluorescent nanosensor for determination of diazepam using molecularly imprinted Mn-doped ZnS quantum dots *Iran J. Pharm. Res.* **21** 1–12

[74] Mahmoudpour M, Ezzati Nazhad Dolatabadi J, Torbati M, Pirpour Tazehkand A, Homayouni-Rad A and de la Guardia M 2019 Nanomaterials and new biorecognition molecules based surface plasmon resonance biosensors for mycotoxin detection *Biosens. Bioelectron.* **143** 111603

[75] Park J H *et al* 2014 A regeneratable, label-free, localized surface plasmon resonance (LSPR) aptasensor for the detection of ochratoxin A *Biosens. Bioelectron.* **59** 321–7

[76] Choi S W, Chang H J, Lee N, Kim J H and Chun H S 2009 Detection of mycoestrogen zearalenone by a molecularly imprinted polypyrrole-based surface plasmon resonance (SPR) sensor *J. Agric. Food Chem.* **57** 1113–8

[77] Jiang W, Wang Z, Nölke G, Zhang J, Niu L and Shen J 2013 Simultaneous determination of aflatoxin B1 and aflatoxin M1 in food matrices by enzyme-linked immunosorbent assay *Food Anal. Methods* **6** 767–74

[78] Jin S, Li D, Feng X and Fu G 2022 Synthesis of carbon dots-based surface protein-imprinted nanoparticles via sandwich-structured template pre-assemble and post-imprinting modification for enhanced fluorescence detection *Microchem. J.* **180** 107611

[79] Anand S K, Sivasankaran U, Jose A R and Kumar K G 2019 Interaction of tetracycline with L-cysteine functionalized CdS quantum dots—fundamentals and sensing application *Spectrochim. Acta* A **213** 410–15

[80] Piloto A M L, Ribeiro D S M, Rodrigues S S M, Santos J L M, Sampaio P and Sales M G F 2022 Cellulose-based hydrogel on quantum dots with molecularly imprinted polymers for the detection of CA19-9 protein cancer biomarker *Microchim. Acta* **189** 134

[81] Piloto A M L, Ribeiro D S M, Rodrigues S S M, Santos J L M and Ferreira Sales M G 2020 Label-free quantum dot conjugates for human protein IL-2 based on molecularly imprinted polymers *Sensors Actuators* B **304** 127343

[82] Zhou X, Gao X, Song F, Wang C, Chu F and Wu S 2017 A sensing approach for dopamine determination by boronic acid-functionalized molecularly imprinted graphene quantum dots composite *Appl. Surf. Sci.* **423** 810–6

[83] Yılmaz G E, Saylan Y, Göktürk I, Yılmaz F and Denizli A 2022 Selective amplification of plasmonic sensor signal for cortisol detection using gold nanoparticles *Biosensors* **12** 482

IOP Publishing

Molecularly Imprinted Polymers for Environmental Monitoring
Fundamentals and applications
Raju Khan and Ayushi Singhal

Chapter 5

Current and emerging techniques for the detection of environmental contaminants

Vijay Vaishampayan, Rashmi Dahake, G K Athira, Mittali Tyagi, Ashish Kapoor and Sarang Gumfekar

The burden to public health of diseases and deaths linked to environmental contamination is evolving as a worldwide challenge. Industries and natural and artificial human activities are the key contributors to generating ecological contaminants in wastewater, microplastics, pathogenic contaminants, electronic waste (e-waste), etc. These contaminants contain a plethora of hazardous organic groups, synthetic chemicals, heavy metals, and so on. Their generation has raised a serious concern in recent years because of their potential threats and continuous output. Often, the level of exposure to these contaminants is unknown and the lack of monitoring makes it difficult to select further treatment methods. The cumulative exposure and long latency period of these materials create hurdles in separating the contaminants from the environment. The existing methods to determine the contaminants quantitatively are sophisticated and bulky. Hence, researchers are developing new generations of sensors for the analytical estimation of these contaminants. This chapter presents a holistic picture of environmental contaminants, their current existing detection methods, and new developing methods. Further, we discuss the integration of these sensing mechanisms with futuristic technologies for better monitoring.

5.1 Introduction

Rapid industrialization and faster growth in developing countries have posed several challenges to communities. One of the biggest challenges faced by any community is environmental contamination. Other than the conventional contaminants, there are numerous emerging contaminants that have been observed in recent years. Electronic waste (e-waste) appears to be a concern due to the release of persistent toxic substances (PTSs) in the environment. E-waste is comprised of a mixture of

doi:10.1088/978-0-7503-4962-8ch5

waste materials, including plastic, heavy metals, etc [1–3]. Technological advancements and never-ending demand to making these technologies more efficient make older devices and apparatus less useful in obtaining the desired results, making them obsolete.

Anthropogenic environmental pollutants are generated by various activities. Some of these pollutants might have been produced for industrial use and, because of their high stability, they are difficult to degrade [4–7]. These toxins might get into the food chain if they are released into the environment. Other environmental pollutants include chemicals that occur naturally; however, industrial activity may increase their circulation and spread, permitting them to enter the food chain at several levels. Environmental pollution is the unintentional alteration of the chemical, physical, and biological properties of water, soil, and air that is detrimental to all living things, including plants and animals. Chemical substances or energy, such as sound, heat, or light, can be pollutants [5, 8–10].

Pollutants are also released into various environments through runoffs, agriculture, wastewater treatment, mining, and metallurgical operations. It has been observed that the anthropogenic processes of heavy metals go beyond the natural fluxes for several elements. Most of the naturally occurring metals in wind-blown dusts come from industrial regions. The combustion of fossil fuels as well as other key anthropogenic sources release nickel, vanadium, mercury, selenium, and tin. The smelting process, pesticides, and automotive exhausts release zinc, arsenic, copper, lead, and other toxic substance [3, 11–13]. Due to the frequent production of goods to satisfy the desires of the huge human population, anthropogenic activities have been contributing more to environmental degradation. As localized point sources, pollutants can enter through direct dumping, channeled waste streams, and piped outflow. As diffuse nonpoint sources, they can enter through runoff and soil percolation to reach different water bodies and groundwater [14, 15]. Nonpoint sources, which include runoff from paved roadways and parking lots, agricultural lots, soil erosion from logging, and atmospheric deposition of acidic or toxic air pollutants, are thought to be substantial contributors to air, water, and soil pollution [7, 13, 16].

After the deadly coronavirus pandemic, biomedical waste has risen sharply. This abundant biomedical waste contaminates bodies of water, soil, the surrounding environment, etc. Bio-contamination is possible during the identification, control, and disposal of such wastage. Current existing sanitary landfills also cause environmental pollution [17, 18]. In various developing nations, waste materials and wastewater are thrown into the rivers, generating enormous water contamination. Recently, India has banned single-use plastic from the country, which may result in curbing the tremendous degree of microplastic pollution in the country.

Apart from these traditional contaminants, various emerging contaminants (ECs) are causing a burden on the global ecology. ECs are potential contaminants of the environments w that have well-known adverse effects on human health cross various generations. They are natural or synthetic chemicals that originated from different regions having been monitored recently and came out as emerging concerns in this climate crisis situation. Talking of the compounds that they include are pharmaceuticals and personal care products (PPCPs), microplastics, antimicrobials, human hormones, detergents [19–21]. Depending on which route they undergo through various transformation reactions their presence in soil, wastewater drains and other sources are causing them to be identified as contaminants. ECs have consequences on the ecosystem and human health as the cycle gets disrupted, causing an imbalance in it. They are being tagged as the emerging concerns in recent situations due to the potential harm they could cause in terms of long and short-term effects on the community's health, including endocrine disruption, immunotoxicity, neurological disorders, and cancer [1, 16]. Hence, there is need to detect the conventional as well as emerging contaminants to monitor their sources, ways and decide the curbing mechanism to limit their spread in the ecology. In this chapter, we are focusing on the current existing techniques as well as some of the emerging detection techniques for the detection of environmental contaminants.

5.2 Existing techniques for the detection of environmental contaminants

The detection of various contaminants can be possible using various methods. These methods can be divided into the traditional methods and emerging methods, as shown in figure 5.1. The traditional methods include gas chromatography (GC), inductively coupled plasma mass spectrometry (ICPMS), atomic absorption spectrometry (AAS), atomic fluorescence spectrometry (AFS), liquid chromatography–mass spectrometry (LC–MS), high-performance liquid chromatography (HPLC), gas chromatography–mass spectrometry (GC–MS), etc. Likewise, the emerging methods are classified based on the applications and techniques of detection, such as electrochemical detection, fluorescent detection, optical detection, colorimetric

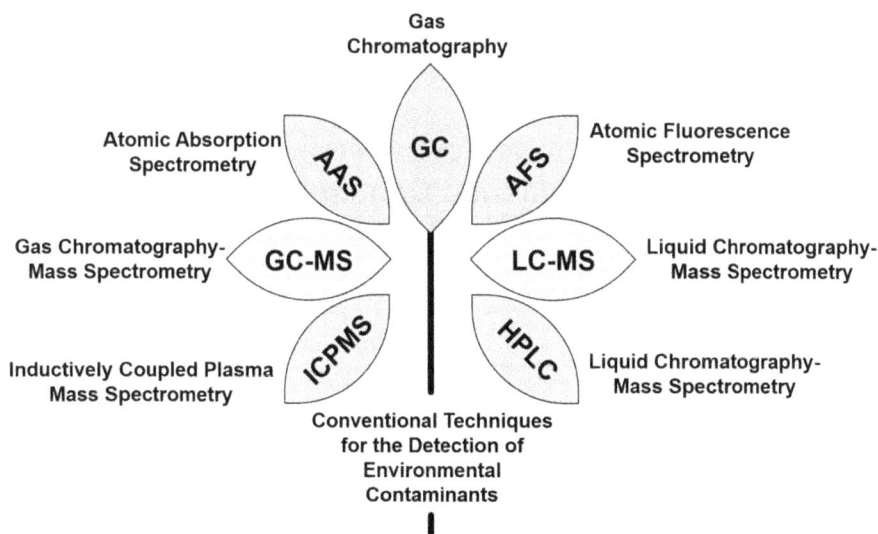

Figure 5.1. Existing conventional techniques for the detection of environmental contaminants.

detection, etc. These methods are more user-friendly, rapid, portable, and affordable compared to conventional methods [4, 22].

Gas chromatography is a primary measurement technique employed for the analysis of numerous families of environmental contaminants such as polychlorinated alkanes (PCAs), pesticides, polychlorinated biphenyls (PCBs), volatile organic compounds (VOCs), etc. The sample preparation stage in GC is labor intensive and consumes a large amount of time. Various schemes have been developed over time to minimize the sample preparation time and utilization of solvent, etc. The analysis of organic contaminants depends on the volatility of the compounds. The VOC analysis and their separation are achieved by the proper selection of the capillary column and stationary phase based on the chemical nature of the analyte sample. Solid samples such as sediments, soils, and biota need the use of a particular extraction methodology to extract the analyte from their matrix phase. Various detectors such as photoionization detectors (PIDs), flame-ionization detectors (FIDs), and mass spectrometry (MS) are more appropriate to analyse VOCs at lower concentrations [23–25].

MS is used widely in the analysis of environmental contaminants with the primary aim to detect organic compounds. MS involves ion generation, separation on the basis of the mass to charge (m/z) ratio, and their detection by their abundance and m/z. It has been used in combination with other techniques such as GC and LC due to their added benefits while analysing a wide spectrum of organic contaminants of different polarities. Many researchers have employed high-resolution mass spectroscopy (HRMS), and tandem mass spectroscopy (MS–MS) with GC–MS for dioxin analysis. Also, HRMS with time-of-flight (TOF) can be used with LC–MS analysis [22, 26–28].

Likewise, the emerging methods are classified based on the applications and techniques of detection, such as electrochemical detection, fluorescent detection, optical detection, colorimetric detection, etc, as shown in figure 5.2. These methods are more user-friendly, rapid, portable, and affordable compared to conventional

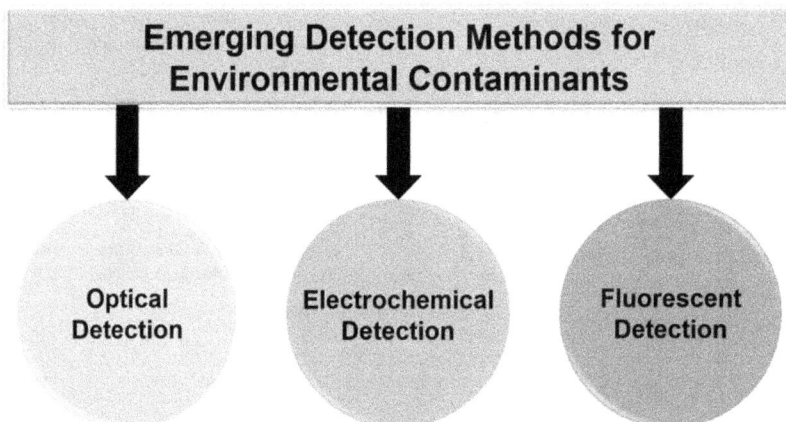

Emerging Detection Methods for Environmental Contaminants

Optical Detection

Electrochemical Detection

Fluorescent Detection

Figure 5.2. Emerging detection techniques for environmental contaminants.

methods [4, 22]. For any sensor, its selectivity, stability, and sensitivity play a crucial role in detecting analytes. Recent research focuses on low-cost, affordable, user-friendly, and sustainable fabrication approaches. The selectivity of sensors is characterized by the signal-to-noise ratio. Selectivity gives the target a specific response for the sensor. It identifies and discriminates between the target molecules and the interfering molecules. Selectivity governs the whole sensing mechanism, particularly in the case of electrochemical sensors [9, 29–32].

5.3 Electrochemical detection of contaminants

In the electrochemical technique, a three-electrode system is applied for detection purposes. The electrodes are named the working, auxiliary/counter, and reference electrodes. The research in this area began with bulk electrode systems in which platinum, carbon, or gold was used as the auxiliary electrodes, Ag/AgCl or calomel as the reference electrodes, and glassy carbon, carbon rod, indium tin oxide (ITO) or fluorine doped tin oxide (FTO) coated glass, stainless steel (SS) plate, metal plates, etc, were used as the working electrodes, which were then modified with the material of interest [33, 34]. Recently, the bulk electrodes were replaced by small-scale chips which are portable and can be applied in the field with ease. There are a variety of options available for small-scale electrodes, of which screen-printed electrodes (SPEs) are a popular choice owing to their advantages, such as bulk production, reproducibility, cost-effectiveness, and small size. SPEs are fabricated using printable conductive inks and non-conductive/insulated substrates. The thickness of the electrodes is only a few microns and can be controlled as per the requirements of the user. Currently researchers are combining microfluidics with the sensor systems, allowing applicability in on-the-spot, field-based analysis [35–37]. A microfluidic electrochemical device includes an electrochemical sensor consisting of a receptor which detects the contaminant, a microfluidic platform, and a potentiostat for signal analysis and processing. Different fabrication techniques such as screen printing, wax patterning, photolithography, 3D printing, inkjet printing, etc, are utilized for the fabrication of microfluidics [38, 39]. A microfluidic works as a transversal technology which combines a variety of scientific disciplines, from biology to fluid control. Microfluidics provides platforms for reagent storage, mixing of a sample with reagents, the flow of the sample, etc. A variety of sensors consisting of electrochemical, optical, and SERS are combined with a microfluidic platform to achieve ease of handling and flow manipulation [40–43]. Recently paper-based microfluidics have caught the attention of many researchers due to the vast interdisciplinary advantages of these devices in the environmental as well as in the medical sectors [44].

Environmental contaminants include inorganic ions such as metals and non-metals, organic pollutants such as organochlorides, organophosphates, etc, and biological contaminants such as bacteria, viruses, etc [45]. Heavy metals pose a severe threat to the biotic and abiotic environments, and human health. Heavy metals are released into the environment from natural as well as anthropogenic activities. Industries such as electroplating, mining, semiconductors, tanneries, etc, use heavy metals and release these toxicants into the water and soil which then also enter into the food chain of living organisms. The toxicity of lead (Pb) can cause neural disorders and kidney failure, Arsenic (As) can cause blood and lung cancer, skin disorders, etc. Cadmium (Cd) is carcinogenic even at lower doses [46]. Therefore, the detection and removal of heavy metals from water, soil, and air are highly essential. The permissible limits of these metals are prescribed by the US Environmental Protection Agency (EPA) and the Occupational Safety and Health Administration (OSHA). Various sensors have been developed for the detection of heavy metals using a disposable lab-on-a-chip platform. A copper-based electro-chemical sensor was fabricated for the detection of Pb (II) in water. Nobel metals such as gold and platinum are very costly and copper is a good option in terms of mass production. A detection limit of 4.4 ppb was achieved by this sensor [47]. The most commonly used materials for electrochemical sensors are carbon and noble metals. Carbon offers a low potential window and good conductivity for electro-chemical studies. Many paper-based sensors use carbon inks [48], carbon cement [49], and carbon cloth [50] for the fabrication of electrodes. Carbon nanotubes (CNTs) provide larger surface areas and they can be functionalized with another nanomaterial to improve their properties. Chemical vapor deposition can be used to grow CNTs on origami paper to achieve better conductivity. For the fabrication of metallic electrodes such as Au, Ag, Cu, and Pt, thin film deposition and metallic ink methods are available. The metallic ink method does not require any high-end instruments and yields a high-quality electrode. Amplification of the electrochemical signal can be achieved by the functionalization of nanoparticles into other materials. Metal NPs can provide a high surface-to-volume ratio with high conductivity. Graphene–polyaniline nanocomposite electrodes were fabricated using different modification methods for simultaneous detection of Zn(II), Cd(II), and Pb(II). Plastic and paper substrates were modified using drop-casting and electrospraying techniques. Good results were obtained on plastic paper and an increased current by three-fold was also observed [51]. Paper-based sensors are effective in the detection of heavy metals due to their easy modification methods and porosity. Paper provides the path for the flow of liquid samples and the solids present in the samples are separated, which demonstrates its use in real water sample analysis [11, 52]. Dings and co-workers discussed the use of paper-based sensors in a liquid sample containing solid impurities and improved the accuracy and sensitivity of the electrode, as shown in the figure 5.3 [53]. The combination of biological material with sensors can greatly increase the sensitivity and selectivity of the sensors. Label-free Ars-3 aptamer was assembled on a screen-printed carbon electrode for the detection of As(III). The absorption of polydiallyldimethylammonium (PDDA) on Ars-3 aptamer occurs due to an electrostatic interaction that repels other cationic species. When As(III) is added to the solution the Ars-3/As(III) complex forms which reduces the absorption of PDDA. The electrochemically active indicator thus produces a turn-on response for the As(III) in a sample [54]. Dual electrochemical as well as colorimetric methods were used for simultaneous determination of Pb(II), Cd (II), and Cu(II). The device consisted of two parts; the first part consisted of an electrochemical detection zone that uses a bismuth modified boron-doped diamond

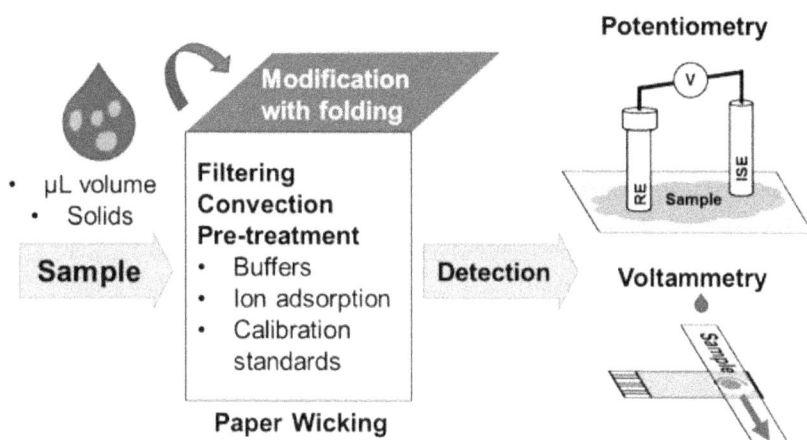

Figure 5.3. Paper as a sampling platform to conduct voltammetry and potentiometry analysis by increasing functionality of the electrodes. (Reproduced with permission from [53]. Copyright 2021 American Chemical Society.)

electrode for the detection of Pb(II) and Cd(II). The second part consists of a colorimetric detection zone prepared by catalytic etching of silver nanoplates by thiosulphate. The addition of copper changes the color of the AgNP from pinkish violet to colorless. The detection limit of 1 ng ml^{-1} was achieved for Pb(II) and Cd(II) and 5 ng ml^{-1} for Cu(II) [55]. Cr(VI) was detected using wax-printed colorimetric paper sensors. A microfluidic PDMS cell was prepared to have four holes for detection and one hole for the sample reservoir. After placing the cell on the filter paper carrying the metal dust sample; acetate buffer solution was added which transferred the ions to the detection zone and a color change to purple occurred [56]. Chrysoidine-G modified microfluidic paper sensors were fabricated using the drop-cast method. The Chrysoidine-G indicator, which is orange in color, was loaded into filter paper and dried. When a sample containing strontium Sr(II) ions comes in contact with an indicator it changes its color from light orange to dark orange thus confirming the presence of aforesaid metal [57]. The combination of an adsorbent and chromogenic reagent was also used for the adsorption of metal ions on the surface and then detection using color change. Zirconium silicate can effectively adsorb Pb(II), hence it was coated on paper in the detection zone followed by sodium rhodizonate, which is a chromatographic reagent. When Pb(II) ions come in contact with the stored reagent a color change from light orange to pink is observed [58]. A Y-shaped paper device was fabricated for the detection of As(III) ions. Au nanoparticles were coated on the paper in the detection zone and the sample, thioctic acid, and thioguanine were passed through the device. Dark bluish black precipitates form which confirm the presence of As(III) [59]. Electrochemical and colorimetric detection methods used in the existing literature for the detection of severe heavy metal ions are summarized in the table 5.1.

Table 5.1. Detection of heavy metal ions using the emerging colorimetric and electrochemical detection methods.

Technique used	Materials used	Detection for	Limit of detection	Reference
Electrochemical detection	Copper	Pb(II)	4.4 ppb	[47]
Electrochemical detection	Graphene–polyaniline (G/PANI) nanocomposite	Zn(II), Cd(II), and Pb(II)	1 ppb, 0.1 ppb, and 0.1 ppb	[51]
Electrochemical detection	Ars-3 aptamer	As(III)	0.15 nM	[54]
Electrochemical detection	Bismuth	Pb(II), Cd(II), and Cu(II)	0.1 ng ml^{-1}	[55]
Colorimetric detection	Chrysoidine-G (CG)	Sr^{2+} ions	200 ppb	[57]
Colorimetric detection	Zirconium silicate	Pb(II)	10 ppb	[58]
Colorimetric detection	Gold nanoparticles	As^{3+} ions	1 ppb	[59]

It is essential to optimize the dimensions of the hydrophobic channel to maintain the capillary action and flow in the device. The detection efficiency is dependent on the type of receptors, reagents, and indicators loaded in the detection zone and their affinity for the paper [41, 60]. The higher the affinity is, the better the sensitivity will be. The thickness of paper also plays an important role; the higher the thickness the lower will be the wicking of solute, resolution, and detection limits. Therefore, proper selection of the type of paper and detection technique is very important in the fabrication of paper-based sensors. The biggest advantage of microfluidic sensors is that they do not need any clean room facilities or tedious sample processing. The detection process involves the simple addition of the sample and the detection of concentration in minimal time. For remote and point-of-care applications, this technique gained enormous attention in analytical sciences as a powerful detection tool [61–63].

The industrial sector is a prime source of growing emissions of greenhouse gases. H_2S, SO_x, NO_x, SF_6, CO, CO_2, etc, are highly hazardous gases that significantly threaten air quality. Over the past several decades, there has been a continuous rise in the greenhouse gases released into the atmosphere due to deforestation, industrialization, burning of conventional fossil fuels, etc [64, 65]. Many industrialized countries have imposed a carbon tax to control emissions and encourage using renewable energy sources. Thus, the monitoring of such toxic gases is necessary [66, 67]. Joshi *et al* prepared SnO_2–Co_3O_4 thin films by depositing a solution on glass substrates using the spin coating technique. The coating was an iterative process to achieve the required thickness of the film. The obtained film was annealed for further use. With an increase in Co_3O_4 concentration, the bandgap and transmittance decrease, which makes these appropriate for CO_2 gas sensing. The porous, spherical morphology of the synthesized material showed the highest sensing response [68]. Hyunsu Kim *et al* synthesized CuO nanostructures and functionalized them with Pd using successive processes of thermal oxidation, dipping, and thermal annealing. The synthesized material was utilized for H_2S sensing at 300 °C. The highly networked Pd functionalized CuO nanostructure showed an excellent response for 100 ppm H_2S gas at 300 °C. The recovery time was shorter in the case of the functionalized CuO nanostructures compared to pristine CuO nanostructures. At lower concentrations of H_2S, the oxidation between H_2S and pre-adsorbed oxygen species occurs. Here, the surface-released electrons combined with the valance band holes led to increased electrical resistance [69]. Likewise, Hu *et al* fabricated CuO nanoneedle arrays on commercial ceramic tubes using magnetron sputtering, chemical etching, and annealing. The fabrication strategy provided a large surface area and avoided the use of binders. The fabricated sensors gave longer stability and selectivity due to a higher number of active adsorption sites and the appearance of p–p nano-homojunctions [70].

5.4 Optical detection

Optical detection includes the monitoring and measurement of properties of light such as absorbance, fluorescence, or luminescence patterns emitted from the sample upon excitation. Most optical sensors are fabricated using paper as it provides ease of handling, is economical, and has high wicking properties, high porosity, stability, non-toxicity, biodegradability, and biocompatibility. Hydrophobic boundaries are prepared on paper to provide a pathway for the sample and reagent zones. The detection areas/zones are already modified with reagents such as organic receptors, dyes, enzymes, polymers, etc. The sample travels the path and when it comes in contact with receptors it changes its color, which can be observed using the naked eye, a camera, or photo-sensors [39, 41]. In the case of fluorimetric sensors, fluorescence is generated which determines the concentration of the analyte. These types of colourimetric/optical lab-on-a-chip devices are very versatile owing to their fast reaction times and outputs [42, 43]. Surface-enhanced Raman spectroscopy (SERS) sensors are prepared using plasmonic nanomaterials as the substrate matrix. SERS produces a vibrational fingerprint of the individual molecule when coupled with nanomaterials. The signals are produced after effective aggregation of the nanoparticles which help in generating the electromagnetic hot spots and localization of Raman probes in hot spots. The use of paper in this platform is very tricky and many functionalization strategies have been exploited by researchers to date [33, 44, 45].

Biosensors are used for measuring chemical substances through biological and chemical reactions by contemplating the proportional concentrations of analytes in the reaction. Biosensors are easy-to-use, low-cost, highly sensitive, and highly selective, contributing to advances in today's generation of sensors. Due to these advantages they are employed in various areas and have applications such as disease monitoring and detection of pollutants, and disease-causing micro-organisms that are indicators of disease in bodily fluids. The increasing amount of harmful

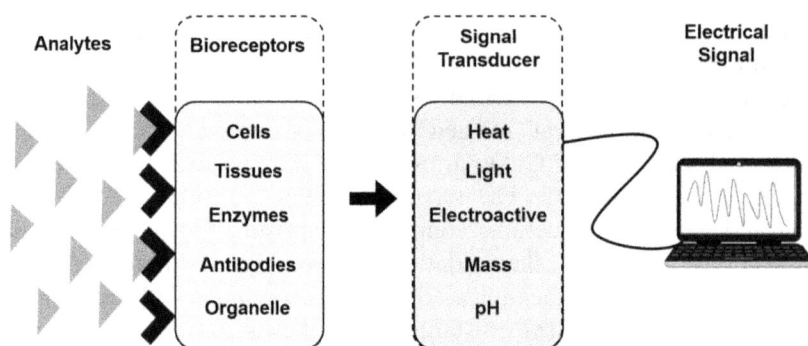

Figure 5.4. Components of biosensors.

pollutants in the environment are motivating the need for devices such as biosensors to be used and enhanced further for the monitoring of the adulterant contaminants. The biosensors employed in this task have analytical devices for detecting and sensing biomaterial or chemical elements [71, 72]. Biosensors for environmental monitoring have several advantages over the previous method of detection, such as portability, miniaturization, and measurement of a pollutant with minimal samples.

Biosensors are developed by integrating various components, such as analytes, receptors, transducers, electronics, and displays as shown in figure 5.4. The specificity and sensitivity play a crucial role in the performance of the biosensors. Aptamers and nanomaterials in the biosensors may increase the lower limit of detection for the biosensors [73, 74].

Pathogens are responsible for numerous water-borne diseases. Their effect depends on the pathogens' size, structure, shape, and composition. These pathogenic contaminants can seriously threaten the environment and water bodies. Hence, their detection is an essential element in monitoring water quality [75, 76]. McConnell *et al* highlighted a recent aptamer-based biosensor for detecting several applications, including heavy metal detection, bacterial pathogen detection, etc. The aptamers' chemical simplicity, stability, and functionality make them valuable and user-friendly compared to antibody-based biosensors. The aptamers-based biosensors can detect bacterial infections, aquatic toxins, and heavy metals. Additionally, these sensors detect antimicrobial resistance (AMR), antibiotics, and pharmaceutical products [9, 77, 78].

5.5 Fluorescent detection

Fluorescence is the emission phenomenon from any substance that absorbs light with a specific wavelength [40, 79]. A Jablonski diagram could better represent the fluorescence phenomena, as shown in figure 5.5. Here, photons with energy above the bandgap are absorbed by the substance, resulting in the excitation of electrons to higher energy levels. These electrons could be either elevated to the first excited level or higher. The molecules in the excited state return to the ground state via the radiative slow relaxation process. It is also known as radiative quenching of excitons [80, 81].

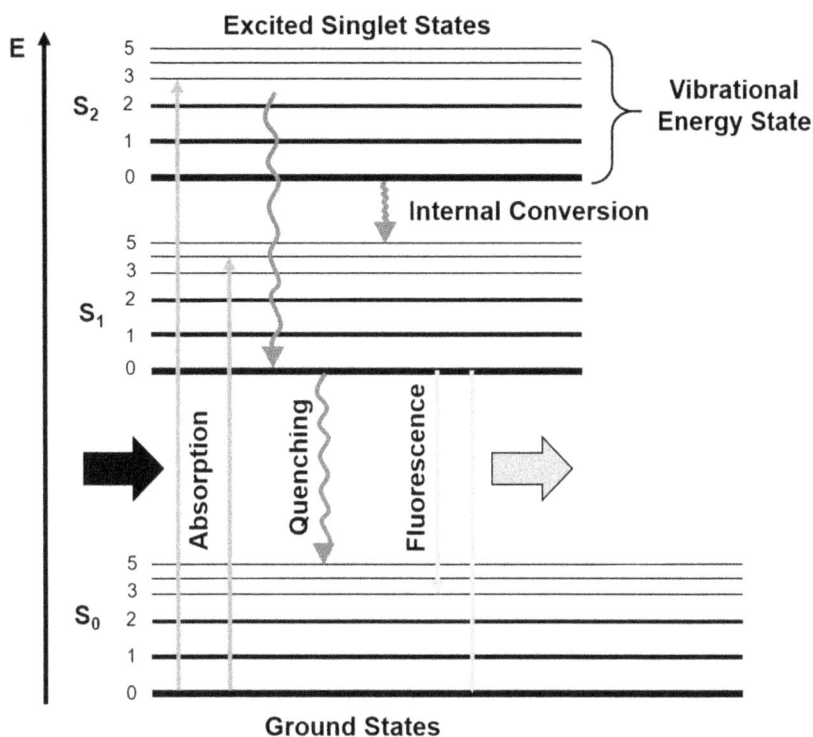

Figure 5.5. Jablonski diagram of fluorescence.

Rotational paper-based microfluidic chips (RPADs) were developed by Qi *et al* using quantum dot molecularly imprinted polymers (QD-MIP) for the detection of phenolic contaminants. This cost-effective, easy, and flexible method was used to analyse 4-nitrophenol (4-NP) and 2,4,6-trinitrophenol (TNP) qualitatively as well as quantitatively. The lower detection limits for 4-NP and TNP were 0.097 and 0.071 mg l^{-1}. The paper-based sensors were fabricated using Whatman chromatography papers. Such reliable methods have tremendous potential and applications related to environmental sensing and point-of-care detection of contaminants [82]. A simple complexation model for the visual detection and quantitative measurement of copper ions was demonstrated by Saleem *et al*. Using an incubating probe and Cu^{2+} ions, confocal fluorescence microscopic images were acquired. A spirolactam ring opening mechanism was proposed for the synthesized probe using a mass spectrum. The proposed colorimetric method was used to determine the concentration at the micromolar level [83]. Lin *et al* developed a fluorescent probe to detect benzenethiols using the thiolysis of a dinitrophenyl ether. The detection of benzenethiols could be performed in aqueous environments, soil, or living cells. It provides superior selectivity with the detection up to 1.8×10^{-9} M. Also, vaporized benzenethiols were detected using aliphatic thiol vapor with a solid support. Such probes would be helpful for the on-site detection of contaminants [84].

A novel fluorescence biosensor was fabricated by Wu *et al* for the rapid quantitative detection of Aflatoxin B1 (AFB1) using MXenes and CRISPR/Cas12a. Dual-AFB1 aptamers blocked the activators, and Cas12a formed inactivated complexes by linking to crRNA. During the process, MXenes adsorbed fluorophore-modified single-stranded DNA (ssDNA-FAM) to quench its fluorescence. Moreover, the fabricated biosensors displayed a linear relationship with the lowest detection limit of 0.92 pg ml^{-1} [85]. Furthermore, Garg *et al* fabricated a fluorescence-based immunosensor to detect ferritin using amine-functionalized graphene quantum dots (afGQDs) and methyl orange as fluorophore–quencher pair. The hydrothermal method was used to synthesize highly fluorescent afGQDs. The described approach eliminated the critical glutaraldehyde chemistries used for the detection as antiferritin antibodies were immobilized directly [86].

5.6 Future perspectives

Advancements in technology are leading to the development of flexible wearable sensors to detect environmental contamination over the conventional laboratory-based methods. The future generation of sensors should be more focused on compact, integrated, and user-friendly approaches with shorter turnaround times for the detection. The multiplexing of various detection tests while detecting numerous analytes could provide an integrated platform in the lab-on-a-chip device form. Embedded circuitry and on-site signal transmission to external devices enhances the usefulness of these environmental contamination monitoring sensors for remote locations. The integration of newer technologies such as artificial intelligence and machine learning may increase the applicability range of flexible sensors for real-time monitoring. Flexible devices are lightweight, stretchable, and bendable to a certain extent, making them perfect candidates for monitoring critical environmental contaminants.

Acknowledgments

The authors are grateful to the management of the Indian Institute of Technology, Ropar, and Harcourt Butler Technical University, Kanpur, for their support throughout this work. The authors are grateful to Dr Balasubramanian Sivasamy for their insightful technical guidance, support, and reviews at various stages of the work.

References

[1] Ravindra K and Mor S 2019 Distribution and health risk assessment of arsenic and selected heavy metals in groundwater of Chandigarh, India *Environ. Pollut.* **250** 820–30

[2] Vasekar S, Kulkarni S and Vaishampayan V 2019 Effect of variation in recycled e-waste reinforcement on mechanical behaviour of polymer matrix composites *AIP Conf. Proc.* **2105** 020015

[3] Leung A, Cai Z W and Wong M H 2006 Environmental contamination from electronic waste recycling at Guiyu, southeast China *J. Mater. Cycles Waste Manag.* **8** 21–33

[4] Beyer J, Jonsson G, Porte C, Krahn M M and Ariese F 2010 Analytical methods for determining metabolites of polycyclic aromatic hydrocarbon (PAH) pollutants in fish bile: a review *Environ. Toxicol. Pharmacol.* 30 224–44

[5] Satheeswaran T *et al* 2019 Assessment of trace metal contamination in the marine sediment, seawater, and bivalves of Parangipettai, southeast coast of India *Mar. Pollut. Bull.* **149** 110499

[6] Dholakia H H and Garg A 2018 Climate change, air pollution and human health in Delhi, India *Climate Change and Air Pollution Springer Climate* (Berlin: Springer) pp 273–88

[7] Tsai W T 2022 Multimedia pollution prevention of mercury-containing waste and articles: case study in Taiwan *Sustainability* **14** 1557

[8] Adimalla N 2020 Spatial distribution, exposure, and potential health risk assessment from nitrate in drinking water from semi-arid region of South India *Hum. Ecol. Risk Assess.* **26** 310–34

[9] Singhal A *et al* 2022 Multifunctional carbon nanomaterials decorated molecularly imprinted hybrid polymers for efficient electrochemical antibiotics sensing *J. Environ. Chem. Eng.* **10** 107703

[10] Ishaq M, Khan M A, Jan F A and Ahmad I 2010 Heavy metals in brick kiln located area using atomic absorption spectrophotometer: a case study from the city of Peshawar, Pakistan *Environ. Monit. Assess.* **166** 409–20

[11] Ghosh R, Vaishampayan V, Mahapatra A, Malhotra R, Balasubramanian S and Kapoor A 2019 Enhancement of limit of detection by inducing coffee-ring effect in water quality monitoring microfluidic paper-based devices *Desalin. Water Treat.* **156** 316–22

[12] Taseidifar M, Makavipour F, Pashley R M and Rahman A F M M 2017 Removal of heavy metal ions from water using ion flotation *Environ. Technol. Innov.* **8** 182–90

[13] Kumar V *et al* 2019 Pollution assessment of heavy metals in soils of India and ecological risk assessment: a state-of-the-art *Chemosphere* **216** 449–62

[14] Kumar V, Sharma A, Pandita S, Bhardwaj R, Thukral A K and Cerda A 2020 A review of ecological risk assessment and associated health risks with heavy metals in sediment from India *Int. J. Sediment Res.* **35** 516–26

[15] Boelee E, Geerling G, van der Zaan B, Blauw A and Vethaak A D 2019 Water and health: from environmental pressures to integrated responses *Acta Trop.* **193** 217–26

[16] Karambelas A *et al* 2018 Urban versus rural health impacts attributable to PM2.5 and O_3 in northern India *Environ. Res. Lett.* **13** 064010

[17] Jain P, Wally J, Townsend T G, Krause M and Tolaymat T 2021 Greenhouse gas reporting data improves understanding of regional climate impact on landfill methane production and collection *PLoS One* **16** e0246334

[18] Sridhar A, Kapoor A and Kumar P S 2021 Conversion of food waste to energy: a focus on sustainability and life cycle assessment *Fuel* **302** 121069

[19] Jang M, Shim W J, Cho Y, Han G M, Song Y K and Hong S H 2020 A close relationship between microplastic contamination and coastal area use pattern *Water Res.* **171** 115400

[20] Lee H, Kunz A, Shim W J and Walther B A 2019 Microplastic contamination of table salts from Taiwan, including a global review *Sci. Rep.* **9** 10145

[21] Kulabhusan P K, Hussain B and Yüce M 2020 Current perspectives on aptamers as diagnostic tools and therapeutic agents *Pharmaceutics* **12** 1–23

[22] Hernández F, Sancho J V, Ibáñez M, Abad E, Portolés T and Mattioli L 2012 Current use of high-resolution mass spectrometry in the environmental sciences *Anal. Bioanal. Chem.* **403** 1251–64

[23] Syage J A, Nies B J, Evans M D and Hanold K A 2001 Field-portable, high-speed GC/TOFMS *J. Am. Soc. Mass Spectrom.* **12** 648–55

[24] Santos F J and Galceran M T 2003 Modern developments in gas chromatography–mass spectrometry-based environmental analysis *J. Chromatogr.* A 1000 125–51

[25] Santos F J and Galceran M T 2002 The application of gas chromatography to environmental analysis *TrAC Trends Anal. Chem.* **21** 672–85

[26] Sivaperumal P, Anand P and Riddhi L 2015 Rapid determination of pesticide residues in fruits and vegetables, using ultra-high-performance liquid chromatography/time-of-flight mass spectrometry *Food Chem.* **168** 356–65

[27] Syage J A, Nies B J, Evans M D and Hanold K A 2001 Field-portable, high-speed GC/TOFMS *J. Am. Soc. Mass Spectrosc.* **12** 648–55

[28] Alaee M, Sergeant D B, Ikonomou M G and Luross J M 2001 A gas chromatography/high-resolution mass spectrometry (GC/HRMS) method for determination of polybrominated diphenyl ethers in fish *Chemosphere* **44** 1489–95

[29] Hu R, Tong X and Zhao Q 2020 Four aspects about solid-state nanopores for protein sensing: fabrication, sensitivity, selectivity, and durability *Adv. Healthcare Mater.* **9** 2000933

[30] Wasik D, Mulchandani A and Yates M 2018 Salivary detection of dengue virus NS1 protein with a label-free immunosensor for early dengue diagnosis *Sensors* **18** 2641

[31] Qazi S and Raza K 2020 Smart biosensors for an efficient point of care (PoC) health management *Smart Biosensors in Medical Care* (Amsterdam: Elsevier) pp 65–85

[32] Syedmoradi L, Norton M L and Omidfar K 2021 Point-of-care cancer diagnostic devices: from academic research to clinical translation *Talanta* **225** 122002

[33] Nie Z *et al* 2010 Electrochemical sensing in paper-based microfluidic devices *Lab Chip* **10** 477–83

[34] Hamedi M M, Ainla A, Güder F, Christodouleas D C, Fernández-Abedul M T and Whitesides G M 2016 Integrating electronics and microfluidics on paper *Adv. Mater.* **28** 5054–63

[35] Camargo J R, Silva T A, Rivas G A and Janegitz B C 2022 Novel eco-friendly water-based conductive ink for the preparation of disposable screen-printed electrodes for sensing and biosensing applications *Electrochim. Acta.* **409** 139968

[36] Zhang S *et al* 2016 Electrochemical immunosensors and their recent nanomaterial-based signal amplification strategies: a review *Electrochim. Acta.* **8** 24995–5014

[37] Syedmoradi L, Daneshpour M, Alvandipour M, Gomez F A, Hajghassem H and Omidfar K 2017 Point of care testing: the impact of nanotechnology *Biosens. Bioelectron.* **87** 373–87

[38] Lu R, Shi W, Jiang L, Qin J and Lin B 2009 Rapid prototyping of paper-based microfluidics with wax for low-cost, portable bioassay *Electrophoresis* **30** 1497–500

[39] Abe K, Kotera K, Suzuki K and Citterio D 2010 Inkjet-printed paperfluidic immuno-chemical sensing device *Anal. Bioanal. Chem.* **398** 885–93

[40] Kalyani N, Goel S and Jaiswal S 2021 On-site sensing of pesticides using point-of-care biosensors: a review *Environ. Chem. Lett.* **19** 345–54

[41] Wang C, Liu M, Wang Z, Li S, Deng Y and He N 2021 Point-of-care diagnostics for infectious diseases: from methods to devices *Nano Today* **37** 101092

[42] Naresh V *et al* 2021 A newly emerging trend of chitosan-based sensing platform for the organophosphate pesticide detection using acetylcholinesterase—a review *Trends Food Sci. Technol.* **19** 345–54

[43] Zarei M 2017 Portable biosensing devices for point-of-care diagnostics: recent developments and applications *TrAC Trends Anal. Chem.* **91** 26–41

[44] Kapoor A, Balasubramanian S, Vaishampayan V and Ghosh R 2018 Lab-on-a-chip: a potential tool for enhancing teaching–learning in developing countries using paper micro-fluidics *Int. Conf. on Transforming Engineering Education, ICTEE 2017* (Piscataway, NJ: IEEE) pp 1–7

[45] Mishra S, Lin Z, Pang S, Zhang W, Bhatt P and Chen S 2021 Recent advanced technologies for the characterization of xenobiotic-degrading microorganisms and microbial communities *Front. Bioeng. Biotechnol.* **9** 632059

[46] Galindo-Miranda J M, Guízar-González C, Becerril-Bravo E J, Moeller-Chávez G, León-Becerril E and Vallejo-Rodríguez R 2019 Occurrence of emerging contaminants in environ-mental surface waters and their analytical methodology—a review *Water Sci. Technol.* **19** 1871–84

[47] Kang W *et al* 2017 Determination of lead with a copper-based electrochemical sensor *Anal. Chem.* **89** 3345–52

[48] Moço A C R *et al* 2021 Carbon ink-based electrodes modified with nanocomposite as a platform for electrochemical detection of HIV RNA *Microchem. J.* **170** 106739

[49] Zhu C C, Bao N and Huo X L 2020 Paper-based electroanalytical devices for stripping analysis of lead and cadmium in children's shoes *RSC Adv.* **10** 41482–7

[50] Liu X, Xu W, Zheng D, Li Z, Zeng Y and Lu X 2020 Carbon cloth as an advanced electrode material for supercapacitors: progress and challenges *J. Mater. Chem.* A **8** 17938–50

[51] Ruecha N, Rodthongkum N, Cate D M, Volckens J, Chailapakul O and Henry C S 2015 Sensitive electrochemical sensor using a graphene–polyaniline nanocomposite for simulta-neous detection of Zn(II), Cd(II), and Pb(II) *Anal. Chim. Acta.* **874** 40–8

[52] Martinez A W, Phillips S T, Wiley B J, Gupta M and Whitesides G M 2008 FLASH: a rapid method for prototyping paper-based microfluidic devices *Lab Chip* **8** 2146–50

[53] Ding R, Cheong Y H, Ahamed A and Lisak G 2021 Heavy metals detection with paper-based electrochemical sensors *Anal. Chem.* **93** 1880–8

[54] Cui L, Wu J and Ju H 2016 Label-free signal-on aptasensor for sensitive electrochemical detection of arsenite *Biosens. Bioelectron.* **79** 861–5

[55] Chaiyo S, Apiluk A, Siangproh W and Chailapakul O 2016 High sensitivity and specificity simultaneous determination of lead, cadmium and copper using μpAD with dual electro-chemical and colorimetric detection *Sensors Actuators* B **233** 540–9

[56] Rattanarat P *et al* 2013 A microfluidic paper-based analytical device for rapid quantification of particulate chromium *Anal. Chim. Acta.* **800** 50–5

[57] Kang S M, Jang S C, Huh Y S, Lee C S and Roh C 2016 A highly facile and selective chemo-paper-sensor (CPS) for detection of strontium *Chemosphere* **152** 39–46

[58] Satarpai T, Shiowatana J and Siripinyanond A 2016 Paper-based analytical device for sampling, on-site preconcentration and detection of ppb lead in water *Talanta* **154** 504–10

[59] Nath P, Arun R K and Chanda N 2014 A paper based microfluidic device for the detection of arsenic using a gold nanosensor *RSC Adv.* **4** 59558–61

[60] Bhattacharya S, Datta A, Berg J M and Gangopadhyay S 2005 Studies on surface wettability of poly(dimethyl) siloxane (PDMS) and glass under oxygen-plasma treatment and correla-tion with bond strength *J. Microelectromech. Syst.* **14** 590–7

[61] Li X, Tian J, Garnier G and Shen W 2010 Fabrication of paper-based microfluidic sensors by printing *Colloids Surf.* B **76** 564–70

[62] Abbas A, Brimer A, Slocik J M, Tian L, Naik R R and Singamaneni S 2013 Multifunctional analytical platform on a paper strip: separation, preconcentration, and subattomolar detection *Anal. Chem.* **85** 3977–83

[63] Nge P N, Rogers C I and Woolley A T 2013 Advances in microfluidic materials, functions, integration, and applications *Chem. Rev.* **113** 2550–83

[64] Gupta M, Hawari H F, Kumar P, Burhanudin Z A and Tansu N 2021 Functionalized reduced graphene oxide thin films for ultrahigh CO_2 gas sensing performance at room temperature *Nanomaterials* **11** 1–18

[65] Siddique S A, Sajid H, Gilani M A, Ahmed E, Arshad M and Mahmood T 2022 Sensing of SO_3, SO_2, H_2S, NO_2 and N_2O toxic gases through aza-macrocycle via DFT calculations *Comput. Theor. Chem.* **1209** 113606

[66] Bareza N J *et al* 2022 Phonon-enhanced mid-infrared CO_2 gas sensing using boron nitride nanoresonators *ACS Photonics* **9** 34–42

[67] Bian Y, Li L, Song H, Su Y and Lv Y 2021 Porous boron nitride: a novel metal-free cataluminescence material for high performance H_2S sensing *Sensors Actuators* B **332** 129512

[68] Joshi G, Rajput J K and Purohit L P 2021 SnO_2–Co_3O_4 pores composites for CO_2 gas sensing at low operating temperature *Microporous Mesoporous Mater.* **326** 111343

[69] Kim H, Jin C, Park S, Kim S and Lee C 2012 H_2S gas sensing properties of bare and Pd-functionalized CuO nanorods *Sensors Actuators* B **161** 594–9

[70] Hu Q *et al* 2021 Binder-free CuO nanoneedle arrays based tube-type sensor for H_2S gas sensing *Sensors Actuators* B **326** 128993

[71] Abubakar Sadique M, Yadav S, Ranjan P, Akram Khan M, Kumar A and Khan R 2021 Rapid detection of SARS-CoV-2 using graphene-based IoT integrated advanced electro-chemical biosensor *Mater. Lett.* **305** 130824

[72] Singhal A *et al* 2022 MXene-modified molecularly imprinted polymer as an artificial bio-recognition platform for efficient electrochemical sensing: progress and perspectives *Phys. Chem. Chem. Phys.* **24** 19164–76

[73] Dixon T A, Williams T C and Pretorius I S 2021 Sensing the future of bio-informational engineering *Nat. Commun.* **12** 1–12

[74] Villalonga A, Pérez-Calabuig A M and Villalonga R 2020 Electrochemical biosensors based on nucleic acid aptamers *Anal. Bioanal. Chem.* **412** 55–72

[75] Kumar N, Hu Y, Singh S and Mizaikoff B 2018 Emerging biosensor platforms for the assessment of water-borne pathogens *Analyst* 143 359–73

[76] Molloy A *et al* 2021 Microfluidics as a novel technique for tuberculosis: from diagnostics to drug discovery *Microorganisms* **9** 1–24

[77] McConnell E M, Nguyen J and Li Y 2020 Aptamer-based biosensors for environmental monitoring *Front. Chem.* **8** 434

[78] Parihar A, Singhal A, Kumar N, Khan R, Khan M A and Srivastava A K 2022 Next-generation intelligent MXene-based electrochemical aptasensors for point-of-care cancer diagnostics *Nano-Micro Lett.* **14** 100

[79] Zhou X *et al* 2004 Determination of SARS-coronavirus by a microfluidic chip system *Electrophoresis* **25** 3032–9

[80] Frackowiak D 1988 The Jablonski diagram *J. Photochem. Photobiol.* B **2** 399

[81] Cai T, Chen B, Han J, Kim M, Yeom E and Kim K C 2022 Effect of excitation duration on phosphorescence decay and analysis of its mechanisms *J. Lumin.* **252** 119423

[82] Qi J, Li B, Wang X, Fu L, Luo L and Chen L 2018 Rotational paper-based microfluidic-chip device for multiplexed and simultaneous fluorescence detection of phenolic pollutants based on a molecular-imprinting technique *Anal. Chem.* **90** 11827–34

[83] Saleem M and Lee K H 2014 Selective fluorescence detection of Cu^{2+} in aqueous solution and living cells *J. Lumin.* **145** 843–8

[84] Lin W, Long L and Tan W 2010 A highly sensitive fluorescent probe for detection of benzenethiols in environmental samples and living cells *Chem. Commun.* **46** 1503–5

[85] Bhardwaj H, Sumana G and Marquette C A 2021 Gold nanobipyramids integrated ultrasensitive optical and electrochemical biosensor for aflatoxin B1 detection *Talanta* **222** 121578

[86] Garg M, Vishwakarma N, Sharma A L and Singh S 2021 Amine-functionalized graphene quantum dots for fluorescence-based immunosensing of ferritin *ACS Appl. Nano Mater.* **4** 7416–25

IOP Publishing

Molecularly Imprinted Polymers for Environmental Monitoring
Fundamentals and applications
Raju Khan and Ayushi Singhal

Chapter 6

The importance of molecularly imprinted polymers in wastewater treatment

Sapna Kumari, Jyoti Singh Chauhan and Vaishnavi Hada

The monitoring of the biological impact of wastewater releases in ecosystems is now significantly relevant for investigating ecological impacts. One of the emerging strategies that gained remarkable importance in the field of wastewater management is molecular imprinting technology. Molecularly imprinted polymers (MIPs) are polymers with precise cavities formed based on the template molecules used in this strategy. MIPs are suitable materials for treating wastewater due to their predefined specificity. MIP-based composites, in particular, have a broad spectrum of applications in the treatment of wastewater. This chapter discusses the significance of MIPs in wastewater treatment, reveals the advancement of MIP-based composites in wastewater by various researchers, and makes future recommendations in the field of MIPs.

6.1 Introduction

The world is reaching new horizons as humans, society, science, and technology evolve, but the price of this advancement may become too high in the near future. This quick expansion has had negative effects on the environment, resulting in a major pollution problem. One of the most important issues is water pollution [1, 2]. Water pollution, produced by the discharge of a variety of pollutants, is a serious environmental concern all over the world. Synthetic or naturally occurring chemicals or microbes that are not often detected in the environment but have the potential to enter the environment and cause known or suspected detrimental ecological or human health impacts are considered as 'emerging contaminants' (ECs) [3, 4].

In recent years there has been a lot of research into these developing pollutants in aqueous media. Various new pollutants have been discovered in wastewater according to research [5]. Prescription and non-prescription medications, personal care products, and other chemical additives that we use on a daily basis are found in municipal wastewater treatment plants (WWTPs) [6]. Emerging pollutants in the environment are thought to be a source of complex difficulties because their impacts are likely to occur at trace amounts. Many studies are currently being conducted to investigate new pollutants in aqueous media [7]. Pharmaceutically active compounds (PACs), endocrine-disrupting chemicals (EDCs), personal care products (PCPs), and heavy metals have all been found in wastewater as per recent studies (figure 6.1) [8–10]. Many studies have found a steady increase in these pollutants in wastewater around the world. EC-rich wastewater can harm metabolic processes, feeding behavior, survival behavior, growth, and the genetic material of aquatic species [11, 12]. Due to their toxicity and carcinogenicity, many of these ECs are hazardous and can cause a significant threat to aquatic life. As a result, prior to being released into the ecosystem they must be treated using a variety of techniques. The techniques which utilize smart materials have gained a lot of interest among researchers [13]. Materials that have been engineered to adapt in a controllable and reversible manner, altering some of their characteristics in response to external stimuli such as mechanical stress or temperature, are referred to as 'smart materials' [14]. Piezoelectric, chromoactive, magnetorheological, shape-memory, and photoactive materials are several types of smart materials available today [15]. Among them, molecularly imprinted polymers (MIPs) are one such material [16].

MIPs can be used in environmental research to prepare samples, and clean up and quantify toxins present in wastewater. They can be used as ideal materials in wastewater treatment because of their predetermined selectivity. These materials have been employed in chromatographic, chemical, and biological sensing

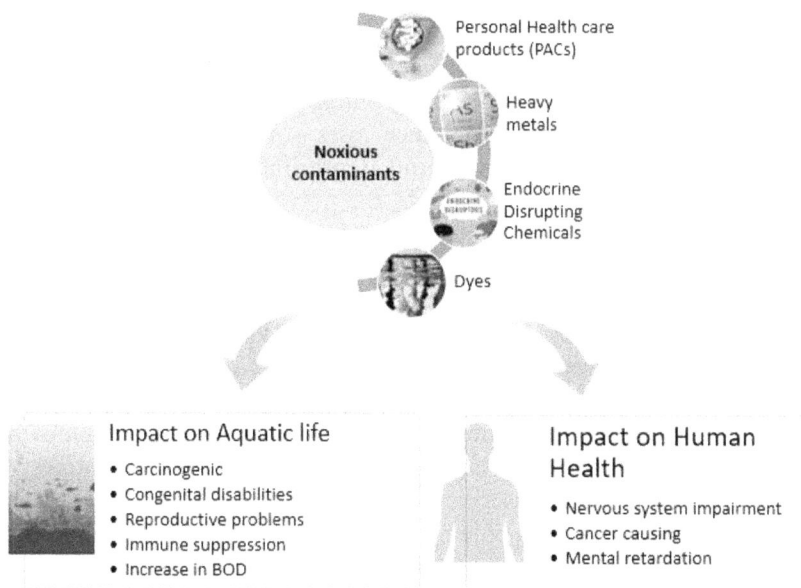

Figure 6.1. The impact of contaminants on various life forms.

applications, with small molecule and protein templates imprinted on them [17]. Because of the ease of preparation, simplicity, and resilience to deterioration in extreme settings of MIPs, the molecular imprinting process is a potential technology with exceptionally high chemical and physical stability of products compared to their biological analogs [18]. MIPs are also less expensive to synthesize and have a long storage life. Furthermore, they have high selectivity and affinity for the target molecule, allowing them to recognize a wide range of biological and chemical compounds [19]. These compounds may include pharmacological substances, amino acids, pollutants, steroid hormones, and metal ions, as well as larger molecules such as peptides or proteins [20]. Because MIPs have high selectivity and affinity for template molecules, they are useful as tailor-made materials for removing low abundance ECs. Nanoscale surface recognition sites for target molecules can be used to construct molecular imprinting, resulting in high binding capacities and rapid mass transfer rates [21]. MIPs can withstand changes in pH, temperature, and the complexity of their surroundings [22].

This chapter provides a basic introduction which shows the potential of MIPs for use in different environmental applications. This article highlights the significance of MIPs in wastewater treatment followed by a literature review in which many different works carried out by researchers worldwide are reported, particularly in the treatment of wastewater. Additionally, a section of the current status and future perspectives is also provided. Finally, the conclusion regarding MIPs provides an overview to open up a new paths in environmental monitoring.

6.2 The significance of MIPs in wastewater treatment

MIPs are synthetic receptors that can retain certain target molecules based on their chemical functionalization and also act as counterparts to natural antigen–antibody systems [23]. The core principle of molecular imprinting is to use covalent or non-covalent bonding to create a polymer by combining functional monomers with a template molecule [24, 25]. The polymer loses solvent solvation as the three-dimensional polymer network grows, and it precipitates in a solid form on the bottom of the reaction vessel. As a result, the polymer only recognizes and binds molecules that have chemical characteristics identical to the template molecule [26]. The molecular recognition events are guided by a polymer matrix. Owing to their outstanding properties such as high sensitivity, low cost, quick adsorption time, and resilience to acids, alkalis, and high temperatures, MIPs are used as adsorbents and solid-phase extraction agents for the concentration and extraction of ECs in wastewater [27, 28].

MIPs are now described in the literature as a potential method for removing trace contaminants from water. MIPs have generated a great deal of interest among scientists working on solid-phase extraction, sensors, catalysts, enzyme mimics, receptors, and antigen–antibody interaction [29]. The pre-concentration and selective elimination of contaminants using the molecular imprinting process has recently been applied to environmental applications [30]. MIPs are ideal for the treatment of trace pollutants because they may be created precisely to eliminate a single or a number of target substances. This is a benefit over general technologies such as activated carbon, which can be eaten while removing significant amounts of non-trace pollutants from the water. Researchers working on MIPs for the treatment of water and wastewater have looked at their isotherms, kinetics, and performance under various conditions, applications, and regeneration [31]. They have so far been used to remove non-steroidal anti-inflammatory medications, antibiotics, antimicrobials, endocrine-disrupting substances, herbicides, phenols, and beta-blockers from contaminated wastewater [32]. Adsorption is a popular and effective technique for purifying water of both inorganic and organic contaminants. The process of adsorption is a surface phenomenon in which adsorbates are drawn to the surface of solid adsorbents and bind there through chemical or physical connections [33]. To date, the development of numerous adsorbents includes materials such as clays, chitosan, and activated carbons. The combination of magnetic separation and selective adsorption offers a very promising method for creating effective adsorbents that can remove target contaminants quickly. The smart MIPs promise to controllably adsorb/detach target pollutants for their high-efficiency enrichment and removal by integrating selective adsorption, a property that is compatible with water, magnetic separation, and temperature management [34].

MIPs have excellent regeneration properties and can be utilized at least 12 times before suffering a substantial reduction in loading capacity. MIPs were suggested as being efficient and selective adsorbents for the quick removal of organic contaminants from contaminated water [35]. MIPs are more selective than other materials and are more likely to be altered or integrated with other materials. MIPs are particularly desirable for applications in several fields owing to their special characteristics [36]. To create unique materials with many functions, MIPs can be coupled with other materials such as metal oxides. MIPs are important for the selective adsorption of organic pollutants, the degradation of pollutants as catalysts, the selective detection of pollutants using MIP-based fluorescence sensors, and the accurate determination of pollutants. High quantities of particular manufacturing-related molecules could be eliminated by using MIPs with high selectivity and great affinity for the target chemicals in industrial wastewater treatment. MIPs are highly useful for the removal of highly toxic organic pollutants (HTOPs) [37]. MIPs are selected materials that show promise for long-term use. MIPs have been used successfully in the preferred photocatalytic destruction of pollutants when combined with high-tech oxidation techniques. Regeneration of MIPs on-site can be accomplished by removing and destroying the pollutants through MIP adsorption, followed by simultaneous extraction and chemical treatment.

In contrast to conventional adsorption materials such as clays and carbonaceous adsorbents, which primarily rely on surface or physical adsorption, MIPs have a high potential for use in wastewater treatment systems because they can bind various classes of pharmaceutical compounds superbly regardless of their structure or pharmacological activity [38]. As a result of daily human activity, thousands of various organic chemicals find their way into water bodies (such as lakes, rivers, oceans, and groundwater). About 90% of these priority pollutants are HTOPs.

Several fairly standardized unit processes are available to remove HTOPs from contaminated water, including chemical degradation (for example, chemical oxidation, electrochemical techniques, and photochemical methods), biological oxidation, physical adsorption, sedimentation, and filtration. But in many real-world water sources, the non-biodegradable HTOPs are present at extremely low concentrations and always coexist alongside more prevalent, but less hazardous and biodegradable pollutants at high levels [39].

A sensor is a device with a recognition component and a transducer that transforms the information from the chemical into discernible signal. MIP-based sensors have benefits including inexpensive production costs, simple storage, a long lifespan, and the capacity to be utilized in urgent situations [40]. For this application, the imprinted polymer is coupled to a transducer to convert the output of the MIP-based sensors into a quantifiable signal. The MIP-based sensors are intended to be used for the analysis of environmental pollutants as well as the detection of some pollutants, such as pesticides, pathogens, explosives, heavy metals, and dyes. They are also intended to be used for the purification of chemical and biological reagent substances [41]. There are potential and promising uses for MIP-based sensors in several fields, including medical, bio-analytical, process control, and environmental applications [42].

Since MIPs were first used as sorbents specifically for solid-phase extraction (SPE), their employment in this field has gained widespread acceptance due to their extremely selective extraction of the target analyte. The term 'molecularly imprinted SPE' (MISPE) has been coined to describe this technique as a result of the broad use of MIPs as SPE sorbents [43]. In MISPE the high selectivity of the sorbent towards a specific structure allows for the selective retention of this structure on the sorbent while the other compounds are not retained, resulting in the desired compound being extracted from the majority of the other compounds present in the sample.

Numerous publications have discussed the use of MIPs as SPE adsorbents, and the number of papers shows exponential growth. The ability to specifically attach to a target in the presence of its structural analogs from a complex matrix is the potential strength of MISPE. Using a very small quantity of imprinted polymer (usually 5–200 mg packed into a cartridge), MIP sorbents can be used to selectively extract the target analyte while also pre-concentrating and cleaning the sample. Environmental fluids, soils, sediments, plant extracts, and soy have all been effectively isolated and pre-concentrated using the effective method known as MISPE. The most sophisticated use of MIPs is in molecular-imprint-based SPE. Off-line use of MIP particles is possible when they are packaged in a cartridge between two frits [44].

Additionally, they can be arranged in a compact column for on-line coupling with liquid chromatography (LC). In recent publications, the off-line and on-line coupling of MIPs with LC has been discussed extensively. The extraction process on an MIP follows the same principles as when using traditional SPE sorbents. The sample is percolated through the MIP after a conditioning stage, and interference-causing substances can be eliminated using a washing step (figure 6.2) [45].

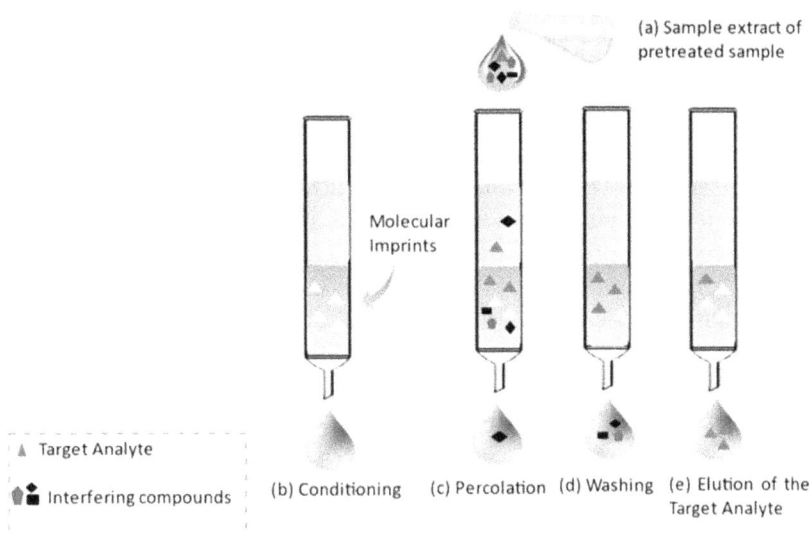

Figure 6.2. Solid-phase extraction process using an MIP sorbent.

6.3 Literature review of MIPs for wastewater treatment

In 2017 Zunngu *et al* produced a specific MIP for ketoprofen and used it as a solid-phase extraction sorbent. Due to its antioxidant properties, ketoprofen is a non-steroidal anti-inflammatory medicine that is frequently used by people. As a functional monomer, cross-linker, initiator, porogenic mixture, and template, respectively, 2-vinylpyridine, ethylene glycol dimethacrylate, 1,1'-azobis (cyclohexanecarbonitrile), toluene/acetonitrile (9:1, v/v), and ketoprofen were used in the synthesis of MIPs. In order to produce a solid monolithic polymer, the polymerization was carried out at 60 °C for 16 h and then raised to 80 °C for 24 h. Similar steps were used to create a non-imprinted polymer (NIP), but without the ketoprofen. For the quantitative analysis of ketoprofen in the Umbilo, Amanzimtoti, and Kingsburgh WWTPs, the SPE technique was optimized and employed using high-performance LC (HPLC). The volume fraction of the washed polymers was around 4% at 40 °C. At 290 °C, the temperature at which the polymeric chains break, significant thermal breakdown of polymers was seen. Additionally, it was shown that the NIP thermally decomposed 100% at 425 °C, although the MIP experienced a 90% mass loss at the same temperature. The thermograms produced by differential scanning calorimetry (DSC) for the MIP and NIP showed similarities and an endothermic peak at 355 °C, which is related to the temperature at which the polymers completely thermally decompose. Relatively similar configurations of the MIPs and NIPs may account for commonalities in the thermograms. In wastewater influent, effluent, and deionized water, the analytical approach provided a limit of detection of 0.23, 0.17, and 0.09 g l^{-1}, respectively. While 114% recovery was achieved for deionized water, only 68% was achieved for wastewater influent and effluent treated with 5 g l^{-1} of ketoprofen. Ketoprofen levels in the influent and effluent samples ranged from 22.5 to 34.0 and 1.1 to 4.33 g l^{-1}, respectively. Throughout the wastewater treatment process, ketoprofen was removed at a rate of 88%–90%. Ultimately, the technique used to analyse the presence of ketoprofen in wastewater was quick, inexpensive, efficient, specific, and precise [46].

In 2018 Mbhele *et al* developed an MIP to remove fenoprofen, a non-steroidal anti-inflammatory drug, obtained mostly from aquatic environmental samples. The assessment of fenoprofen from wastewater was then performed, accompanied by chromatographic evaluation, using the imprinted polymer as the chosen sorbate. Additionally, the methodology described in this paper was used to analyze fenoprofen quantitatively in South African WWTPs. The research methodology produced a 99.6% efficiency and a 0.64 ng ml^{-1} limit of detection. Fenoprofen was found at high proportions in wastewater samples, along with effluent, indicating an uncontrolled discharge of this drug into the surface and groundwater. Between 24 and 58 ng ml^{-1} of fenoprofen was found in influent and effluent samples from two WWTPs. Due to the disparity in concentration levels in the influent and effluent samples, it was discovered that the treatment systems had 41% removal efficiency for fenoprofen throughout the wastewater processing. Ultimately, this research investigates a straightforward way for consistently removing fenoprofen from WWTPs using MIPs, which may also be helpful for surface waterways [39].

In order to adsorb the target substance from water, in 2018 Madikizela *et al* developed an MIP that is specific to ketoprofen. Ketoprofen, 2-vinylpyridine, ethylene glycol dimethacrylate, toluene, and 1,1′-azobis(cyclohexanecarbonitrile) were utilized as the template, functional monomer, cross-linker, porogen, and initiator, respectively, in the bulk polymerization technique used to create the MIP at high temperatures (60 °C–80 °C). In a manner comparable to that of MIPs, but without the use of ketoprofen, NIPs were produced. According to molecular dynamics simulation, hydrogen bonds were discovered to form the basis of the associations among the template and the functional monomer. This was confirmed scientifically, where a higher removal efficiency of about 90% was attained under acidic circumstances (pH 5) because of the protonation of ketoprofen. The highest amount of ketoprofen could be absorbed from 10 ml of contaminated water using 8 mg of the adsorbent during an exposure time of 45 min. Additionally, computational evidence supported the use of 2-vinylpyridine as the appropriate functional monomer in the production of MIPs for ketoprofen. SEM was used to characterize the synthesized MIPs and revealed that they had a rougher surface than the NIPs. Whilst the MIPs and NIPs showed commonalities in their NMR and FTIR data, selectivity tests revealed that the MIPs were more specific to ketoprofen in the context of fenoprofen and gemfibrozil because of the imprinting effects. In comparison to the NIPs, specificity for ketoprofen was shown to be 8 and 2 times higher in the presence of fenoprofen and gemfibrozil, respectively. When utilizing MIPs, the removal rates for ketoprofen were around 70% and 100%. By attaining a comparative selectivity coefficient of 7.7 for ketoprofen in the presence of medicines with a similar structure, MIPs demonstrated better binding affinity than NIPs. Ketoprofen was followed by fenoprofen and then gem brozil in the degree of adsorption onto the MIPs from water. The study showed the enormous potential of MIPs in distinguishing ketoprofen from closely comparable chemicals in wastewater [47].

One of the anti-retroviral (ARV) medications frequently used to treat human immunodeficiency virus is efavirenz. Surface water and wastewater have been reported to contain such anti-retroviral medications. The limitation of specificity in the widely viable SPE adsorbent, however, appears to make investigation times longer. In order to specifically recognize and remove efavirenz from wastewater and surface water samples, Mtolo et al created an MIP in 2019. SPE is used to isolate and pre-concentrate anti-retroviral medicines before their chromatographic examination, depending on the complexity of the environmental materials. This polymer was then used as an SPE sorbent. Efavirenz was used as the template, 2-vinyl-pyridine as the functional monomer, 1,1′-azobis-(cyclohexanecarbonitrile) as the initiator, ethylene glycol dimethacrylate as the cross-linker, and toluene:acetonitrile (9:1, v/v) as the porogenic solvent mixture to create the MIPs and NIPs. The findings of the characterization revealed structural similarities between the two polymers (MIPs and NIPs), with the MIPs exhibiting greater surface area and surface roughness. In order to effectively remove efavirenz from wastewater influent and effluent as well as surface water, the SPE with the MIP sorbent was adjusted. It was possible to remove 120% in distilled water and 81% in wastewater. Although the amounts in surface water were around 0.975 and 2.88 g l^{-1}, those in wastewater varied from 2.79 to 120.7 g l^{-1}. Since the key emphasis of this work was on water analysis and efavirenz was found in some samples, the next research step should examine solid samples, such as sewage sludge, aquatic plants, and biota. The findings of this investigation showed that efavirenz could not be effectively removed from water by the processing technique utilized in the analysed WWTPs. Overall, this work has shown that MIPs may be utilized to analyze specific ARV medications in water samples [48].

In 2020 Gornik et al carried research to see if MIPs might be used as adsorbents to remove selective serotonin re-uptake inhibitors (SSRIs) from water. SSRIs are anti-depressants frequently found in the environment. The objective of the work was to create an MIP that could be used to remove not only sertraline (SER—the template they specifically targeted), but also the entire class of SSRIs. The synthesized MIPs were assessed for their reactivity, capability, and specificity for SER, and the strongest compounds were selected. The additional characterization included cross-reactivity with other anti-depressants such as fluoxetine (FLU), paroxetine (PXT), escitalopram (ESC), bupropion (BUP), two SER metabolites—norsertraline (NS) and sertraline ketone (SEK), and the structurally similar substance bupivacaine (BUC), as shown in figure 6.3.

The work suggested that SSRIs and their derivatives can be eliminated by MIPs with the basic form of SER. Based on the capability, reactivity, and specificity of the synthesized MIPs, the functionality of the monomer and porogen had a significant effect. The specific MIPs showed cross-reactivity with SSRIs and the metabolite norsertraline, but were less bound to the BUP and BUC. Further, the existence of salt ions had a substantial impact on both the imprinted and non-imprinted materials' efficiency, enhancing it in wastewater. MIP functionality remained

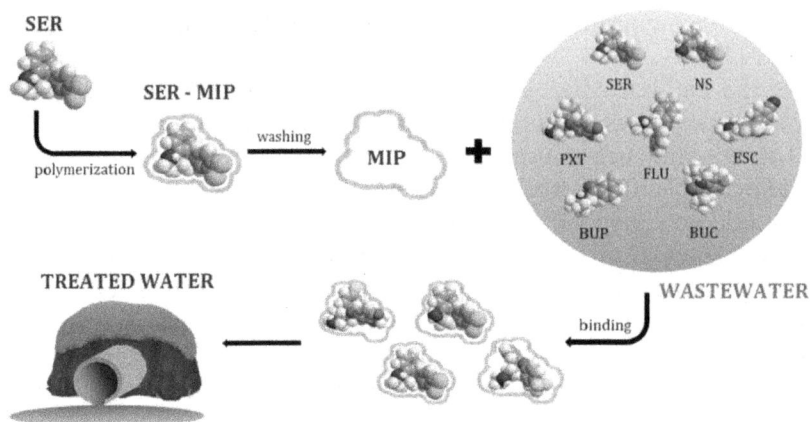

Figure 6.3. Figure representing MIPs and SER as the chosen template for wastewater treatment. (Reproduced from [49]. Copyright 2020 the authors. CC BY 4.0.)

constant over the pH range of 6–8, which is essential for wastewater. The study found a maximal imprinting value of 3.7 and the highest MIP for sertraline in water of 72.6 mg g^{-1}. Although they had less contact area (approximately 27.4 and 193.8 m^2.g^{-1}) than activated carbon (1400 m^2.g^{-1}), their adsorption properties in waste-waters were ideal. The MIPs with greater porosity had more non-specific associations with the targets significantly increasing the total sorption capacity, making it suitable for application in treating wastewater [49].

For effective 4-nitrophenol (4-NP) adsorption and removal from water, a new sorbent depending on surface MIPs (SMIPs) was developed by Liang *et al* in 2020. p-nitrophenol (PNP) was used as template. The surface molecularly imprinted polymer of Fe$_3$O$_4$@SiO$_2$@PNP-SMIP was produced using microwave-assisted surface imprinting technology with 4-NP serving as the template molecule, ureido propyl trimethoxysilane as the functional monomer, and Fe$_3$O$_4$@SiO$_2$ as the base material. The adsorbent quantities of Fe$_3$O$_4$@SiO$_2$@PNP-SMIP are 134.23 mg g^{-1} and 64.07 mg g^{-1}, respectively, and the removal rate of the SMIP to 4-NP can achieve 99% when the recommended dose is 0.01 g, the initial concentration of 4-NP is 6.5 g ml^{-1}, the pH of the solution is 7, the adsorption time is 60 min, and the temperature is 30 °C. Fe$_3$O$_4$@SiO$_2$@PNP-SMIP demonstrated am excellent adsorption property towards 4-NP and it could preferentially adsorb 4-NP in a binary mixed solution when compared to a non-imprinted polymer of Fe$_3$O$_4$@SiO$_2$@PNP-SNIP. Additionally, even in the presence of several molecular moieties of PNP, the specific SPE effectiveness reached 99.2%. The greatest 4-NP removal rate in the previously reviewed literature is 97.3%. As a result, the new material Fe$_3$O$_4$@SiO$_2$@PNP-SMIP is an outstanding adsorbent and eluent to extricate 4-NP from water samples and it offers a novel method for humans to track and identify the presence of 4-NP in water [50].

One of the primary routes by which antibiotics enter the environment is through urban wastewater. To determine the amount of antibiotics released into the environment by urban wastewater treatment systems the levels of antibiotics in wastewater must be monitored. For the *in situ* detection of two common antibiotics, fluoroquinolones (FQs) and sulfonamides (SAs), in urban wastewater, in 2020 Cui *et al* revealed a unique diffusive gradient in thin films (DGT) method based on MIPs. When MIP-DGT is deployed, the two target antibiotics are selectively taken up because MIPs exhibit unique affinity against their templates and their structural analogs. The interpretation of the research of the MIP-DGTs' absorption capability revealed that it was mostly unaffected by the solution's pH (4.0–9.0), ionic strength (1–750 mmol l^{-1}), and dissolved organic matter (DOM, 0–20 mg l^{-1}). Three SA (sulfamethoxazole, sulfapyridine, and trimethoprim) and one FQ (ofloxacin) antibiotics were found with concentrations ranging from 25.50 to 117.58 ng l^{-1} in the effluent of urban WWTPs during experimental studies using MIP-DGT samplers. These findings are in accordance with those obtained using random sampling. The processing plant's overall antibiotic elimination efficacy was 80.1%. The findings suggest that MIP-DGT is a useful method for *in situ* assessment of residual antibiotics in sophisticated urban wastewaters [51].

One of the issues that is currently emerging is the occurrence of pharmaceutical compounds including carbamazepine (CBZ) in environmental wastewater. In 2021 Elmasry *et al* developed up-conversion particles coated with new molecular imprinted polymer (UCNPs@MIP) through *in situ* photo-polymerization by utilizing the internal green light of UCNPs upon photo-excitation at 980 nm. This has been utilized for ultrasensitive detection through fluorescence quenching of UCNP emission by CBZ. The detection limits for CBZ with two different UCNPs@MIPs were 28.5 and 40.0 pM (*S/N* = 3) with wide linear ranges and good correlation coefficients. Furthermore, the practical application of the sensors for the CBZ determination in wastewater and serum matrix was successfully investigated with acceptable recoveries of 98.2%–104.2% and an RSD less than 5%. What is noteworthy is that the UCNP emissions exhibited in the visible wavelength range allowed us to demonstrate the functional utility of our architecture in the form of turn-off sensors. These sensors are capable of detecting CBZ in picomolar concentrations and possess sensitivity at least 300 times lower than the other reported methods. Hydrogen bonds between the hydroxyl groups and the amide, and the π–π conjugation of the benzene rings, both participated in the specific recognition. Remarkably, UCNPs@MIPs could be also blended with chitosan (CS) to form CS–MIP composite film as a simple removal technique of CBZ from wastewater samples with an adsorption capacity of 2.23 mg g^{-1} and removal percentage of 99.7% after 18 h. Furthermore, this UCNP-based sensing method can be used for detecting CBZ in environmental water samples and be further applied to human serum samples showing noticeable sensitivity with good recovery results. This research proposes a reliable and portable strategy for the detection and removal of CBZ from water, which is foreseen to open new avenues for its application in the treatment of wastewater [52].

In 2021 Li *et al* carried out a study in which Methyl red (MR) was used as a target molecule to create a new MIP with altered cellulose carbon microspheres as its core. In order to build a high-sensitivity sensor for MR identification, an SPE column was loaded with a mixture of MR-MIP and cobalt-doped iron carbide (Co–FeC) nanoparticles as the electrode material. Based on MR-MIP@Co–FeC, which merges adsorbent screening and electrochemical monitoring of the residue quantity, an integrative adsorption–detection platform for MR in printing and dyeing waste-water was developed. Utilizing cellulose as the carbon source and MR as the template, the distillation–precipitation method was used to produce microspheres with molecularly imprinted surfaces. The electrochemical sensor made from the MIP and iron paste electrode, co-modified with nitrogen-doped carbonized iron nano-particles, has excellent sensitivity for MR identification in addition to having acceptable adsorption efficiency for the target molecule MR with responsive adsorption efficiency up to 264.2 mg g^1. Additionally, the MR-MIP@Co–FeC/CPE sensor showed good reliability in a number of experiments (as shown in figure 6.4); residual recognition as well as the remediation of actual industrial printing and dyeing wastewater were also successfully completed. Owing to the versatility of the technology based on molecularly imprinted microspheres, it is simple to adapt it to other interesting molecules. The environmental monitoring, food safety testing, and even drug residue testing industries all hold considerable promise for the sensor system described in this paper. The sensing system does present possible difficulties, however. The electrodes, for instance, cannot be

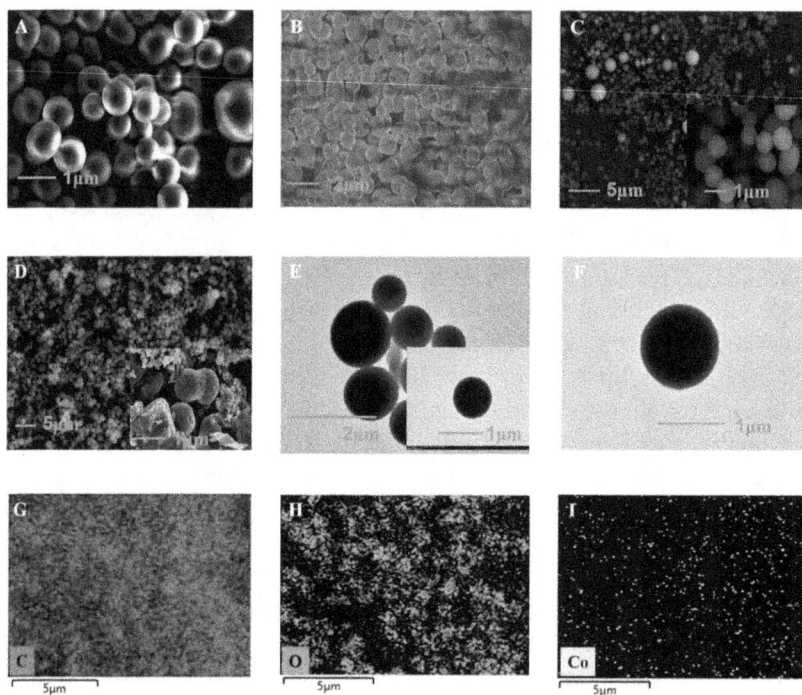

Figure 6.4. SEM images of (A) cellulose carbon microspheres, (B) modified cellulose carbon microspheres, (C) non-imprinted polymer microspheres, and (D) molecularly imprinted polymer microspheres; TEM images of (E) cellulose carbon microspheres and (F) molecularly imprinted polymer microspheres; EDS images of (G) carbon, (H) oxygen, and (I) cobalt in cellulose carbon microspheres. (Reproduced with permission from [53]. Copyright 2021 Elsevier.)

Table 6.1. Different works carried out by various researchers to develop MIPs and MIP composites.

S. No.	Developed MIP/ composite	Methodology	Motive of the work	Reference
1	MIP	SPE, HPLC	Removal of ketoprofen from wastewater	[46]
2	MIP	—	Removal of fenoprofen from wastewater	[39]
3	MIP	Bulk polymerization technique	Removal of ketoprofen from wastewater	[47]
4	MIP	SPE	Removal of efavirenz from wastewater	[48]
5	MIP	—	To remove selective serotonin re-uptake inhibitors (SSRIs) from water	[49]
6	Fe_3O_4@SiO_2@PNP-SMIP	Microwave-assisted surface imprinting technology	4-nitrophenol (4-NP) adsorption and removal from water	[50]
7	MIP-DGT	—	*In situ* detection of two common antibiotics, fluoroquinolones (FQs) and sulfonamides (SAs), in urban wastewater	[51]
8	UCNPs@MIP	Photo-polymerization	Detection and removal of CBZ from water	[52]
9	MR-MIP@Co–FeC	SPE	MR identification for environmental monitoring	[53]

precisely the same after each refining; therefore, deviations are unavoidable and may restrict the use of the developed sensor. A micro-fluidic electrochemical sensing platform incorporating MIP@Co–FeN and other carbon-modified electrodes will be described in a subsequent work to solve this issue [53].

Table 6.1 describes the developed composites of MIPs, the methodology, and the motives of the works carried out by various researchers worldwide. It also gives an idea of how MIPs are useful in wastewater treatment by removing hazardous chemicals which cause water pollution. Through this, an overview regarding MIPs used for treating wastewater can also be revealed and can be implemented in the future for many more advancements in environmental monitoring.

6.4 Current status and future prospects

To remove toxins from wastewater, a variety of methods have been used in the past, including conventional adsorption, coagulation, chemical oxidation, and biological treatments. The majority of these approaches, however, have drawbacks such as poor processing, convoluted operation, high cost, and challenging experimental settings [54]. It is indeed impossible to determine how often contamination is still present in the wastewater after it has been treated. Furthermore, the functional preceding process requires high levels of sample handling and enrichment, as well as very sensitive detection to achieve test findings [55].

Technology for imprinting water-compatible MIPs is still necessary [56]. Traditional polymers are frequently manufactured in organic solvents, and when utilized in an aqueous solution, they have a distinct swelling effect. The MIPs are unable to recognize the target from water samples due to their apparent altered structure after swelling [57]. Water molecules, on the other hand, will compete with the template, weakening or breaking the non-covalent bond between the template and the functional monomer. Despite this, some progress in the creation of water-compatible MIPs has been achieved. The insertion of hydrophilic characteristics into the polymer is a new way for making water-compatible MIPs [58].

MIPs have a particularly high potential for wastewater treatment applications due to their great selectivity and affinity for target compounds. MIPs have been shown to be a highly beneficial approach for selective SPE and removal of a certain analyte or a set of structurally comparable chemicals as excellent sorbents [59]. MIPs play a key role in pollutant degradation when used as catalysts. Furthermore, an MIP-based fluorescence sensor gives an option for special detection and assessment of contaminants from real-world samples. MIP-based composite materials, in particular, have a wide range of potential applications in wastewater treatment [60]. Despite the vast number of patents registered on MIPs throughout the world, the technology is primarily used in the tertiary sector. The translation of MIP technology from the lab to the final product has been hampered by technological obstacles in two areas (i) device design and fabrication and (ii) manufacturing process scaling up [61].

A functional device should not only capture the target compound precisely from a variety of different and complicated matrices, which is where the majority of current research efforts have been focused, but also offer the user a mechanism to retrieve and store the measured data. The system should be compact, ideally portable, user-friendly, quick, and inexpensive [54]. A completely developed and calibrated MIP-based system is currently missing, and its ultimate development would need considerable research expenditure. The manufacturing process, particularly the scale-up from the academic laboratory to large quantities of MIP materials, is the biggest barrier to commercialization [62]. The biggest disadvantage is the lack of repeatability of MIPs between batches, both in terms of morphological and chemical binding capabilities. The problem appears to be caused by a lack of control over manufacturing parameters, which is made even more difficult by the complicated synthesis of some of the suggested sensors [63]. Furthermore, when moving to large-scale manufacturing in an industrial setting, academic laboratory protocols for manufacturing must usually be completely re-engineered, requiring additional research and development tasks before a successful prototype can be obtained and requiring large investments in the early stages of product development [64]. Finally, bulk manufacture of MIPs will need a large quantity of templates, which may be unavailable or uneconomical. In order to reuse the target molecule in different batches, a procedure for target recovery and purification following elution from newly produced MIPs is required [65]. The pursuit for improved sensitivity and selectivity has sparked a lot of study in materials science and engineering, and several sensors using various nanomaterials and advanced production techniques have been described. The novel materials have been thoroughly described and calibrated in laboratory-made solutions, including some real-world examples of tests [66].

While these initiatives may increase analytical performance, they potentially hinder commercialization efforts due to the high cost and complexity of production. MIPs are now being commercialized in specialist sectors such as biotechnology, analytical chemistry, and separation chemistry. The technology is now being commercialized by a number of start-ups originating from tertiary institutions [67].

Several start-ups derived from academic laboratories are currently commercializing the technology. Semorex (Fanwood, NJ, US) specializes in protein-imprinted polymers for the elimination of specific proteins from the gastrointestinal tract in the therapy of Crohn's disease. MIP Technologies AB (Lund, Sweden) offers tailored purification resins. AFFINISEP (Petit Couronne, France) has developed a range of solid-phase-extraction phases used in food and environment analysis, life sciences, and pharmaceuticals. MIP Diagnostics Ltd (Bedford, UK) commercializes different types of tailor-made MIPs for *in vitro* diagnostics. Biotage (Cardiff, UK) designs resins for the removal of low-level contaminants, or extraction of high-value desirables, from any process, particularly for the food, beverage, flavor, and fragrance industries. In addition, the life science technologies and specialty chemicals company Sigma-Aldrich (St Louis, MO, USA) offers solid-phase-extraction materials based on MIP technology. The consumer market for MIP-based sensors has significant challenges [68].

First, structural analog compound interference with the target molecule is a well-known issue in the literature, and some of the complicated materials and synthesis procedures described to prevent this problem are not feasible for large-scale production due to added costs and manufacturing constraints. In most situations, converting laboratory bench protocols to industrial production processes would necessitate the re-engineering of fabrication methods [69]. The evaluations show a number of instances of MIP-based sensors that matched the LOD and linear range requirements for biological and environmental applications, however, the bulk of these sensors are made using expensive and time-consuming processes [70]. The optimization of the production methods is critical in order for these materials to reach the consumer and be mass-produced efficiently [71].

Second, while most of the examined studies involve real-sample testing, they are restricted to only a few promising outcomes. The most prevalent matrices in environmental sensing are natural fluids and wastewaters [72]. Their pH, dissolved solids concentration, and organic matter content may all vary significantly. Clinical studies using MIP-based sensors are required before a biomedical device can be approved by the FDA [73]. Large-scale testing is an expensive and time-consuming process that is one of the most significant impediments to technological innovation. Despite the hurdles, MIP technology continues to attract a large number of application-oriented researchers who are seeking to realize its full potential [74].

6.5 Conclusion

MIPs have notable promise in treating wastewater due to their great efficiency and strong affinity toward target compounds. MIP priorities and potential improvement prospects have been highlighted in order to advance MIP research. MIPs have been demonstrated to be an extremely useful method for specific SPE and elimination of a certain analyte or a set of molecules of similar structure. MIPs contribute significantly to the breakdown of contaminants when used as catalysts. Additionally, an MIP-based fluorescence sensor offers a substitute for the specific identification and quantification of contaminants from actual samples. In particular, a diverse array of potential outcomes in the treatment of wastewater is offered by MIP-based composite materials. MIPs can retain their strong specificity and potentially produce additional performance, such as outstanding charge separation–transportation efficiency and strong adsorption capability, with the use of magnetic particles, nanoparticles, or quantum dots. Molecularly imprinted technology and biological technology are not well understood; future research might concentrate on identifying possible feedstocks that could be combined with MIPs. All of these studies imply that MIPs' prospective applicability for cleaning water is almost certainly true. The production of MIPs that are suitable with water and the probable release of remaining motif are among the remaining difficulties. We persist optimistic for the future possibilities of MIPs as selective adsorbents for treating wastewater as well as a range of other applications after considering these factors into account.

Funding statement

This work received no specific grant from any funding agency.

Conflicts of interest

The authors declare no conflicts of interest.

References

[1] Singhal A, Parihar A, Kumar N and Khan R 2021 High throughput molecularly imprinted polymers based electrochemical nanosensors for point-of-care diagnostics of COVID-19 *Mater. Lett.* **306** 130898

[2] Inyinbor Adejumoke A, Adebesin Babatunde O, Oluyori Abimbola P, Adelani Akande Tabitha A, Dada Adewumi O and Oreofe Toyin A 2018 Water pollution: effects, prevention, and climatic impact *Water Challenges of an Urbanizing World* (London: IntechOpen) pp 33–47

[3] Singhal A, Sadique M A, Kumar N, Yadav S, Ranjan P, Parihar A, Khan R and Kaushik A K 2022 Multifunctional carbon nanomaterials decorated molecularly imprinted hybrid polymers for efficient electrochemical antibiotics sensing *J. Environ. Chem. Eng.* **10** 107703

[4] Khan S, Naushad M, Govarthanan M, Iqbal J and Alfadul S M 2022 Emerging contaminants of high concern for the environment: current trends and future research *Environ. Res.* **207** 112609

[5] Crini G and Lichtfouse E 2019 Advantages and disadvantages of techniques used for wastewater treatment *Environ. Chem. Lett.* **17** 145–55

[6] Obotey Ezugbe E and Rathilal S 2020 Membrane technologies in wastewater treatment: a review *Membranes* **10** 89

 [7] Khan N A, Khan S U, Ahmed S, Farooqi I H, Yousefi M, Mohammadi A A and Changani F 2020 Recent trends in disposal and treatment technologies of emerging-pollutants—a critical review *TrAC Trends Anal. Chem.* **122** 115744

 [8] Haq I and Raj A 2019 Endocrine-disrupting pollutants in industrial wastewater and their degradation and detoxification approaches *Emerging and Eco-Friendly Approaches for Waste Management* (Singapore: Springer) pp 121–42

 [9] Zamri M F, Bahru R, Pramanik S K and Fattah I M 2021 Treatment strategies for enhancing the removal of endocrine-disrupting chemicals in water and wastewater systems *J. Water Process Eng.* **41** 102017

[10] Mahesh N, Balakumar S, Danya U, Shyamalagowri S, Babu P S, Aravind J, Kamaraj M and Govarthanan M 2022 A review on mitigation of emerging contaminants in an aqueous environment using microbial bio-machines as sustainable tools: progress and limitations *J. Water Process Eng.* **47** 102712

[11] Chen L, Fu W, Tan Y and Zhang X 2021 Emerging organic contaminants and odorous compounds in secondary effluent wastewater: identification and advanced treatment *J. Hazard. Mater.* **408** 124817

[12] Rout P R, Zhang T C, Bhunia P and Surampalli R Y 2021 Treatment technologies for emerging contaminants in wastewater treatment plants: a review *Sci. Total Environ.* **753** 141990

[13] Surana D, Gupta J, Sharma S, Kumar S and Ghosh P 2022 A review on advances in removal of endocrine disrupting compounds from aquatic matrices: future perspectives on utilization of agri-waste based adsorbents *Sci. Total Environ.* **826** 154129

[14] Mrinalini M and Prasanthkumar S 2019 Recent advances on stimuli-responsive smart materials and their applications *Chem. Plus. Chem.* **84** 1103–21

[15] Kamel N A 2022 Bio-piezoelectricity: fundamentals and applications in tissue engineering and regenerative medicine *Biophys. Rev.* **14** 717–33

[16] BelBruno J J 2018 Molecularly imprinted polymers *Chem. Rev.* **119** 94–119

[17] Ansari S and Masoum S 2019 Molecularly imprinted polymers for capturing and sensing proteins: current progress and future implications *TrAC Trends Anal. Chem.* **114** 29–47

[18] Zhou T, Ding L, Che G, Jiang W and Sang L 2019 Recent advances and trends of molecularly imprinted polymers for specific recognition in aqueous matrix: preparation and application in sample pretreatment *TrAC Trends Anal. Chem.* **114** 11–28

[19] Dong C, Shi H, Han Y, Yang Y, Wang R and Men J 2021 Molecularly imprinted polymers by the surface imprinting technique *Eur. Polym. J.* **145** 110231

[20] Crapnell R D, Hudson A, Foster C W, Eersels K, Grinsven B V, Cleij T J, Banks C E and Peeters M 2019 Recent advances in electrosynthesized molecularly imprinted polymer sensing platforms for bioanalyte detection *Sensors* **19** 1204

[21] Piletsky S, Canfarotta F, Poma A, Bossi A M and Piletsky S 2020 Molecularly imprinted polymers for cell recognition *Trends Biotechnol.* **38** 368–87

[22] Musarurwa H and Tavengwa N T 2022 Stimuli-responsive molecularly imprinted polymers as adsorbents of analytes in complex matrices *Microchem. J.* **181** 107750

[23] Parisi O I, Francomano F, Dattilo M, Patitucci F, Prete S, Amone F and Puoci F 2022 The evolution of molecular recognition: from antibodies to molecularly imprinted polymers (MIPs) as artificial counterpart *J. Funct. Biomater.* **13** 12

[24] Huang Y J, Chang R and Zhu Q J 2018 Synthesis and characterization of a molecularly imprinted polymer of spermidine and the exploration of its molecular recognition properties *Polymers* **10** 1389

[25] Morsi S M, Abd El-Aziz M E and Mohamed H A 2022 Smart polymers as molecular imprinted polymers for recognition of target molecules *Int. J. Polym. Mater. Polym. Biomater.* **72** 612–35

[26] Reville E, Sylvester E, Benware S, Negi S and Berda E B 2022 Customizable molecular recognition: advancements in design, synthesis, and application of molecularly imprinted polymers *Polym. Chem.* **13** 3387–411

[27] Zare E N *et al* 2022 Remediation of pharmaceuticals from contaminated water by molecularly imprinted polymers: a review *Environ. Chem. Lett.* **20** 2629–64

[28] Kamaruzaman S, Nasir N M, Mohd Faudzi S M, Yahaya N, Mohamad Hanapi N S and Wan Ibrahim W N 2021 Solid-phase extraction of active compounds from natural products by molecularly imprinted polymers: synthesis and extraction parameters *Polymers* **13** 3780

[29] Cantarella M, Carroccio S C, Dattilo S, Avolio R, Castaldo R, Puglisi C and Privitera V 2019 Molecularly imprinted polymer for selective adsorption of diclofenac from contaminated water *Chem. Eng. J.* **367** 180–8

[30] Metwally M G, Benhawy A H, Khalifa R M, El Nashar R M and Trojanowicz M 2021 Application of molecularly imprinted polymers in the analysis of waters and wastewaters *Molecules* **26** 6515

[31] Murray A and Örmeci B 2012 Application of molecularly imprinted and non-imprinted polymers for removal of emerging contaminants in water and wastewater treatment: a review *Env. Sci. Pollut. Res.* **19** 3820–30

[32] Gornik T *et al* 2020 Molecularly imprinted polymers for the removal of antide-pressants from contaminated wastewater *Polymers* **13** 120

[33] Huang D L, Wang R Z, Liu Y G, Zeng G M, Lai C, Xu P, Lu B A, Xu J J, Wang C and Huang C 2015 Application of molecularly imprinted polymers in wastewater treatment: a review *Environ. Sci. Pollut. Res.* **22** 963–77

[34] Guan G, Pan J H and Li Z 2021 Innovative utilization of molecular imprinting technology for selective adsorption and (photo) catalytic eradication of organic pollutants *Chemosphere* **265** 129077

[35] Forner A *et al* 2015 Lack of arterial hypervascularity at contrast-enhanced ultrasound should not define the priority for diagnostic work-up of nodules <2 cm *J. Hepatol.* **62** 150–5

[36] Links D A 2011 Recent advances in molecular imprinting technology: current status, challenges and highlighted applications *Chem. Soc. Rev.* **40** 2922–42

[37] Shen X, Zhu L, Wang N and Tang H 2012 Molecular imprinting for removing highly toxic organic pollutants *Chem. Commun* **48** 788–98

[38] Hussain M 2015 Molecular imprinting as multidisciplinary material science: today and tomorrow *Int. J. Adv. Mater. Res.* **1** 132–54

[39] Mbhele Z E, Ncube S and Madikizela L M 2018 Synthesis of a molecularly imprinted polymer and its application in selective extraction of fenoprofen from wastewater *Environ. Sci. Pollut. Res.* **25** 36724–35

[40] Haupt K 2003 Peer reviewed: molecularly imprinted polymers: the next generation *Anal. Chem.* **75** 376A–83A

[41] Whitcombe M J, Kirsch N and Nicholls I A 2014 Molecular imprinting science and technology: a survey of the literature for the years 2004–2011 *J. Mol. Recognit.* **27** 297–401

[42] Lopes R P, Reyes R C, Romero-González R, Frenich A G and Vidal J L 2012 Development and validation of a multiclass method for the determination of veterinary drug residues in

chicken by ultra high performance liquid chromatography–tandem mass spectrometry *Talanta* **89** 201–8

[43] Beltran A, Borrull F, Marcé R M and Cormack P A 2010 Molecularly-imprinted polymers: useful sorbents for selective extractions *TrAC Trends Anal. Chem.* **29** 1363–75

[44] Jandera P 2009 Molecularly imprinted polymers and their application in solid phase extraction *J. Sep. Sci.* 799–812

[45] Chapuis-Hugon F 2008 Role of molecularly imprinted polymers for selective determination of environmental pollutants—a review *Anal. Chim. Acta* **2** 48–61

[46] Zunngu S S, Madikizela L M, Chimuka L and Mdluli P S 2017 Synthesis and application of a molecularly imprinted polymer in the solid-phase extraction of ketoprofen from wastewater *C. R. Chim.* **20** 585–91

[47] Madikizela L M, Zunngu S S, Mlunguza N Y, Tavengwa N T, Mdluli P S and Chimuka L 2018 Application of molecularly imprinted polymer designed for the selective extraction of ketoprofen from wastewater *Water SA* **44** 406–18

[48] Mtolo S P, Mahlambi P N and Madikizela L M 2019 Synthesis and application of a molecularly imprinted polymer in selective solid-phase extraction of efavirenz from water *Water Sci. Technol.* **79** 356–65

[49] Gornik T, Shinde S, Lamovsek L, Koblar M, Heath E, Sellergren B and Kosjek T 2020 Molecularly imprinted polymers for the removal of antide-pressants from contaminated wastewater *Polymers* **13** 120

[50] Liang W, Lu Y, Li N, Li H and Zhu F 2020 Microwave-assisted synthesis of magnetic surface molecular imprinted polymer for adsorption and solid phase extraction of 4-nitrophenol in wastewater *Microchem. J.* **159** 105316

[51] Cui Y, Tan F, Wang Y, Ren S and Chen J 2020 Diffusive gradients in thin films using molecularly imprinted polymer binding gels for *in situ* measurements of antibiotics in urban wastewaters *Front. Environ. Sci. Eng.* **14** 1–2

[52] Elmasry M R, Tawfik S M, Kattaev N and Lee Y I 2021 Ultrasensitive detection and removal of carbamazepine in wastewater using UCNPs functionalized with thin-shell MIPs *Microchem. J.* **170** 106674

[53] Li X, Yu P, Feng Y, Yang Q, Li Y and Ye B C 2021 Specific adsorption and highly sensitive detection of methyl red in wastewater using an iron paste electrode modified with a molecularly imprinted polymer *Electrochem. Commun.* **132** 107144

[54] Mahmoudpour M, Torbati M, Mousavi M M, de la Guardia M and Dolatabadi J E 2020 Nanomaterial-based molecularly imprinted polymers for pesticides detection: recent trends and future prospects *TrAC Trends Anal. Chem.* **129** 115943

[55] Janczura M, Luliński P and Sobiech M 2021 Imprinting technology for effective sorbent fabrication: current state-of-art and future prospects *Materials* **14** 1850

[56] Zhu G, Cheng G, Wang P, Li W, Wang Y and Fan J 2019 Water compatible imprinted polymer prepared in water for selective solid phase extraction and determination of ciprofloxacin in real samples *Talanta* **200** 307–15

[57] Horemans F, Weustenraed A, Spivak D and Cleij T J 2012 Towards water compatible MIPs for sensing in aqueous media *J. Mol. Recognit.* **25** 344–51

[58] Benito-Peña E, Martins S, Orellana G and Moreno-Bondi M C 2009 Water-compatible molecularly imprinted polymer for the selective recognition of fluoroquinolone antibiotics in biological samples *Anal. Bioanal. Chem.* **393** 235–45

[59] Marć M, Kupka T, Wieczorek P P and Namieśnik J 2018 Computational modeling of molecularly imprinted polymers as a green approach to the development of novel analytical sorbents *TrAC Trends Anal. Chem.* **98** 64–78

[60] Yusof N A, Ab. Rahman S K, Hussein M Z and Ibrahim N A 2013 Preparation and characterization of molecularly imprinted polymer as SPE sorbent for melamine isolation *Polymers* **5** 1215–28

[61] Wang Y, Cottman M and Schiffman J D 2012 Molecular inversion probes: a novel microarray technology and its application in cancer research *Cancer Genet.* **205** 341–55

[62] Wan Q, Liu H, Deng Z, Bu J, Li T, Yang Y and Zhong S 2021 A critical review of molecularly imprinted solid phase extraction technology *J. Polym. Res.* **28** 1–6

[63] Ozcelikay G, Kaya S I, Ozkan E, Cetinkaya A, Nemutlu E M, Kır S and Ozkan S A 2022 Sensor-based MIP technologies for targeted metabolomics analysis *TrAC Trends Anal. Chem.* **146** 116487

[64] Lowdon J W, Diliën H, Singla P, Peeters M, Cleij T J, van Grinsven B and Eersels K 2020 MIPs for commercial application in low-cost sensors and assays—an overview of the current status quo *Sensors Actuators* B **325** 128973

[65] Li R, Feng Y, Pan G and Liu L 2019 Advances in molecularly imprinting technology for bioanalytical applications *Sensors* **19** 177

[66] Gao M, Gao Y, Chen G, Huang X, Xu X, Lv J, Wang J, Xu D and Liu G 2020 Recent advances and future trends in the detection of contaminants by molecularly imprinted polymers in food samples *Front. Chem.* **8** 616326

[67] Tong P, Li M, Meng Y and Li J 2021 Molecularly imprinted polymer composites in biological analysis *Molecularly Imprinted Polymer Composites* (Cambridge: Woodhead) pp 143–72

[68] Kadhem A J, Gentile G J and Fidalgo de Cortalezzi M M 2021 Molecularly imprinted polymers (MIPs) in sensors for environmental and biomedical applications: a review *Molecules* **26** 6233

[69] Patel K D, Kim H W, Knowles J C and Poma A 2020 Molecularly imprinted polymers and electrospinning: manufacturing convergence for next-level applications *Adv. Funct. Mater.* **30** 2001955

[70] Sargazi S, Fatima I, Kiani M H, Mohammadzadeh V, Arshad R, Bilal M, Rahdar A, Díez-Pascual A M and Behzadmehr R 2022 Fluorescent-based nanosensors for selective detection of a wide range of biological macromolecules: a comprehensive review *Int. J. Biol. Macromol.* **206** 115–47

[71] Ansari S and Masoum S 2021 Recent advances and future trends on molecularly imprinted polymer-based fluorescence sensors with luminescent carbon dots *Talanta* **223** 121411

[72] Ayankojo A G, Reut J, Nguyen V B, Boroznjak R and Syritski V 2022 Advances in detection of antibiotic pollutants in aqueous media using molecular imprinting technique—a review *Biosensors* **12** 441

[73] Alberti G, Zanoni C, Losi V, Magnaghi L R and Biesuz R 2021 Current trends in polymer based sensors *Chemosensors* **9** 108

[74] Srivastava A, Gupta S, Quamara M, Chaudhary P and Aski V J 2020 Future IoT-enabled threats and vulnerabilities: state of the art, challenges, and future prospects *Int. J. Commun. Syst.* **33** e4443

IOP Publishing

Molecularly Imprinted Polymers for Environmental Monitoring
Fundamentals and applications
Raju Khan and Ayushi Singhal

Chapter 7

Molecularly imprinted polymers for the detection of heavy metals

Yasmin Bano

Heavy metal contamination causes adverse effects on the environment and its inhabitants, and the remediation of contaminates is important for people to live healthy lives in healthy surroundings. Molecular recognition through molecularly imprinted polymers (MIPs) has been found to be an excellent form of heavy metal determination and they have important functions in biological contexts. MIPs are an attractive technique and have set a benchmark in polymer science, basically working on the lock and key principle. Molecular imprinting techniques (MITs) are used in the design of MIPs by incorporating self-assembled monomers and templates in the presence of cross-linkers. The aim of the present review is to highlight the various MIP-based techniques that are used in the detection and elimination of heavy metals and their ions to achieve a sustainable healthy and hearty environment.

7.1 Introduction

7.1.1 Heavy metals

Humanity's war against human, animal, and environmental safety and health is an ambitious and gruelling conflict that is yet to be won and needs enhanced/transformed tools to succeed according to the challenges. Several types of pollution are threats worldwide and of them heavy metal (HM) pollution is a major cause of danger for plant, animal, and human health, as well as the quality of the environment. The term HMs is used for a group of elements showing metallic and/or metalloid properties, with a high atomic number (>20), and with relatively higher atomic density (>5 g cm^{-3}) in comparison to water [1]. Arsenic (As), mercury (Hg), lead (Pb), copper (Cu), cobalt (Co), iron (Fe), chromium (Cr), cadmium (Cd), manganese (Mn), nickel (Ni), zinc (Zn), and selenium (Se) are the names of some common HMs. Small amounts of certain HMs, such as Fe, Co, Mo, Ni, Zn, and Cu are essential for health, working as cofactors for different enzymes that crucially maintain the metabolism processes of organisms, but after a certain limit they pose considerable negative impacts, and at higher concentrations could be life-threatening. Other HMs and metalloids such as As, Pb, Cd, and Hg do not have any beneficial effects and are listed as non-essential, and are highly toxic even at low exposure levels [2]. HMs are ubiquitous and naturally found elements that exist in the Earth's crust and persist through the phenomenon of never being destroyed or degraded. These HMs and metalloids may slowly enter into plants, animals, and humans via various processes through contaminated air and water. Further, because of the progression up the food chain over time and the tendency toward bioaccumulation (a gradual increase in the amount of a chemical substance in a biological organism due faster accumulation than excretion), these HMs can cross the threshold level of safety [3]. Ions of these HMs have been known to interact with almost all cell components (proteins and DNA) and change their conformations or damage DNA, or generate reactive oxygen species which alter many vital processes that lead to carcinogenesis or apoptosis and ultimately death [4]. The detection and removal of these HM ions is strongly recommended to achieve a healthy ecosystem.

7.1.2 Biomimetics: methods of detection

For a long time, natural biological resources have been used for the detection of different molecules, e.g. antibody based detection. However, such types of processes have limitations in terms of storage, stability, shelf life, cost, ethical approval, and more importantly notable batch-to-batch variation. Reports demonstrate that around 75% of total available antibodies performed well and show no validation below the mark or perform their intended role adequately [5]. Notwithstanding improved validation ability, there remains the need for a large number of animals for antibody production. It is estimated that in Europe alone around one million animals per year are required, and this knowledge is enough to drive research to focus toward finding antibody alternatives [6]. Several different methods based on antibody mimicry have been used for the detection of molecules. One method is 'antibody mimics', which are robust proteins engineered to be more stable and smaller counterparts used for detection. These antibody mimics are similar in function but structurally different to antibodies, for example affibodies with three α-helices and a molecular mass of ~6 kD in comparison to ~150 kD for typical monoclonal antibodies. The extreme mass reduction makes them able to resist an extremely high range of pH and temperature. The Food and Drug Administration (FDA) approved affibodies for therapeutic and diagnostic uses but they are expensive and show limited market viability due to a lack of purification techniques [7]. Another example of antibody mimics is the 'single-chain fragment variable' (scFv) antibody, a class of fusion proteins engineered by fusing immunoglobulin's heavy (V_H) and light (V_L) variable regions linked via a short polypeptide linker. Thus this tiny fragment of an immunoglobulin with molar masses of ~50 kD preserves the typical antigen-binding specificity [8]. Another key feature of using scFv antibodies is their easy and economical mass production by using straightforward expression systems such as *Escherichia coli*, whereas monoclonal antibody production requires intricate mammalian systems for articulation and massive post-translational modification. Fragment of antigen-binding (Fab) regions are comprised of the entire light chain (V_L and C_L) and V_H of an immunoglobulin (molecular mass ~28 kD). The enhancement in phase display methods has made it easier to find suitable scFv and Fab; these multifaceted *in vitro* methods are capable of fast assortment of strong affinity and highly specific antibodies or antibody fragments. The basic motivation for constructing these antibodies was therapeutic applications, and further extended via the bio-sensing community as recognition elements. These fragments have the advantages of low-cost and easy production processes with improved sensitivity, but a major limitation is their denaturation at the time of immobilization upon the surface of sensors [9].

Aptamer based mimicry also provides an option for detection. Aptamers are a newly emerged class of oligonucleotide-based molecular recognition elements conflicting antibody based methods, which can fold into a desire geometry to bind specific and selectively targeted molecules. The systematic evolution of ligands by exponential enrichment (SELEX) method has been used to generate aptamers from the large random-sequence libraries [10]. In this process, generally an oligonucleotide molecule is incorporated in a mixture of targets, the bonded oligonucleotides are separated from the unbonded pool, then they are amplified to generate a new population of molecules with identical properties. Selecting sensitivity is further increased, first by performing the counter selection, a process of adding similar molecular structures to the analyte, and second by removing extra base-pairs bonded to non-targeted targets [11]. Basically, six different methods are outlined in the literature for SELEX and all were compared by Zhang's group in 2019 [12]. Is is considered to be a low-cost synthesis method and improved technology has sustained an upsurge in the popularity of SELEX. To facilitate surface immobilization modification, i.e. the addition of groups or labelling, it can be done at the 3' or 5' ends, but carefully, because the added group can interact and hinder the analyte-biding capability since aptamer–target binding is based on conformational changes. Comparatively, DNA aptamers provide more stability than antibodies, as they are sensitive to nuclease degradation and lead to the synthesis of more stable hybrid-systems via integrating polymeric elements through chemical alterations to form electropolymerized polymers on all sides the aptamers to fix them in place [13].

7.1.3 Polymers to molecular-imprinted polymers

Since 2010, as aptamers were starting to grow in popularity and applications to measure proteins, molecularly imprinted polymers (MIPs) have also followed a trend of growth and improvement, and there has been an exponential increase in the number of publications on these materials as polymeric recognition elements. Structurally, MIPs are a completely different concept, depending on novel polymerization technology to create specific identification sites for a target in a polymeric matrix. Polymerization is the process of merging monomers chemically and produces long chains of polymers into 3D forms through various mechanisms [14]. MIPs are an augmented polymerization technology where cavity-based polymeric templates are used to construct highly specific binding sites that can bind to selective molecular targets. Initially these systems were used in chromatographical methods but have further expanded to clinical applications such as antibody development and ligand/target binding [15]. Biochemical revolutions have accelerated the opportunities for using MIPs in various ways, such as purification, identification, and separation of pollutants, pesticides, toxic elements, pharmaceuticals residues, toxic anions, HM ions, sensing of microorganisms, and environmental monitoring [16]. A number of polymerization techniques have been used for the creation of MIPs. A few examples are free-radical (co)polymerization (FRCP), atom transfer radical polymerization (ATRP), solution-phase polymerization, and reversible addition fragmentation chain-transfer polymerization (RAFT). Amid all these approaches FRCP is a widely used method because of its attractive properties of ease and rapidity, its lower requirements for sophisticated instruments, as well as its suitability for mass scale production [17]. Based on the same principle, ATRP is a catalyst driven method used for the synthesis of ultrathin films from a solid substrate. Due to zero production of solution-phase radical species, this method is free from solution-phase polymerization.

In addition, template based polymerization of functional monomers is a key factor for designing highly selective MIPs. Principally, the synthesizing technology of MIPs comprises three stages—(a) distribution of a (non-)covalent conjugate between the template molecule and functional monomer, (b) polymerization or aggregation of the conjugated monomer–template composite, and (c) removal of the template. Removal can be performed via any appropriate method, such as chemical cleavage or executing solvent extraction. Five fundamental variables are needed to design an MIP: the template (a fundamental component that leads functional groups to the functional monomers), functional monomers (they provide the requisite functional groups so the template complex can form by non-/covalent bonds), crosslinkers (they help to control the morphology and stability of the polymer matrices and encompass recognition sites as well), initiators (these chemicals react initially with a single monomer and yield a compound linked with a large number of other monomers to form a polymeric matrix), and porogens (porogenic solvents carry all the mentioned components and aid in producing porous structures in macroporous polymers) [18] (figure 7.1).

7.2 Molecularly imprinted polymers

Advancements in science and technology have surged on many fronts, through the improvement of techniques which help to improve life and MIPs are one such technique. Based on imprinting matrices, MIPs provide robust molecular detection and are able to mimic natural entities, namely antibodies and biological receptors, because the qualities (functional and structural) of the building blocks (monomers) are imprinted on polymers that are similar to natural amino acids. Instead of having of vast structural dissimilarity to antibodies, one can consider the similarity between the binding sites of MIPs and the active sites of enzymes [19]. It is clear that the molecular imprinting approach facilitates creating very specific and selective artificial receptors/entities of preselected materials that can surpass some other synthetic recognition

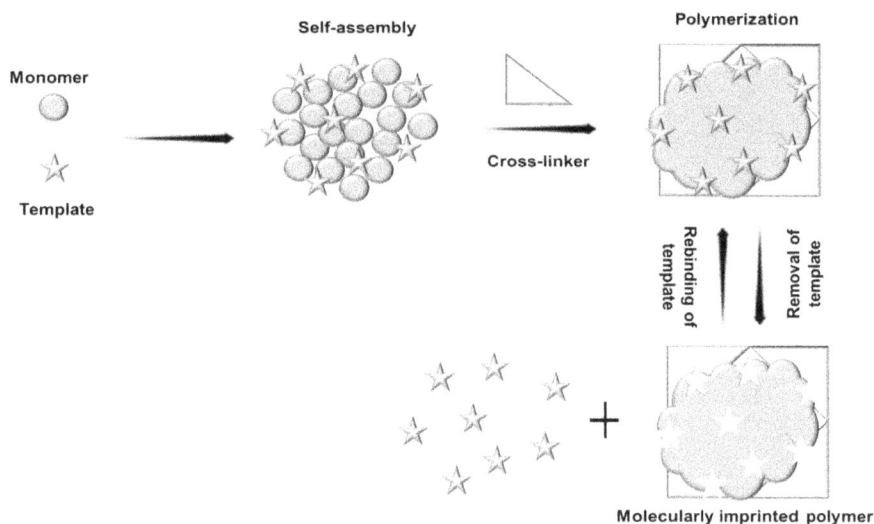

Figure 7.1. Synthetic steps to prepare molecularly imprinted polymers.

materials in terms of cost, stability, and durability (e.g. can be resist from nuclease enzyme degradation) and that can be used as an ideal material in various developing approaches [20]. The main advantages of using MIPs is that they are commercially viable and the market competency make the things easily reachable while on the other hand MIPs allow virtual and the market competency make the things easily reachable while on the other hand MIPs allow virtual detection of targeted targets [21], for example, Aspira Biosystems sells MIPs to capture microorganisms.

The next section described the techniques based on MIPs for the significant detection and removal of HMs by means of specific templates for a material. A great deal of research is focusing on this issue because of the challenging degradation and easy bioaccumulation properties of HMs. Imprinted polymers have an outstanding ability to identify the HM ions by absorbing them on polymers [22]. This MIP-based sensing of HMs can be achieved using electrochemical as well as optical detection methods. Electrochemical sensing is based on assessing the changes in signals generated at the time of interaction between a chemical and a sensing surface. Three different electrochemical techniques utilized in sensing relays on MIPs are described in the literature (figure 7.2). (i) The first is for redox-active analytes and for this direct determination is used in the form of Faradaic current generated due to direct electron movement between the target and electrode surface. (ii) In the second technique, an indirect quantification can be used, as in the case of enzymatic targets where the production of redox-active products at the electrode surface indicates the presence of the intended targets, and (iii) third, in another indirect quantification method the flux of a redox marker is used, such as ferri/ferrocyanide, and on binding of the targets signals are modulated. Electrochemical detection with great specificity, simpler and cheap investigational aspects, as well as the potential for incorporation into portable devices, is gaining attention [23].

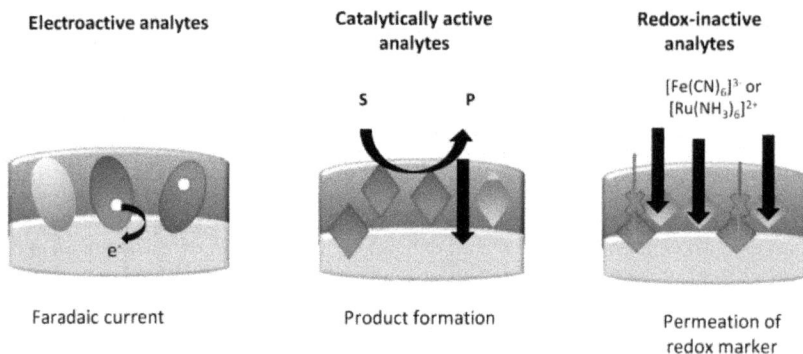

Figure 7.2. Illustrations of the three main detection approaches used in electrochemical readouts of MIP-based sensors. (Reproduced with permission from [23]. Copyright 2020 the authors. CC BY 4.0.)

7.3 MIP-based sensing and removal of heavy metals

This section is focused on the detection of HMs or HM ions as these substances are considered one of the most challenging environmental hazards. Many separation and purification based scientific studies have described the well-established utilization of MIPs in the removal of HM ions from various environmental and synthetic sources. Compared to existing platforms such as membrane separation and ion exchange mechanisms, MIPs have gained advantage in terms of excellent recognition, selectivity, and specificity [24]. The principle behind imprint polymerization is primarily the co-polymerization of a monomer and template complex via amalgamation with a cross-linker to form a 3D polymer complex. The template is subsequently removed, yielding a highly selective polymer matrix with respect to the template molecules with active binding entities. The features of any MIP depend upon the desired application [25]. The monomer–template complex is a fundamental factor that plays a key role in the effectiveness of an MIP because the identification of the target is mainly influenced by the extent of complexation. Many MIPs have been manufactured for the adsorption and uptake of toxic HM ions for their sensing and removal. Furthermore, imprinted polymers can be interfused with different transducing elements, such as microgravimetric [26], optical [27], thermal [28], and electrochemical [29]. Broadly, electrochemical sensing technology encompasses two components, one is the recognition element which provides the signal upon binding with the targeted analyte, if present, and the other is the transducer which converts the interacting signals into a readable or analytical signal [30]. Furthermore, if the polymer matrices are prepared selectively for the recognition of ions, such an imprinting method is termed as ion imprinting polymerization (IIP) technology [31], and this technology has drawn substantial marketable interest as it is low-cost and compatible with mass-manufacturing.

7.3.1 Mercury (II) ions

Mercury (Hg), also known as quicksilver, exists naturally in the environment and leaches through geothermal and anthropogenic activities. It is reportedly the most common metal pollutant and approximately 3400 t/yr Hg is flushed out into the environment, about 95% of this dwells in terrestrial soil, and 3% and 2% reside in the surface ocean water and atmosphere, respectively. Anthropogenic activities account for 70% of Hg contamination worldwide. The occurrence of Hg ions in water bodies presents a great threat to ecological balance [32]. In the environment Hg occurs in many oxidative stages, including inorganic salts and organic complexes. Released Hg ions generate toxicity as they bind with cell components (proteins and enzymes) and change either their morphology or block the functions, eventually their destructive action leads to cell death. Chemical properties drive the mercuric soil contamination and methylation changes it into its most toxic form. Methylmercury has a great potential of binding with the sulfhydryl ligands of protein monomers (amino acids), which lessens the changes in protein folding and restricts their functioning [33]. The World Health Organization (WHO) set a a 6 μg l^{-1} (30 nM) threshold amount for Hg in drinking water [34].

Singh *et al* [35] described the design of MIPs in detail, for which they used methacrylic acid (MAA), ethylene glycol dimethacrylate (EGDMA), azobisisobutyronitrile (AIBN), and cyclohexanol as the monomer, cross-linker, initiator, and porogenic solvent, respectively, and the recorded detectable adsorption ability was 125 μM g^{-1}. The process of polymerization used to design the MIP is shown in figure 7.3.

Another study was performed by Zhang *et al* [36] who proposed a Hg^{2+}–IIP sorbent array on dithizone–Hg^{2+} chelation for the specific detection of Hg^{2+} in aqueous, biological, and environmental samples. In the sol–gel process (a wet chemical technique) the strong chelation compound dithizone was used as the template material that made a complex with Hg^{2+} ions due to its chelating property. The template complex was induced to couple non-covalently with a functional

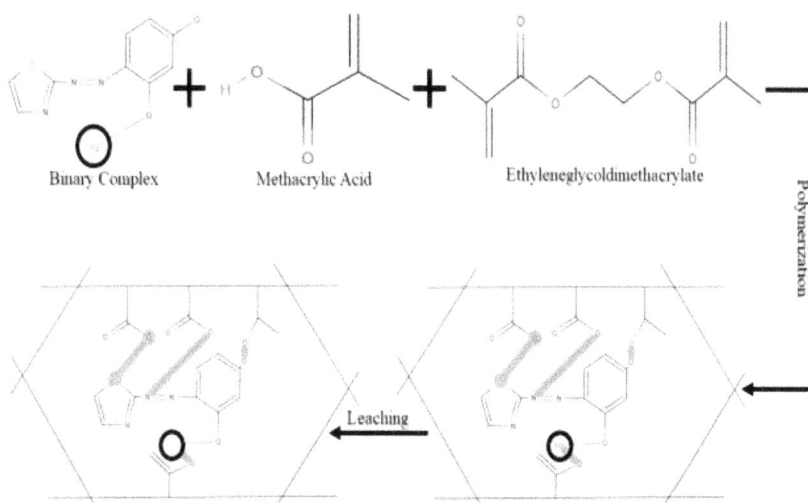

Figure 7.3. The polymerization mechanism for the Hg^{2+}–MIP designed by [35]. (Reproduced with permission from [35]. Copyright 2010 Elsevier.)

monomeric group called 3-amino-propyltriethoxysilane (APTES), as a silane or silicon hydride coupling agent is usually used in the sol–gel platform because its hydrophobicity. Hg^{2+} ions were eliminated from the polymeric matrix which left imprints of Hg^{2+} ions on the polymer surface that could specifically recognize Hg^{2+} in the samples. The chelating method demonstrated IIPs as an ideal sorbent compound for Hg^{2+} ions when compared to inductively coupled plasma mass spectrometry (ICP-MS) and provide a simpler and more robust environmentally favourable method in contrast to conventional methods.

Very limited literature is available that reports IIP-based electrochemical sensing of Hg^{2+} detection. However, Alizadeh et al [37] defined an easy and effective IIP-based sensor based on a modified carbon paste electrode for the determination of Hg^{2+} ions in natural water samples. In this technique vinylpyridine was the functional monomer and a complexing agent and initiator promoted the interaction of Hg^{2+} with the vinylpyridine. The Hg^{2+} sensing selectivity can be improved further by adding itaconic acid monomers or by making changes in the precipitation polymerization methodology [38]. A voltametric sensing method can also be used for the detection of Hg^{2+} through carbon ionic liquid paste electrodes (CILEs) impregnated with Hg^{2+}–IIP nanobeads [39]. In such a case, for the improvement of the conductivity of the electrode, in the process of synthesis of the carbon paste the non-conductive organic binder was substituted with carbon ionic liquid, making it able to sense the presence of a small amount, from 1 nM, to a higher range of up to 2000 nM of Hg^{2+} in a complex mixture. Recently, gold electrodes were introduced for the construction of an IIP electrochemical sensor, with a 1 pM limit of detection (LOD) concentration of metal ions. Modified surfaces of electrodes are being developed using gold with diazonium salt or via developing ZnO nanorods [40] followed by electropolymerization of pyrrole, while Hg^{2+} and L-cysteine are available as the template and cross-linker, respectively. Yasinzai et al [41] found the presence of Hg^{2+} ions in an aqueous environment via interdigital electrodes coated with IIP receptors with the help of three polymeric systems—styrene, N-vinylpyrrolidone, and styrene-co-vinylpyrrolidone. All three polymer based systems were capable of recognizing the analyte, even at very minute concentrations such as 10 ppm.

7.3.2 Lead (II) ions

The Agency for Toxic Substances and Disease Registry listed lead (Pb) in the second position in the list of most toxic HMs [42]. There are many ways in which Pb leaches into the environment, for example, the most common source is ores of metals (Zn, Cu, Ag, etc) or through industries (paints, plastic, cables, pipelines, etc), but among all the sources fuel combustion is the most dangerous anthropogenic cause of Pb^{2+} ions introduction [43], while plumbing increases the contamination in drinking water. The acute accumulation of Pb causes severe poisoning and harmful cognitive developmental effects on children [44]. To minimize the Pb induced poisoning, the WHO has consecutively reduced the threshold value of Pb from 100 $\mu g\ l^{-1}$, to 50 $\mu g\ l^{-1}$, to 10 $\mu g\ l^{-1}$ in drinking water in the past two decades [34].

In the literature the detection of Pb^{2+} ions is proposed via hydrophilic Pb^{2+}–MIPs based on a precipitation polymerization technique, where the active monomer 2-(allyl sulphur) nicotinic acid was cross-linked with the template through the cross-linker EGDMA [45]. $Pb(NO_3)_2$ and the functional monomer 2-(allyl sulphur) niacin were mixed in methanol or triple-distilled water and reflux treated with nitrogen. It was further refluxed by adding methanol containing EGDMA and azobisisobutyr-onitrile (AIBN). The obtained sample was rinsed with methanol and the synthesized template was removed, as illustrated in figure 7.4. In another study a novel IIP was synthesized for the adsorption of Pb^{2+} ions by using two active monomers to fabricate Pb^{2+}–IIP driven ionic interactions. To obtain the IIPs, suspension polymerization was used. $Pb(NO_3)_2$ was mixed with two monomers, namely MAA and 4-vinylpyridine, one by one in the presence of ultrapure water. Toluene and hydroxyethyl cellulose were mixed in the same container and then after the adding of EGDMA and AIBN it was degassed. The degassed solution was mixed with nitrogen to remove unbounded particles. The filtered polymerized particles we purified repeatedly with acetone mixed water to eliminate the components of monomers and cross-linkers and were placed into a vacuum drier. Subsequently, the ions were removed from the resultant IIPs by reacting with HNO_3 and again the polymers were washed and dried to obtain suspended imprinted polymers [46]. With their excellent qualities of fast kinetics, stability, and specific binding capability of Pb^{2+}, the IIPs present a high significant adsorption ability (8.35 mg g^{-1}). In addition, the Pb^{2+}–IIP can be reused for up to 10 cycles of adsorption with minimal loss of function, and the Pb^{2+}–IIP not only proved to be a potential candidate for the analysis of Pb in water samples, but also a potential candidate for rapid removal from aqueous solutions with a recovery mass of up to approximately 105%.

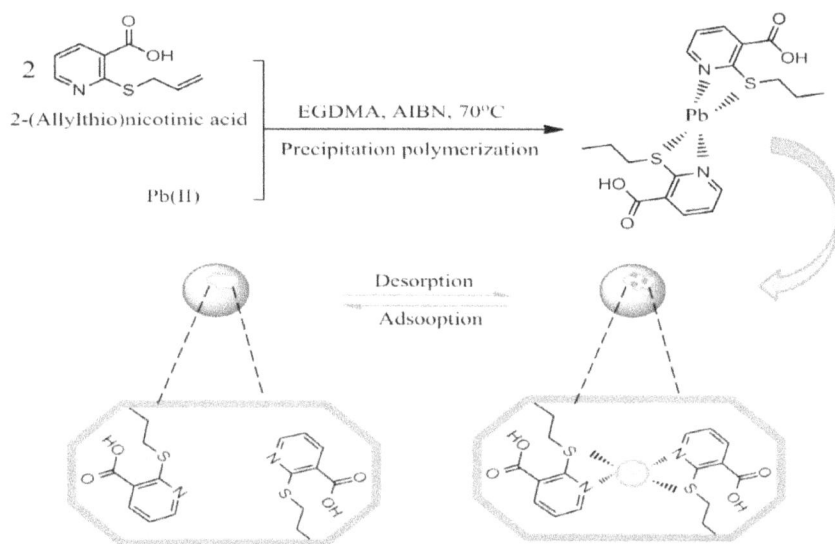

Figure 7.4. Illustration of the synthesis and polymerization mechanism for designing Pb^{2+}–IIP/MIPs. (Reproduced from [45]. Copyright 2023 SAGE Publications Ltd. CC BY 4.0.)

In the literature electrochemical sensors based on IIPs also play an important role for environmental HM identification. The creation of electrochemical sensors has been inspired by the above described methods, for example precipitation polymerization was adopted to create a novel nano-structured IIP by using complexes of Pb^{2+}, MAA, EGDMA, and modified CPEs that lead to identifying the presence of Pb^{2+} ions in different water models and edible salts, with an LOD in the 0.6 nM range [47]. Another group of researchers used the precipitation polymerization process for Pb^{2+} ions with complexes of Pb^{2+}, 4-vinylpyridine, EGDMA, AIBN, and 4-(2-pyridylazo) resorcinol as the template, functional monomer, cross-linker, initiator, and Pb-binding ligand, respectively, in acetonitrile solution. The sensor exhibited a higher selectivity (the LOD was 30 pM) towards Pb^{2+} ions in a mixture of HM ions, and thus was useful for tracking trace amounts of Pb^{2+} ions in polluted natural bodies of water [48]. Hu and colleagues [49] designed a Pb^{2+}–IIP with the help of methyl methacrylate (functional monomer) that can detect Pb^{2+} contamination in the waste pool water and rice with interference from other HMs above 60% with an LOD of 0.01 μM. Furthermore, for enhancing the analysis of Pb^{2+} ions researchers modified the electrode design by expanding the active surface area via integrating some additional nano-material supplies for the intensification of faster electrons transfer. For example, recently Dahaghin *et al* [50] obtained pre-polymerized IIP for Pb^{2+} by co-polymerizing EGDMA with an itaconic acid–Pb^{2+} nexus. The formed IIP material was further infused with trace amounts of multi-walled carbon nanotubes (MWCNTs, 6% w/w) that transformed the identification capability in watery samples (with an LOD of 3.8 pM). However, increased amounts of interfering HMs, such as 50 fold concentration of Fe^{2+} and Zn^{2+} and 40 fold concentration of Cu^{2+}, had an adverse effect on the sensor activity [51]. Dahaghin *et al* [50] also designed another IIP-based electrochemical sensors by embedding Fe_3O_4@SiO_2 nanomaterials as a consequence of polymerization of 2-(2-aminophenyl)-1*H*-benzimidazole (the ligand) and 4-vinylpyridine (the functional monomer). The synthesized polymer can be used for modification in glassy carbon electrodes (GCEs) and sensing Pb^{2+} ions in fruit juices and water samples with an LOD of 0.24 nM.

7.3.3 Cadmium (II) ions

Cadmium (Cd) is a carcinogenic metal that causes toxicity when a person breathes in contaminated air or eats or drinks things containing high levels of cadmium. It has the propensity to bio-accumulate in the cells and long-term exposure leads to severe health effects, in particular kidney failure, bone damage, and lung cancer [52]. The WHO limits the threshold value of Cd in drinking water to below 10 μg l^{-1} (the safe value is 3 μg l^{-1}) [34, 53]. Cd contamination is commonly combined with Zn and is released into the environment from Zn ores during the processing of Zn. Industrial anthropogenic activities also result in Cd toxicity due to its critical applications in manufacturing industries (Ni–Cd batteries, electroplating, alloys, pesticide, etc) from which it is leaked directly into the air and water.

In a noteworthy study, Xi *et al* [54] showed the design of a novel Cd–MIP to detect Cd^{2+} ions in wastewater composites using of different components—a

Cd–dithizone complex, MAA (functional monomer), AIBN (initiator), and EGDMA (cross-linker). In its preparation protocol, dithizone was dissolved in NH_4OH and a solution of the Cd supplement ($CdCl_2 \cdot 2.5H_2O$) was added drop by drop. The obtained precipitate was filtered before washing and placing into a desiccator. The Cd–dithizone complex, MAA, and ethyl acetate solution were mixed in an aqueous medium and a nitrogenous purge was apply. The polymer mixture was left for some time to allow the self-assembly of the monomer and template. For the eradication of porogens basic washing was preferred with water and methanol, while the Cd^{2+} ion template was removed by acid washing to obtain pure fabricated MIPs. The measured adsorption of the MIP was found to be 44.6 mg g^{-1}. Yang's group [55] further extended this technique through using nanotube structures. Their group designed a Cd^{2+}–IIP by means of halloysite nanotubes (HNTs) to analyse the presence of Cd^{2+} ions in environmental samples. The HNTs offered a safe, secure, non-dangerous, nontoxicy, low budget, and strong biologically portable mechanism for MIP synthesis. Moreover, batch experiments also provide proof of safe, effective, and specific absorption and removal of Cd^{2+} ions from water bodies, as HNTs are cost-effective and widely available in China at a very cheap cost [56].

Similarly to Pb^{2+} and Hg^{2+}, limited literature has focused on the electrochemical sensing of Cd^{2+} ions through IIP/MIP-based detection. For example, a decade ago Alizadeh and his team established the IIP-based detection of Cd^{2+} ions by applying modified CPEs on electrodes (the LOD was 0.52 nM). That sensor had a highly specific selectivity and, despite the presence of 500-times as much of various alkaline or earth alkaline cationic elements, was able to sense 50 nM Cd^{2+} ions in real water samples [57]. Later, in 2018, Mousavi *et al* patented a design for of Cd^{2+}–IIP via a co-precipitation polymerization mechanism through the use of some varieties of monomers in acetonitrile that were consequently attached to GCEs. Most importantly, the presence of the cations Pb^{2+}, K^+, Ag^+, Zn^{2+}, Cr^{3+}, Hg^{2+}, and Ni^{2+} in amounts more than 100 times larger did not affect the sensing of 0.03 μM Cd^{2+} ions (the LOD was 0.1 nM) in various water samples such as river, tap, and wastewater models [58]. Recently, Wu *et al* [59] used the natural polymer chitosan in the form of the base substrate for the design of Cd^{2+}–IIPs. However, the sensing of the obtained Cd^{2+} (at 0.09 μM) sensor could be reduced if the interference by other cations crossed the 20-fold limit. The results exhibit a linear voltammetry response in the presence of Cd^{2+} ions (the LOD was 17 nM) in a range of 0.1 μM to 0.9 μM concentration.

7.3.4 Chromium (III) ions

The key sources of chromium (Cr) released into the environment are the industrial wastes released in the form of water or smog. Industries such as leather, textiles, cement-producing plants, photography, electro-painting, metal cleaning, and chromate mines, continue to utilize Cr for commercial purposes and dispose of waste with high amounts of toxic metals directly into nearby water bodies or into the air. The increasing concentration in the environment is alarming because it is not only toxic to humans but also to microorganisms, plants, and other animals, as it a mutagenic as well as a carcinogenic metal that can hamper cell growth and consequentially lead to death [60]. Cr is found in many oxidative stages in the environment. Cr^{3+} is mostly reported naturally but can be produced through the reduction of Cr^{4+}. Comparatively, the Cr^{3+} ionic form is less toxic, long persisting, and a substantial essential nutritive element for humans as it plays a role in metabolic processes (carbohydrates, insulin, and lipids), but an excess is dangerous, while the Cr^{4+} ionic stage is the most toxic and carcinogenic. A concentration of Cr^{4+} above 0.1 mg l^{-1} is considered mutagenic and toxic, and it is advisable to remove this HM before use as guided by the US-EPA authority [61].

For the removal of Cr^{3+} ions by the mechanism of molecularly imprinting technology, primarily Birlik's group optimized a monomer–template complex, i.e. Cr^{3+}–methacryloyl histidine. A polymerized Cr^{3+}-imprinted EGDMA-N-(L)-methacryloyl histidine molecule was produced, with a fast (within 30 min) determination ability and a maximum adsorption limit of 69.28 mg l^{-1}. The fabrication mechanism is described in figure 7.5, which is based on the polymerization of poly(EDMA-MAH/Cr^{3+}) where methacryloylamidohistidine is the monomer [62].

In another approach, Ren *et al* [63] develop a method to remove Cr^{4+} ions via using AIBN, EGDMA, and acetone, playing the roles of initiator, cross-linker, and porogenic solvent, respectively. In this experiment they described eight different functional monomers depending on different pH agents, of them 4-VP yielded greater efficiency of adsorption. The rate of adsorption was very quick and it took only 3 min to reach the saturation point of about 339 mg g^{-1}. The primed Cr^{4+}–MIPs displayed adequate specificity, reusability, constancy, selectivity, and discriminating qualities. This proved a good method for the effective clearance of Cr^{4+} ions in less than 3 min, which was significantly higher than the previous available MIP methods as well as other conventional approaches developed for the removal of Cr^{4+} ions [56].

7.3.5 Copper (II) ions

Copper is a heavy metal with the symbol Cu, and is in the category of essential substance for human growth and development in trace amounts; for adults it is

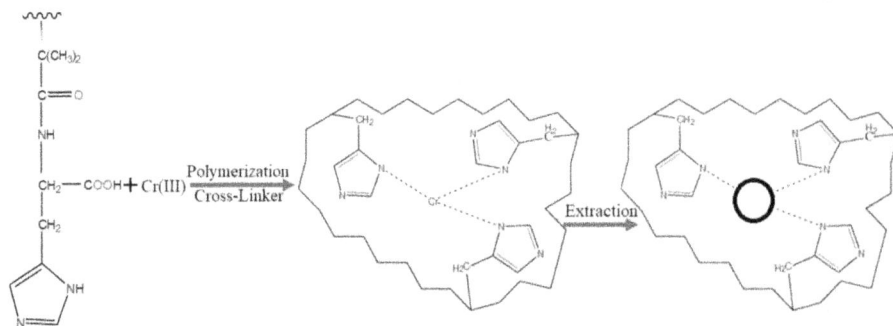

Figure 7.5. Reaction mechanism of the polymerization of Cr^{3+}–MIP {poly(EDMA-MAH/Cr(III))}. (Reproduced with permission from [62]. Copyright 2007 Elsevier.)

required in amounts of 1–100 mg/day. It helps in gene expression regulation, blood vessel development, metabolizing of iron, connective tissue formation, immune system functioning, and many more processes. Both excess and deficiency impair the biochemical processes and arrangements of the human body. A level of Cu in plasma below 40 μg dl^{-1} and in plasma ceruloplasmin below 13 mg dl^{-1} is considered as Cu deficiency, which can cause hypochromic anaemia and osteoporosis. In contrast, an excessive amount above the level of 140 mcg dl^{-1} in blood produces adverse effects on the liver and kidneys, and also generates immunotoxicity. There are significant toxic effects from large amounts of Cu on the environment and other animals, for example it changes the chemosensory abilities of aquatic organisms, producing difficulty in interacting with other organisms and the environment [64]. In every case, it is essential to regulate the Cu level, whether it is in body fluids or in environmental models. It is found in many different oxidative states (Cu^0, Cu^+, Cu^{2+}, Cu^{3+}, and Cu^{4+}) in the environment, depending on the compound formed, among which Cu^{2+} is the most toxic with a safe limit of 2 mg l^{-1} in drinking water (according to the WHO [34]).

In a recent study Wang's team demonstrated an economical, user-friendly, and green mechanism to achieve the design of green IIPs to eliminate Cu^{2+} ions from different aqueous solutions. It combined the approaches of using the green energy of three diverse functional monomers gelatin (G), 8-hydroquinone (HQ), and chitosan (C), thus they were called G-HQ-C IIPs. With a maximum adsorption magnitude of about 112 mg g^{-1} at room temperature, the IIPS produced by this study were found to be the most significant and selective method compared to other types of adsorbents reported in the literature. The most prominent application is reusability, as they can be reused for more than ten cycles without hampering adsorption capacity [65]. Furthermore, promising results were obtained when this concept was extended to the detection and elimination of other HM ions by changing the Cu^{2+} ions to Cd^{2+}, Hg^{2+}, and Pb^{2+} ions. In addition, this novel multi-monomer based study was based on the concept of multifunctional IIPs, where multiple HM ions can be removed using a single type of IIP [56]. Cu^{2+}–MIPs are another developed imprint design for the recovery of Cu^{2+} ions from various liquid solutions (the calculated adsorption limit was 14.8 mg l^{-1}). EGDMA (cross-linker) and AIBN (free-radical initiator) were placed in itaconic acid (a functional monomer) containing methanol and in the presence of Cu^{2+} ions, and the reaction was allow to take place to grow a solid polymercrumpled polymer. Cu^{2+} ion residues were washed out (HCl-EDTA washing) before vacuum drying. Finally the prepared MIPs were used for Cu^{2+} ion extraction [66]. Figure 7.6 is presents the process of polymerization of Cu^{2+} MIPs. Ren et al [67] proposed a modified method for Cu^{2+}–MIPs by bringing together graphene oxide ($C_{140}H_{42}O_{20}$), EGDMA, acrylamide, and Cu^{2+} ions. This widely recognized method exhibited a good imprinting factor and Cu^{2+} ions removal ability with the excellent capacity of more than 132 mg g^{-1} in favourable conditions. In the same year Ban and colleagues [68] patented an assay to develop Cu^{2+}–MIPs. Their innovative approach used surface imprinted polymers of Cu^{2+} ions reinforced by a polystyrene synthetic polymer and synthesized MIPs that not only worked as solid phase, double-site eliminators of Cu^{2+} ions from aqueous samples, but this

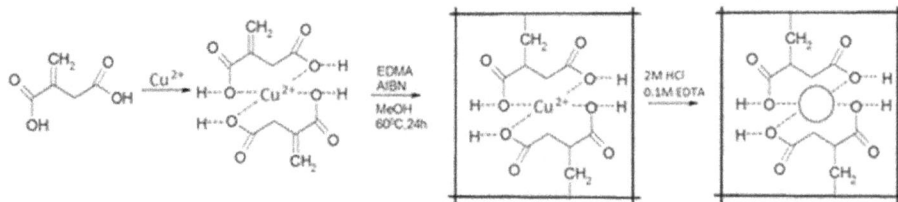

Figure 7.6. Polymerization mechanism for designing Cu(II)-ion MIPs. (Reproduced with permission from [66]. Copyright 2015 Springer. CC BY 4.0.)

selective adsorption and extraction technique also had the ability of simultaneous removal of counter-anions. In addition to these applications, an additional key advantage of this MIP is the specific extraction of chelate and Cu together with counter-ions once the elution step is completed. This innovation is also able to maintain the stability between anions and cations throughout the procedure of extraction, make these MIPS a most appropriate candidate for efficient polymer materials. Recently, Di Masi's group fabricated a sensitive and highly selective novel voltammetric sensor based on electro-synthesized IIP film to determine the presence of Cu^{2+} ions in aqueous models via an electropolymerisation process. Electrochemical polymerization was achieved through the availability of p-phenylenediamine and a template of Cu^{2+} ions, which were combined with screen-printed technology to create stable and robust films. In addition, the developed sensor represented a device that can remove Cu^{2+} ions from polluted water with an excellent recovery performance of around 95% to 105%. The sensor also offered an alternative platform to investigate Cu^{2+} ions in biological fluids [69].

7.3.6 Arsenic ions

Arsenic is at the top of the list of toxic HMs [42] and is a ubiquitous and naturally occurring metalloid that is employed extensively in the fields of medicines, agriculture (pesticides, fungicides), steel, and rubber manufacturing. These industries release excess amounts of As into the environment which indirectly contaminates vegetation and water. Finally, when humans and other animals come into contact with these vegetative crops and water, the As will bio-accumulate into the different life forms as a hazardous substance [70]. Short term exposure can induced nausea and/or diarrhoea, whereas long-term exposure leads to skin lesions, cancer, and multi-organ failure. Most commonly arsenic is found in four oxidative state in nature, As^{3-}, As^{0}, As^{3+}, and As^{5+}, of which As^{3+} and As^{5+} are toxic in anaerobic and aerobic conditions, respectively, while As^{3-} (arsine) is considered a very toxic gaseous compound.

To mitigate the amount of arsenic, an IIP-based approach was developed by Mafu's group [71]. Their arsenic–IIP polymeric compounds play two roles—one is to recognize the element and the other is to be able to adsorb 568 μg g^{-1} arsenic from various wastewater samples with fair reusability. The process of forming arsenic–IIP is, in short, as follows. $Na_2HAsO_4 \cdot 7H_2O$ was mixed with C_8H_8 (styrene) in a reaction chamber and to dissolve this mixture acetic acid and methanol were used. In the same chamber EGDMA and AIBN were mixed and allow to freeze (liquid nitrogen) so that the polymerization reaction could be initiated. Grounding, sieving, and washing were carried out thoroughly to remove unbound particles. Before the final washing, HNO_3 treatment was carried out to remove the template. The study suggested that the polymerized arsenic–IIPs showed six month stability; after six months the rate of adsorption was reduced because of damage to the adsorption site which ultimately affected the rate of adsorption which dropped by 12 $\mu g\ m^{-1}$. The rate of adsorption was also affected by the repetition of cycles and harsh solvent chemical operations. In addition, comparatively, the column format captured more analytes than the batch format system, however, adsorption was reported to be much higher in the batch system. Moreover, to draw the arsenic-IIPs reusability test graph, HNO_3:HCl (1:1, v/v) were preferred and results conclusively reported that HNO_3:HCl recovered higher amount of arsenic ions with a higher recovery rate (97%) from the aquas samples even after six adsorption-desorption cycles.

7.3.7 Selenium ions

Selenium (Se) occurs naturally in the environment, appearing in bodies of water and some food stuffs. It falls into the trace element category as its small amount in the body plays roles in essential metabolic bioprocesses and it also has antibiotic properties which protect cells from damage. However, like Cu, the excessive use of Se in different manufacturing industries and mining process has released it into the human environment, and consumption of excess and toxic amounts of Se directly or indirectly may be fatal because of its bioaccumulation property [70, 72].

Filip et al [73] suggested that to remove Se from the environment three aspects should be accounted for in developing high-performance sensing (electrochemical) electrodes. The first is the mechanism via which Se reacts on the sensor prior to the electrochemical process; the second is the oxidative state of the Se detecting species; and the third is the transformation of the electrode surface to improve the affinity of the sensor towards Se. It is currently a necessity to remove the excess amount of Se and reduce its ingestion from drinking water [74]. Keeping this in mind, Mafu et al [71] suggested a potent research approach via synthesizing an IIP-based adsorbent for specific recognition of Se. The primary process of Se–IIPs construction is quite similar to the above discussed method of arsenic–IIPs with a few modifications. For creating Se–IIPs, complexes of SeO_2, methanol, 4,7-dichloroquinoline, and EGDMA were used to prepare the template and 2,2′-azobis(cyclohexyl-carbonitrile) was added into this homogeneous mixture. Further, following the same processes as carried out to form the arsenic–IIPs, the adsorption capacity of the Se–IIPs was found to be little smaller (530 $\mu g\ g^{-1}$) in comparison to the arsenic–IIPs whereas they showed the same reusability and capability of removing toxic metal ions from wastewater samples. They also exhibited the same features as arsenic–IIPs, such as adsorption being found to be higher in column experiments. More than a decade ago Khajeh et al prepare the MIP-particles primarily for the adequate extraction and detection of Se from aqueous media. Polymeric printed molecules were achieved by combining SeO_2, AIBN, EDMA, 2-vinylpyridine, and o-phenylenediamine in appropriate amounts [75].

7.3.8 Iron (II) ions

Iron (Fe) is a very well-known micronutrient and an important HM in nature A as well (32.1% of the Earth's mass is iron). It is most commonly dissolved in water bodies in either its Fe^{2+} (ferrous) and Fe^{3+} (ferric) oxidative forms, and is also notable for a high redox potential [76]. Although it is prime component of blood and many metabolic activities, a high concentration of Fe in the blood can produce an enormous amount of dangerous free radicals (super ions, hydroxyl ions) from peroxidase, which are main causes of damage to the body such as DNA and proteins, and too much iron can ultimately lead to shock, a fatal coma, multi-organ failure, and even death. The set value for the daily intake of Fe is 45 mg/day for young people and 40 mg/day in the case of children (up to the age of 14 years) [77].

Many conventional methods for Fe extraction have been described in various sources in the literature, mainly focused on neutral extractants (for example ketones and ethers). These neutral extractants have some limitations such as slow kinetics and the requirement for a high volumes of mineral acid to extract large quantities of Fe. To resolve the problem Ara *et al* [78] validated an IIP technique based on co-polymerization of a three compound multiplex, including Fe^{3+} ions, MAA, and divinyl benzene. The core part of this study was to approach the complete removal of Fe^{3+} ions from the targeted module solutions and this method demonstrated excellent adsorption of Fe^{3+} ions from different closed and open ecological samples compared to methods reported previously in the literature [56]. In the preparatory phase of Fe^{3+}–MIPs, MAA and divinyl benzene were used as monomer and cross-linker, respectively. These two compounds were further hybridized in a glass tube with Fe(III)8-hydroxyquinolone, and this step is co-polymerization. In the next step extraction was performed in a chloroform solution and purging was achieved via liquid nitrogen. In the next step washing was performed to collect the purged molecules and untreated particles were steeped out. The process of synthesis for Fe^{3+}–MIPs is described in figure 7.7. Behind the creation of Fe^{3+}–MIPs, the chemistry is such that electro-negative N atoms present at the 2′ and 4′ positions of the pyridine ring of the compound complex (Fe(III)8-hydroxyquinolone) incorporate an electron (e^-) reduction, so the nucleophilic addition of MAA occurs at the 2′ position rather than the 3′ position of the pyridine ring. The incorporated molecules are fixed in place when the MAA monomer is cross-linked with the cross-linker divinyl benzene. At this phase the obtained polymer retained the immobilized Fe^{3+} particles/ions. After removing the Fe^{3+} ion templates, the entire polymerized polymer was washed out and only then used for further experiments. The reusability experiment showed that the produced MIPs were ideal candidates for the removal of Fe^{3+}ions with an excellent recovery rate of approximately 98% even after being recycled for 15 cycles without a change in the quality. They tested the efficiency of the MIPs in different water bodies and found that the recovery rate was around 98% from reservoir water, 97% from stream water, 95% from tube well water, and 99% from spring water samples when 100 mg l^{-1} Fe^{3+} ions were added in all respective samples, and the highest adsorption was 170 μmol g^{-1}.

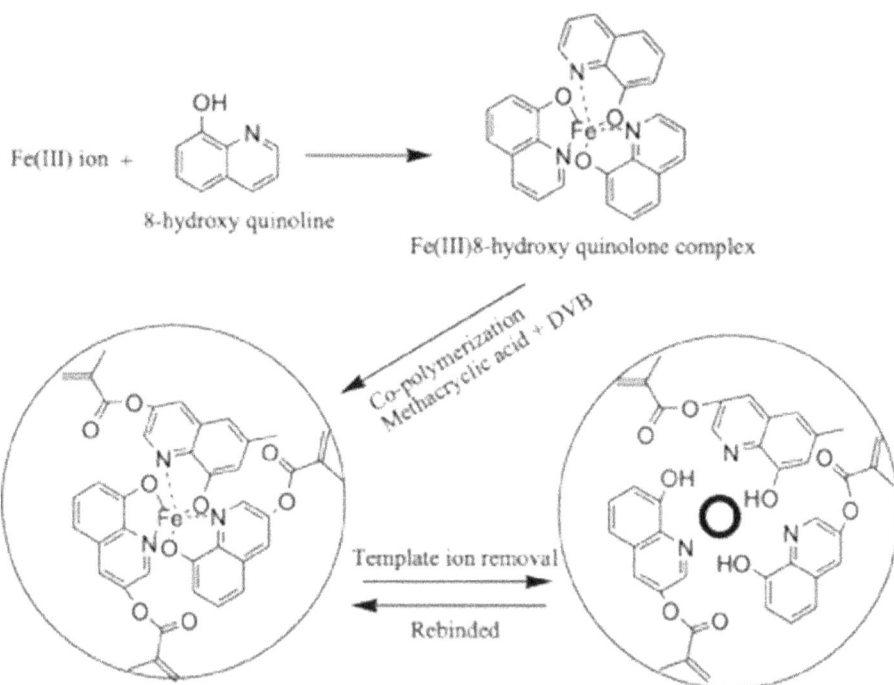

Figure 7.7. The systematic removal of Fe^{3+}–IIP using an 8-hydroxyquinoline complex. (Reproduced with permission from [78]. Copyright 2018 Springer. CC BY 4.0.)

7.3.9 Nickel (II) ions

Of the HMs discussed in the chapter, after Fe, nickel (Ni) is the most abundant element on Earth. It also comes under the essential micronutrient category as it is a necessary element in lipid metabolism and hormonal activities. The hazardous form is all man-made and the occupationally exposed groups of people, those exposed at their workplaces such as in the manufacturing industries (e.g. batteries manufacturing or electrochemical-planting), are the most common victims of Ni toxicity. Apart of this, wide-scale environmental changes due to uncontrolled transportation, industrial development, urbanization, and the misuse of fertilizers in agriculture and related industries have caused physical threat to all individuals through the intake of contaminated materials in form of air and food [79]. Ni exposure toxicity depends on the mode of its exposure and the dissolubility of the compound like other HMs, and it has direct effects on the immune system, neurology, reproductive system, and development, and can cause carcinogenesis, lung fibrosis, kidney diseases, cardiovascular diseases, etc [80].

One group of researchers introduced an IIP-based polymer, which they called super-macroporous cryogels synthesized from poly(hydroxyethyl methacrylate) (PHEMA) with the help of N-methacryloyl-histidine methyl ester (MAH) used as the functional monomer [81]. The monomer–template complex was synthesized through the polymerization of Ni^{+2} ions and MAH in two different molar concentrations—1:1 and 1:2 of Ni^{+2}:MAH. HEMA (the active monomer), PEDGA (the cross-linker cross-linker), and polymerized Ni^{+2}:MAH in a ratio of 1:1 and 2:1 in two differently synthesized assays for the preparation of the super-macroporous cryogels MIP1 and MIP2, respectively. They were places in plastic syringes for polymerization and consequent washing with an excess of water to remove untreated monomer residues. Furthermore, washing to remove the templates was carried out using EDTA and then HNO_3 solution. Finally the synthesized MIP1 and MIP2 could be stored for a long time at 4 °C in 0.02% sodium azide solution. The relative adsorption powers were found to be around ~2 and ~6 mg g^{-1} for MIP1 and MIP2, respectively. These imprinted cryogels exhibited sophisticated adsorption intake of Ni^{+2} ions when compared to previously published Ni^{+2}-imprinted microparticles, polypyrrole films, and silica gel methods for the removal of Ni^{+2} ions from various sources [82]. In addition to this, reusability was retained for up to ten rounds of adsorption and desorption with the capability of 92% removal of Ni^{+2} ions (for both MIP1 and MIP2). There is a most economically important benefit to using these cryogel techniques, including their notable and specificity over others method, as it suggests low diffusion resistance for Ni^{+2} ion uptake. In addition, the interconnected ultra-macro-porosity of the Ni^{+2}-imprinted MIP-based cryogels revealed as a potential platform for the removal of Ni^{+2} ions from water models. In recent years Zhou *et al* synthesized a Ni^{+2}-imprinted polymeric substance using a bulk polymerization assay. The combinations of several functional monomers, cross-linkers, templates (different molar ratios), and solvents were explored to discover the best adsorber for Ni^{+2} ions. After a significant amount of experimental characterization, they reported a Ni^{+2}-imprinted polymeric candidate with the highest adsorption capacity of 86.3 mg g^{-1} when the initial amount of Ni was 500 mg l^{-1}, while the selectivity coefficients found more than one for Ni^{+2}/interfering ions [83]. Before this study, Ersöz's research group, using solid-phase extraction polymerization, proposed an excellent extraction polymeric column to isolate and pre-concentrate Ni^{+2} ions from aqueous samples with the limit of detection of 0.3 ng ml^{-1} for flame atomic absorption spectrometry. An important outcome of this study was that Ni^{+2} ions can be identified well even in the presence of a complex medium such as an ocean water sample [84]. In 2019 Aravind and Mathew, for the fast sensing of Ni^{+2} ions, produced an electrochemical sensor fabricated pt/MWCNT–IIP by forming a nano-layer on the surface of a platinum sensor, which made an excellent adsorbent of Ni^{+2} ions. This novel and cost-effective approach to sensing allowed the sensing of the presence of a low concentration of HM ions ranging between 1 and 5 ppm with an LOD value of 0.028 μM [85].

7.4 Transformed MIP-based detection

In addition to the above discussed MIP-based sensing of HMs, researches continue to find improved methods for better, faster, and more cost-effective and sensitive detection and removal of target molecules. Among the most direct approaches to MIP-based sensing of elements is the capability of converting a bonding reaction at the surface of the MIPs into a visualized/optical result that can be readable with the naked eye or via basic analytical instruments such spectrophotometers. Optical outcomes can be obtained through several molecular techniques such as fluorescence quenching, redox reactions, and dye displacement [86]. HMs and their ions can be detected using different types of optical approaches, such as plasmon resonance or flame atomic absorption spectroscopy [87]. Very recently, Meza Lopéz et al [88] proposed an optical approach for the identification of Pb^{+2} ions where they used an optical surface imprinted fibre with 2-acrylamido-2-methylpropane sulphonic acid. The spectrophotometric coupled optical fibre detected ultra-trace amounts of Pb^{+2} ions in real aqueous samples with a limit of 85 ng l^{-1}. Another direct approach is colorimetric detection using either dye based templating or structurally mimicking molecules. The presence of the target molecules released dye from the dye-conjugated MIP into the medium which confirmed the presence of the target.

To achieve the aim of enhancing the detection properties and alternative approach employing fluorescent dyes has been used and nowadays, encountering nanoparticles is one of the most commonly used phenomena to enhance the fluorescence individuality of any MIPs. Specifically, quantum dots introduced as core–shell molecules are manufactured using metals such as Zn and Cu, and also C and graphenes, and are able to emit light of a specific wavelength that can be adjusted as desired [89]. The covering layer of MIP remains thin, which allows the emitted light to pass through when the target is absent, but emission is halted in the presence of the target molecules because of the binding to the respective entities [9].

Electrochemical sensing based on screen-printing is a way to determine and analyse the molecule of interest (examples are discussed in section 7.3.1 and 7.3.3). Developing and optimizing screen-printed electrodes in the form of transducers in the electroanalytical stages has further enriched their application from the sensing point of view because they maintain excellent reusability, reproducibility, cost-effectiveness, sensitivity, and chemical stability for bulk synthesis [90]. Versatile applications of using screen-printed electrodes as transducers for the detection of environmental contamination, in particular in terms of the detection and identi-fication of HM ions, can be found in the literatures [91].

Potentiometric methods are another potent approach and work on a zero current approach where the resultant record data are the voltages across a membrane due to the electromotive force, which directly indicates the composition of a sample [92], and usually depends on the determination of the concentration of the targeted ion rather than its activity. Malon et al [93] describe direct potentiometry at trace levels. Potentiometric methods are highly sensitive and capable of detecting up to the femtomolar (fM) limit of detection for Ca^{2+}, Pb^{2+}, and Ag^{2+} HM ions, but it is still a challenge to detect unknown concentrations in real samples. You's group introduce the μISE (an ion-selective electrode) application relay on a microfabrication assay to fabricate a multiplexed micro-array encompassing a plethora of HM ionophore targets such as AsO_2^-, Hg^{2+}, Pb^{2+}, and Cd^{2+} which can be detected at ppb concentrations in drinking water [94].

7.5 Future prospects of MIPs in toxicity remediation

The most important advantage of using MIPs is the chromatographic identification that also permits the involvement of other supporting schemes. The production of restricted binding sites in polymeric compounds is highly useful for detection and environmental monitoring with regard to toxic pollutants. Although MIP-based detection has shown remarkable results, some refinements are still needed to make improvements in terms of finding promising well-organized and low toxicity template materials, enhancing the kinetic properties, improving their relevance in supramolecular chemistry, etc. Moreover, MIPs have limited applications in the field of clinical research, as MIP drug delivery is experiencing issues due to exhibiting low safety and a high chance of toxicity taking place while interacting with body tissues. MIPs have a bright future in terms of addressing toxicity related concerns, such as cancer toxicity induced by HM ions. MIPs can support the creation of a novel drug-delivery system in conditions where only a specific analogous template is manufactured without creating any type of toxicity itself. In the last decade, electrochemical sensors have been in demand for wide-scale research, in particular regarding environmental and food toxicity analysis. In addition, modulated selective MIPs for the adsorption of toxic elements are needed in the future to establish a healthy and protected environment by fabricating highly specific biochemical sensing electrodes that have the capability of sensing and measuring the hazardous effects and amounts of substances, weather it would be for HM ions, toxic dyes, or pollutants [95].

Moreover, a core challenge is the reuse of the MIPs, and the need for them to be desorbed after completing a single cycle, but this is quite difficult to manage separately on an industrial platform. Thus the new concept of 'self-desorbing MIPs' has been introduced, which have the self-desorbed property and elute properly after every individual adsorption–desorption cycle [56]. Thus, from the perspective of reusability and economy, the construction of MIPs/IIPs makes them favourable for utilization in toxic metal ion remediation.

The hybridization of MIPs with nanotechnology has set a benchmark to develop an MIP–aptamer system, called 'AptaMIPs', to enhance the exclusive performance of MIPs and encourage their adoption at a commercial level. They also exhibit improved stability and affinity [96]. The production of AptaMIPs generally involves the formation of a polymerizable aptamer, and then this polymerizable aptamer is incorporated into a monomeric mixture which can further be used in the imprinting process. This concept was utilized to design a hybrid affinity substance to recognize moxifloxacin antibiotic by Sullivan and his group [97]. In 2022 Ali and Omer [98] published a review of the literature and detailed the process of synthesizing AptaMIP platform together with their benefits for the identification of different entities, including metal ions, microorganisms (bacteria and viruses), proteins, etc. However, the hybrid of these two techniques is in its very initial stages and still required experimental confirmation. Questions to investigate remain, such as, in generating the polymerizable aptamer, what length for the linker and what position of the linker would be best within the aptamer itself?

7.6 Conclusion

The main features of MIPs are their low cost, specific selectivity, and robust affinity towards a ligand, offering an exceptional approach for the examination, quantification, and elimination of HM ions in distinctive complicated environments with different pH and temperatures, and help to establish an ecological balance in the environment. This review has focused on developmental studies of MIPs and IIPs for use in the elimination of well-known and toxic HMs and their ions from the natural and spiked sample models, as these approaches have great promise for combination into portable and commercial devices as well. In recent times researchers have been focusing on the electrochemical, colorimetric, and optical MIP-based sensing strategies, but their production for future marketable uses must be optimized. In particular, the affinities of the polymerized MIPs to specific targets in applications in real and complex models must be studied and improved. After a comprehensive literature review, it is clear that nanomaterials promise to be important factors which will accelerate the analytical power of sensors, but the way in which they are synthesized should be simple and suitable for bulk production. The review of published articles also indicates that electrochemical sensor based identification is more promising from the commercial point of view because they allow good sensitivity, simplicity, and economic benefits. Moreover, screen-printed analysis methods also offer a cost-effective bulk-production approach for sensing electrodes. Colorimetric sensing is preferred in areas where sensitivity is not considered such an important factor. Overall, this array of determinations is profoundly appropriate for testing in fields which required no expensive instruments. In the coming era, revolutions in nano-biotechnology along with different forms of MIPs (electro-biochemical, screen-printing etc) will pave the way for the commercialization of MIP-based sensing technology.

The science of MIPs based on MIT not only allows the sensing of chemical and biological molecules (microorganisms, HMs, pollutants, etc) but also facilitates drug-delivery systems to purification mechanisms. The ultimate goal of the present article is to promote knowledge about MIPs and move towards their commercialization for the improvement of life. Advancements in science and technology have been developed in many dimensions through these techniques and we need many more to win the battle for the betterment of a healthy environment.

References

[1] Abbas A, Al-Amer A M, Laoui T, Al-Marri M J, Nasser M S, Khraisheh M and Atieh M A 2016 Heavy metal removal from aqueous solution by advanced carbon nanotubes: critical review of adsorption applications *Sep. Purif. Technol.* **157** 141–61

[2] Bhalara P D, Punetha D and Balasubramanian K 2014 A review of potential remediation techniques for uranium (VI) ion retrieval from contaminated aqueous environment *J. Environ. Chem. Eng.* **2** 1621–34

Bano Y, Khan R, Sharma J and Shrivastav A 2019 Changes in activities of nitrogen metabolism enzymes in arsenic stressed *Phaseolus vulgaris Int. J. Sci. Res. Rev.* **8** 336–41

Bano Y, Shrivastava S, Jatav S K, Jayant S K, Sharma J and Shrivastav A 2018 Physiological and molecular responses underlying differential arsenic tolerance in *Phaseolus vulgaris Int. J. Creat. Res. Thoughts* **6** 2010–7

[3] Morsi R E, Al-Sabagh A M, Moustafa Y M, ElKholy S G and Sayed M S 2018 Polythiophene modified chitosan/magnetite nanocomposites for heavy metals and selective mercury removal *Egypt. J. Pet.* **27** 1077–85

[4] Hashemian S, Saffari H and Ragabion S 2015 Adsorption of cobalt (II) from aqueous solutions by Fe₃O₄/bentonite nanocomposite *Water Air Soil Pollut.* **226** 1–10

Jalilzadeh M, Uzun L, Şenel S and Denizli A 2016 Specific heavy metal ion recovery with ion-imprinted cryogels *J. Appl. Polym. Sci.* **133** 10

Wang S and Shi X 2001 Molecular mechanisms of metal toxicity and carcinogenesis *Mol. Cell. Biochem.* **222** 3–9

Beyersmann D and Hartwig A 2008 Carcinogenic metal compounds: recent insight into molecular and cellular mechanisms *Arch. Toxicol.* **82** 493–512

Bano Y, Jayant S K, Sharma J and Shrivastava A 2017 Arsenic effect on morphology and rate limiting step of nitrate assimilation pathway of *Phaseolus vulgaris Int. J. Adv. Res. Dev.* **2** 341–4

[5] Berglund L *et al* 2008 A genecentric human protein atlas for expression profiles based on antibodies *Mol. Cell. Proteomics* **7** 2019–27

[6] Gray A C, Bradbury A R, Knappik A, Plückthun A, Borrebaeck C A and Dübel S 2020 Animal-derived-antibody generation faces strict reform in accordance with European Union policy on animal use *Nat. Methods* **17** 755–6

[7] Ahlgren S *et al* 2009 Targeting of HER2-expressing tumors with a site-specifically 99mTc-labeled recombinant affibody molecule, ZHER2: 2395, with C-terminally engineered cysteine *J. Nucl. Med.* **50** 781–9

Barozzi A, Lavoie R A, Day K N, Prodromou R and Menegatti S 2020 Affibody-binding ligands *Int. J. Mol. Sci.* **21** 3769

[8] Satheeshkumar P K 2020 Expression of single chain variable fragment (scFv) molecules in plants: a comprehensive update *Mol. Biotechnol.* **62** 151–67

de Aguiar R B, da Silva T D, Costa B A, Machado M F, Yamada R Y, Braggion C, Perez K R, Mori M A, Oliveira V and de Moraes J Z 2021 Generation and functional characterization of a single-chain variable fragment (scFv) of the anti-FGF2 3F12E7 monoclonal antibody *Sci. Rep.* **11** 1432

[9] Tchekwagep P M *et al* 2022 A critical review on the use of molecular imprinting for trace heavy metal and micropollutant detection *Chemosensors* **10** 296

Crivianu-Gaita V and Thompson M 2016 Aptamers, antibody scFv, and antibody Fab' fragments: an overview and comparison of three of the most versatile biosensor biorecognition elements *Biosens. Bioelectron.* **85** 32–45

[10] Ellington A D and Szostak J W 1990 *In vitro* selection of RNA molecules that bind specific ligands *Nature* **346** 818–22

[11] Levine H A and Nilsen-Hamilton M 2007 A mathematical analysis of SELEX *Comput. Biol. Chem.* **31** 11–35

Prante M, Segal E, Scheper T, Bahnemann J and Walter J 2020 Aptasensors for point-of-care detection of small molecules *Biosensors* **10** 108

[12] Zhang Y, Lai B S and Juhas M 2019 Recent advances in aptamer discovery and applications *Molecules* **24** 941

[13] Keum J W and Bermudez H 2009 Enhanced resistance of DNA nanostructures to enzymatic digestion *Chem. Commun.* **45** 7036–8

Poma A, Brahmbhatt H, Pendergraff H M, Watts J K and Turner N W 2015 Generation of novel hybrid aptamer–molecularly imprinted polymeric nanoparticles *Adv. Mater.* **27** 750–8

Jolly P, Tamboli V, Harniman R L, Estrela P, Allender C J and Bowen J L 2016 Aptamer–MIP hybrid receptor for highly sensitive electrochemical detection of prostate specific antigen *Biosens. Bioelectron.* **75** 188–95

[14] Menger M, Yarman A, Erdőssy J, Yildiz H B, Gyurcsányi R E and Scheller F W 2016 MIPs and aptamers for recognition of proteins in biomimetic sensing *Biosensors* **6** 35

Arreguin-Campos R, Jiménez-Monroy K L, Diliën H, Cleij T J, van Grinsven B and Eersels K 2021 Imprinted polymers as synthetic receptors in sensors for food safety *Biosensors* **11** 46

Aslıyüce S, Bereli N, Uzun L, Onur M A, Say R and Denizli A 2010 Ion-imprinted supermacroporous cryogel, for *in vitro* removal of iron out of human plasma with beta thalassemia *Sep. Purif. Technol.* **73** 243–9

Soni D, Trivedi M and Ameta R 2014 Polymerization *Microwave-Assisted Organic Synthesis. A Green Chemical Approach* (Oakville, ON: Apple Academic Press) vol 8 pp 311–37

[15] Trotta F, Biasizzo M and Caldera F 2012 Molecularly imprinted membranes *Membranes* **2** 440–77

[16] Pichon V and Chapuis-Hugon F 2008 Role of molecularly imprinted polymers for selective determination of environmental pollutants—a review *Anal. Chim. Acta* **622** 48–61

Lasáková M and Jandera P 2009 Molecularly imprinted polymers and their application in solid phase extraction *J. Sep. Sci.* **32** 799–812

Li W and Li S 2007 Molecular imprinting: a versatile tool for separation, sensors and catalysis *Adv. Polym. Sci.* **206** 191–210

Jia M, Zhang Z, Li J, Ma X, Chen L and Yang X 2018 Molecular imprinting technology for microorganism analysis *TrAC Trends Anal. Chem.* **106** 190–201

Luliński P 2017 Molecularly imprinted polymers based drug delivery devices: a way to application in modern pharmacotherapy: a review *Mater. Sci. Eng.* C **76** 1344–53

Gupta P, Lapalikar V, Kundu R and Balasubramanian K 2016 Recent advances in membrane based waste water treatment technology: a review *Energy Environ. Focus* **5** 241–67

[17] Vasapollo G, Sole R D, Mergola L, Lazzoi M R, Scardino A, Scorrano S and Mele G 2011 Molecularly imprinted polymers: present and future prospective *Int. J. Mol. Sci.* **12** 5908–45

Zaidi S A 2017 Molecular imprinting polymers and their composites: a promising material for diverse applications *Biomater. Sci.* **5** 388–402

Speltini A, Scalabrini A, Maraschi F, Sturini M and Profumo A 2017 Newest applications of molecularly imprinted polymers for extraction of contaminants from environmental and food matrices: a review *Anal. Chim. Acta* **974** 1–26

[18] Chen L, Xu S and Li J 2011 Recent advances in molecular imprinting technology: current status, challenges and highlighted applications *Chem. Soc. Rev.* **40** 2922–42

Martín-Esteban A 2013 Molecularly-imprinted polymers as a versatile, highly selective tool in sample preparation *TrAC Trends Anal. Chem.* **45** 169–81

Yan H and Row K H 2006 Characteristic and synthetic approach of molecularly imprinted polymer *Int. J. Mol. Sci.* **7** 155–78

Santora B P, Gagné M R, Moloy K G and Radu N S 2001 Porogen and cross-linking effects on the surface area, pore volume distribution, and morphology of macroporous polymers obtained by bulk polymerization *Macromolecules* **34** 658–61

[19] Whitcombe M J, Kirsch N and Nicholls I A 2014 Molecular imprinting science and technology: a survey of the literature for the years 2004–2011 *J. Mol. Recognit.* **27** 297–401

Canfarotta F, Poma A, Guerreiro A and Piletsky S 2016 Solid-phase synthesis of molecularly imprinted nanoparticles *Nat. Protoc.* **11** 443–55

[20] Vasapollo G, Sole R D, Mergola L, Lazzoi M R, Scardino A, Scorrano S and Mele G 2011 Molecularly imprinted polymers: present and future prospective *Int. J. Mol. Sci.* **12** 5908–45

Anantha-Iyengar G, Shanmugasundaram K, Nallal M, Lee K P, Whitcombe M J, Lakshmi D and Sai-Anand G 2019 Functionalized conjugated polymers for sensing and molecular imprinting applications *Prog. Polym. Sci.* **88** 1–29

Malik A A, Nantasenamat C and Piacham T 2017 Molecularly imprinted polymer for human viral pathogen detection *Mater. Sci. Eng.* C **77** 1341–8

[21] Lowdon J W, Diliën H, Singla P, Peeters M, Cleij T J, van Grinsven B and Eersels K 2020 MIPs for commercial application in low-cost sensors and assays—an overview of the current status quo *Sensors Actuators* B **325** 128973

[22] Chen L, Xu S and Li J 2011 Recent advances in molecular imprinting technology: current status, challenges and highlighted applications *Chem. Soc. Rev.* **40** 2922–42

[23] Yarman A and Scheller F W 2020 How reliable is the electrochemical readout of MIP sensors? *Sensors* **20** 2677

[24] Jalilzadeh M, Uzun L, Şenel S and Denizli A 2016 Specific heavy metal ion recovery with ion-imprinted cryogels *J. Appl. Polym. Sci.* **133** 43095

[25] Aslıyüce S, Bereli N, Uzun L, Onur M A, Say R and Denizli A 2010 Ion-imprinted supermacroporous cryogel, for *in vitro* removal of iron out of human plasma with beta thalassemia *Sep. Purif. Technol.* **73** 243–9

Ye L 2015 Synthetic strategies in molecular imprinting *Molecularly Imprinted Polymers in Biotechnology* (Berlin: Springer) pp 1–24

Mujahid A, Mustafa G and Dickert F L 2018 Label-free bioanalyte detection from nanometer to micrometer dimensions—molecular imprinting and QCMs *Biosensors* **8** 52

[26] Schirhagl R, Qian J and Dickert F L 2012 Immunosensing with artificial antibodies in organic solvents or complex matrices *Sensors Actuators* B **173** 585–90

Phan N V, Sussitz H F, Ladenhauf E, Pum D and Lieberzeit P A 2018 Combined layer/ particle approaches in surface molecular imprinting of proteins: signal enhancement and competition *Sensors* **18** 180

Zhao X, He Y, Wang Y, Wang S and Wang J 2020 Hollow molecularly imprinted polymer based quartz crystal microbalance sensor for rapid detection of methimazole in food samples *Food Chem.* **309** 125787

[27] Zhang Z, Guan Y, Li M, Zhao A, Ren J and Qu X 2015 Highly stable and reusable imprinted artificial antibody used for *in situ* detection and disinfection of pathogens *Chem. Sci.* **6** 2822–6

Cai W, Li H H, Lu Z X and Collinson M M 2018 Bacteria assisted protein imprinting in sol–gel derived films *Analyst* **143** 555–63

[28] van Grinsven B *et al* 2018 SIP-based thermal detection platform for the direct detection of bacteria obtained from a contaminated surface *Phys. Status Solidi* A **215** 1700777

Peeters M M, Van Grinsven B, Foster C W, Cleij T J and Banks C E 2016 Introducing thermal wave transport analysis (TWTA): a thermal technique for dopamine detection by screen-printed electrodes functionalized with molecularly imprinted polymer (MIP) particles *Molecules* **21** 552

Steen Redeker E *et al* 2017 Biomimetic bacterial identification platform based on thermal wave transport analysis (TWTA) through surface-imprinted polymers *ACS Infect. Dis.* **3** 388–97

[29] Ramanaviciene A and Ramanavicius A 2004 Molecularly imprinted polypyrrole-based synthetic receptor for direct detection of bovine leukemia virus glycoproteins *Biosens. Bioelectron.* **20** 1076–82

Golabi M, Kuralay F, Jager E W, Beni V and Turner A P 2017 Electrochemical bacterial detection using poly (3-aminophenylboronic acid)-based imprinted polymer *Biosens. Bioelectron* **93** 87–93

Zhang L *et al* 2018 Chirality detection of amino acid enantiomers by organic electrochemical transistor *Biosens. Bioelectron.* **105** 121–8

[30] Eersels K, Lieberzeit P and Wagner P 2016 A review on synthetic receptors for bioparticle detection created by surface-imprinting techniques from principles to applications *ACS Sens.* **1** 1171–87

Iskierko Z, Sharma P S, Bartold K, Pietrzyk-Le A, Noworyta K and Kutner W 2016 Molecularly imprinted polymers for separating and sensing of macromolecular compounds and microorganisms *Biotechnol. Adv.* **34** 30–46

[31] Rahangdale D, Kumar A, Archana G and Dhodapkar R S 2018 Ion cum molecularly dual imprinted polymer for simultaneous removal of cadmium and salicylic acid *J. Mol. Recognit.* **31** e2630

Fu J, Chen L, Li J and Zhang Z 2015 Current status and challenges of ion imprinting *J. Mater. Chem.* A **3** 13598–627

Wei X, Li X and Husson S M 2005 Surface molecular imprinting by atom transfer radical polymerization *Biomacromolecules* **6** 1113–21

[32] Marrugo-Negrete J, Enamorado-Montes G, Durango-Hernández J, Pinedo-Hernández J and Díez S 2017 Removal of mercury from gold mine effluents using *Limnocharis flava* in constructed wetlands *Chemosphere* **167** 188–92

Erdem Ö, Saylan Y, Andaç M and Denizli A 2018 Molecularly imprinted polymers for removal of metal ions: an alternative treatment method *Biomimetics* **3** 38

Frossard A, Donhauser J, Mestrot A, Gygax S, Bååth E and Frey B 2018 Long- and short-term effects of mercury pollution on the soil microbiome *Soil Biol. Biochem.* **120** 191–9

Mason R P, Fitzgerald W F and Morel F M 1994 The biogeochemical cycling of elemental mercury: anthropogenic influences *Geochim. Cosmochim. Acta* **58** 3191–8

Mustafai F, Balouch A, Bhanger M, Abdullah A, Rajar K, Panah P, Ahmed B, Shah T and Kumar A 2018 Synthesis of molecularly imprinted polymer for the selective removal of mercury *Eurasian J. Anal. Chem.* **13** em61

Arora R, Singh N, Balasubramanian K and Alegaonkar P 2014 Electroless nickel coated nano-clay for electrolytic removal of Hg (II) ions *RSC Adv.* **4** 50614–23

[33] Broussard L A, Hammett-Stabler C A, Winecker R E and Ropero-Miller J D 2002 The toxicology of mercury *Lab. Med.* **33** 614–25

Sahetya T J, Dixit F and Balasubramanian K 2015 Waste citrus fruit peels for removal of Hg (II) ions *Desalin. Water Treat.* **53** 1404–16

Gonte R R, Balasubramanian K and Mumbrekar J D 2013 Porous and cross-linked cellulose beads for toxic metal ion removal: Hg(II) ions *J. Polym.* **2013** 1–9

[34] World Health Organization, WHO 2004 *Guidelines for Drinking-Water Quality* (Geneva: World Health Organization)

[35] Singh D K and Mishra S 2010 Synthesis and characterization of Hg (II)-ion-imprinted polymer: kinetic and isotherm studies *Desalination* **257** 177–83

[36] Zhang Z, Li J, Song X, Ma J and Chen L 2014 Hg^{2+} ion-imprinted polymers sorbents based on dithizone–Hg^{2+} chelation for mercury speciation analysis in environmental and biological samples *RSC Adv.* **4** 46444–53

[37] Alizadeh T, Ganjali M R and Zare M 2011 Application of an Hg^{2+} selective imprinted polymer as a new modifying agent for the preparation of a novel highly selective and sensitive electrochemical sensor for the determination of ultratrace mercury ions *Anal. Chim. Acta* **689** 52–9

[38] Alizadeh T, Hamidi N, Ganjali M R and Rafiei F 2018 Determination of subnanomolar levels of mercury (II) by using a graphite paste electrode modified with MWCNTs and Hg (II)-imprinted polymer nanoparticles *Microchim. Acta* **185** 1–9

[39] Bahrami A, Besharati-Seidani A, Abbaspour A and Shamsipur M 2015 A highly selective voltammetric sensor for nanomolar detection of mercury ions using a carbon ionic liquid paste electrode impregnated with novel ion imprinted polymeric nanobeads *Mater. Sci. Eng. C* **48** 205–12

[40] Ait-Touchente Z, Sakhraoui H E, Fourati N, Zerrouki C, Maouche N, Yaakoubi N, Touzani R and Chehimi M M 2020 High performance zinc oxide nanorod-doped ion imprinted polypyrrole for the selective electrosensing of mercury II ions *Appl. Sci.* **10** 7010

[41] Yasinzai M, Mustafa G, Asghar N, Ullah I, Zahid M, Lieberzeit P A, Han D and Latif U 2018 Ion-imprinted polymer-based receptors for sensitive and selective detection of mercury ions in aqueous environment *J. Sens.* **2018** 8972549

[42] ATSDR 2015 *Priority List of Hazardous Substances* (Atlanta, GA: Agency for Toxic Substances and Disease Registry, Division of Toxicology and Human Health Sciences) pp 1–9

[43] Frontalini F, Curzi D, Giordano F M, Bernhard J M, Falcieri E and Coccioni R 2015 Effects of lead pollution on *Ammonia parkinsoniana* (foraminifera): ultrastructural and microanalytical approaches *Eur. J. Histochem.* **59** 2460
Malar S, Shivendra Vikram S, Favas J C, Perumal P and Lead V 2016 Heavy metal toxicity induced changes on growth and antioxidative enzymes level in water hyacinths [*Eichhornia crassipes* (Mart.)] *Bot. Stud.* **55** 1–11

[44] Ara A and Usmani J A 2015 Lead toxicity: a review *Interdiscip. Toxicol.* **8** 55
Lanphear B P *et al* 2005 Low-level environmental lead exposure and children's intellectual function: an international pooled analysis *Environ. Health Perspect.* **113** 894–9

[45] Ao X and Guan H 2018 Preparation of Pb (II) ion-imprinted polymers and their application in selective removal from wastewater *Adsorp. Sci. Technol.* **36** 774–87

[46] Cai X Q, Li J H, Zhang Z, Yang F F, Dong R C and Chen L X 2014 Novel Pb^{2+} ion imprinted polymers based on ionic interaction via synergy of dual functional monomers for selective solid-phase extraction of Pb^{2+} in water samples *ACS Appl. Mater. Interfaces* **6** 305–13

[47] Alizadeh T and Amjadi S 2011 Preparation of nano-sized Pb^{2+} imprinted polymer and its application as the chemical interface of an electrochemical sensor for toxic lead determination in different real samples *J. Hazard. Mater.* **190** 451–9

[48] Bojdi M K, Mashhadizadeh M H, Behbahani M, Farahani A, Davarani S S and Bagheri A 2014 Synthesis, characterization and application of novel lead imprinted polymer nanoparticles as a high selective electrochemical sensor for ultra-trace determination of lead ions in complex matrixes *Electrochim. Acta* **136** 59–65

[49] Hu S, Xiong X, Huang S and Lai X 2016 Preparation of Pb (II) ion imprinted polymer and its application as the interface of an electrochemical sensor for trace lead determination *Anal. Sci.* **32** 975–80

[50] Dahaghin Z, Kilmartin P A and Mousavi H Z 2020 Novel ion imprinted polymer electrochemical sensor for the selective detection of lead (II) *Food Chem.* **303** 125374

[51] Alizadeh T, Hamidi N, Ganjali M R and Rafiei F 2017 An extraordinarily sensitive voltammetric sensor with picomolar detection limit for Pb^{2+} determination based on carbon paste electrode impregnated with nano-sized imprinted polymer and multi-walled carbon nanotubes *J. Environ. Chem. Eng.* **5** 4327–36

[52] Liu Y, Xiao T, Perkins R B, Zhu J, Zhu Z, Xiong Y and Ning Z 2017 Geogenic cadmium pollution and potential health risks, with emphasis on black shale *J. Geochem. Explor.* **176** 42–9

Genchi G, Sinicropi M S, Lauria G, Carocci A and Catalano A 2020 The effects of cadmium toxicity *Int. J. Environ. Res. Public Health* **17** 3782

[53] Li M, Feng C, Li M, Zeng Q, Gan Q and Yang H 2015 Synthesis and characterization of a surface-grafted Cd (II) ion-imprinted polymer for selective separation of Cd (II) ion from aqueous solution *Appl. Surf. Sci.* **332** 463–72

[54] Xi Y, Luo Y, Luo J and Luo X 2015 Removal of cadmium (II) from wastewater using novel cadmium ion-imprinted polymers *J. Chem. Eng. Data* **60** 3253–61

[55] Yang B, Yang L Z, Xue L H, Wu D and Yu D Y 2018 Preparation method and application of cadmium ion imprinted halloysite nanotubes CN107715849A, China https://worldwide. espacenet.com/publicationDetails/biblio?CC=CN&NR=107715849A&KC=A&FT=D&DB= EPODOC&locale=en_EP&date=20180223&rss=true

[56] Sharma G and Kandasubramanian B 2020 Molecularly imprinted polymers for selective recognition and extraction of heavy metal ions and toxic dyes *J. Chem. Eng. Data* **65** 396–418

[57] Alizadeh T, Ganjali M R, Nourozi P, Zare M and Hoseini M 2011 A carbon paste electrode impregnated with Cd^{2+} imprinted polymer as a new and high selective electrochemical sensor for determination of ultra-trace Cd^{2+} in water samples *J. Electroanal. Chem.* **657** 98–106

[58] Dahaghin Z, Kilmartin P A and Mousavi H Z 2018 Determination of cadmium (II) using a glassy carbon electrode modified with a Cd-ion imprinted polymer *J. Electroanal. Chem.* **810** 185–90

[59] Wu S, Li K, Dai X, Zhang Z, Ding F and Li S 2020 An ultrasensitive electrochemical platform based on imprinted chitosan/gold nanoparticles/graphene nanocomposite for sensing cadmium (II) ions *Microchem. J.* **155** 104710

[60] Bhalara P D, Balasubramanian K and Banerjee B S 2015 Spider–web textured electrospun composite of graphene for sorption of Hg (II) ions *Mater. Focus* **4** 154–63

Agrafioti E, Kalderis D and Diamadopoulos E 2014 Arsenic and chromium removal from water using biochars derived from rice husk, organic solid wastes and sewage sludge *J. Environ. Manage.* **133** 309–14

[61] Mertz W 1998 Interaction of chromium with insulin: a progress report *Nutr. Rev.* **56** 174
EFSA Panel on Dietetic Products, Nutrition and Allergies 2014 Scientific opinion on dietary reference values for chromium *EFSA J.* **12** 3845
Chen Q Y, Murphy A, Sun H and Costa M 2019 Molecular and epigenetic mechanisms of Cr (VI)-induced carcinogenesis *Toxicol. Appl. Pharmacol.* **377** 114636
Dima J B, Sequeiros C and Zaritzky N E 2015 Hexavalent chromium removal in contaminated water using reticulated chitosan micro/nanoparticles from seafood processing wastes *Chemosphere.* **141** 100–11

[62] Birlik E, Ersöz A, Açıkkalp E, Denizli A and Say R 2007 Cr (III)-imprinted polymeric beads: sorption and preconcentration studies *J. Hazard. Mater.* **140** 110–6

[63] Ren Z, Kong D, Wang K and Zhang W 2014 Preparation and adsorption characteristics of an imprinted polymer for selective removal of Cr (VI) ions from aqueous solutions *J. Mater. Chem.* A **2** 17952–61

[64] Bennett B L and Pierce M C 2011 Bone health and development *Child abuse and neglect* (Amsterdam: Elsevier) 260–74

[65] Wang L, Li J, Wang J, Guo X, Wang X, Choo J and Chen L 2019 Green multi-functional monomer based ion imprinted polymers for selective removal of copper ions from aqueous solution *J. Colloid Interface Sci.* **541** 376–86

[66] Kuras M J and Więckowska E 2015 Synthesis and characterization of a new copper (II) ion-imprinted polymer *Polym. Bull.* **72** 3227–40

[67] Ren Z, Zhu X, Du J, Kong D, Wang N, Wang Z, Wang Q, Liu W, Li Q and Zhou Z 2018 Facile and green preparation of novel adsorption materials by combining sol–gel with ion imprinting technology for selective removal of Cu (II) ions from aqueous solution *Appl. Surf. Sci.* **435** 574–84

[68] Ban X X, Fan B H, Chen N, Zhang P W and Zhang T L 2018 Copper (II) ion surface imprinted polymer and preparation method CN107573462A, China https://worldwide.espacenet.com/publicationDetails/biblio?CC=CN&NR=107573462A&KC=A&FT=D&DB=EPODOC&locale=en_EP&date=20180112

[69] Di Masi S, Pennetta A, Guerreiro A, Canfarotta F, De Benedetto G E and Malitesta C 2020 Sensor based on electrosynthesised imprinted polymeric film for rapid and trace detection of copper (II) ions *Sensors Actuators* B **307** 127648

[70] Hung D Q, Nekrassova O and Compton R G 2004 Analytical methods for inorganic arsenic in water: a review *Talanta* **64** 269–77
Gore P, Khraisheh M and Kandasubramanian B 2018 Nanofibers of resorcinol–form aldehyde for effective adsorption of As (III) ions from mimicked effluents *Environ. Sci. Pollut. Res.* **25** 11729–45

[71] Mafu L D, Mamba B B and Msagati T A 2016 Synthesis and characterization of ion imprinted polymeric adsorbents for the selective recognition and removal of arsenic and selenium in wastewater samples *J. Saudi Chem. Soc.* **20** 594–605

[72] Morgan Griffin R 2021 Selenium *WebMD* https://webmd.com/a-to-z-guides/supplement-guide-selenium
Ostovar M, Saberi N and Ghiassi R 2022 Selenium contamination in water; analytical and removal methods: a comprehensive review *Sep. Sci. Technol.* **57** 1–21

[73] Filip J, Vinter Š, Čechová E and Sotolářová J 2021 Materials interacting with inorganic selenium from the perspective of electrochemical sensing *Analyst* **146** 6394–415

[74] Ratnaike R N 2003 Acute and chronic arsenic toxicity *Postgrad. Med. J.* **79** 391–6
Sharma S, Balasubramanian K and Arora R 2016 Adsorption of arsenic (V) ions onto cellulosic–ferric oxide system: kinetics and isotherm studies *Desalin. Water Treat.* **57** 9420–36

[75] Khajeh M, Yamini Y, Ghasemi E, Fasihi J and Shamsipur M 2007 Imprinted polymer particles for selenium uptake: synthesis, characterization and analytical applications *Anal. Chim. Acta* **581** 208–13

[76] Frykman E, Bystrom M, Jansson U, Edberg A and Hansen T 1994 Side effects of iron supplements in blood donors: superior tolerance of heme iron *J. Lab. Clin. Med.* **123** 561–4

[77] Pośpiech B and Walkowiak W 2010 Studies on iron (III) removal from chloride aqueous solutions by solvent extraction and transport through polymer inclusion membranes with D2EHPA *Physicochem. Probl. Miner. Process.* **44** 195–204

[78] Ara B, Muhammad M, Salman M, Ahmad R and Islam N 2018 Preparation of microspheric Fe (III)-ion imprinted polymer for selective solid phase extraction *Appl. Water Sci.* **8** 1–4

[79] Chakraborty P, Gopalapillai Y, Murimboh J, Fasfous I I and Chakrabarti C L 2006 Kinetic speciation of nickel in mining and municipal effluents *Anal. Bioanal. Chem.* **386** 1803–13

[80] Das K K and Büchner V 2007 Effect of nickel exposure on peripheral tissues: role of oxidative stress in toxicity and possible protection by ascorbic acid *Rev. Environ. Health* **22** 157–73
Matović V, Bulat Z P, Djukić-Ćosić D and Soldatović D 2010 Antagonism between cadmium and magnesium: a possible role of magnesium in therapy of cadmium intoxication *Magnesium Res.* **23** 19–26
Das K K, Das S N and Dhundasi S A 2010 Nickel: molecular diversity, application, essentiality and toxicity in human health *Biometals, Molecular Structures, Binding Properties and Applications* (New York: Nova Science) pp 33–58
Bal W, Kozłowski H and Kasprzak K S 2000 Molecular models in nickel carcinogenesis *J. Inorg. Biochem.* **79** 213–8
Cangul H, Broday L, Salnikow K, Sutherland J, Peng W, Zhang Q, Poltaratsky V, Yee H, Zoroddu M A and Costa M 2002 Molecular mechanisms of nickel carcinogenesis *Toxicol. Lett.* **127** 69–75

[81] Tamahkar E, Bakhshpour M, Andaç M and Denizli A 2017 Ion imprinted cryogels for selective removal of Ni (II) ions from aqueous solutions *Sep. Purif. Technol.* **179** 36–44
Say R, Garipcan B, Emir S, Patır S and Denizli A 2002 Preparation of poly (hydroxyethyl methacrylate-co-methacrylamidohistidine) beads and its design as a affinity adsorbent for Cu (II) removal from aqueous solutions *Colloids Surf.* A **196** 199–207

[82] Du X, Zhang H, Hao X, Guan G and Abudula A 2014 Facile preparation of ion-imprinted composite film for selective electrochemical removal of nickel (II) ions *ACS Appl. Mater. Interfaces* **6** 9543–9
Nacano L R, Segatelli M G and Tarley C R 2010 Selective sorbent enrichment of nickel ions from aqueous solutions using a hierarchically hybrid organic–inorganic polymer based on double imprinting concept *J. Braz. Chem. Soc.* **21** 419–30
Jiang N, Chang X, Zheng H, He Q and Hu Z 2006 Selective solid-phase extraction of nickel (II) using a surface-imprinted silica gel sorbent *Anal. Chim. Acta* **577** 225–31

[83] Zhou Z, Kong D, Zhu H, Wang N, Wang Z, Wang Q, Liu W, Li Q, Zhang W and Ren Z 2018 Preparation and adsorption characteristics of an ion-imprinted polymer for fast removal of Ni (II) ions from aqueous solution *J. Hazard. Mater.* **341** 355–64

[84] Ersöz A, Say R and Denizli A 2004 Ni (II) ion-imprinted solid-phase extraction and preconcentration in aqueous solutions by packed-bed columns *Anal. Chim. Acta* **502** 91–7

[85] Aravind A and Mathew B 2019 An electrochemical sensor and sorbent based on mutiwalled carbon nanotube supported ion imprinting technique for Ni (II) ion from electroplating and steel industries *SN Appl. Sci.* **1** 1–11

[86] Lowdon J W, Diliën H, Singla P, Peeters M, Cleij T J, van Grinsven B and Eersels K 2020 MIPs for commercial application in low-cost sensors and assays—an overview of the current status quo *Sensors Actuators* B **325** 128973

Piletsky S A, Piletskaya E V, El'Skaya A V, Levi R, Yano K and Karube I 1997 Optical detection system for triazine based on molecularly-imprinted polymers *Anal. Lett.* **30** 445–55

Vlatakis G, Andersson L I, Müller R and Mosbach K 1993 Drug assay using antibody mimics made by molecular imprinting *Nature* **361** 645–7

Shariati R, Rezaei B, Jamei H R and Ensafi A A 2019 Application of coated green source carbon dots with silica molecularly imprinted polymers as a fluorescence probe for selective and sensitive determination of phenobarbital *Talanta* **194** 143–9

Chmangui A, Driss M R, Touil S, Bermejo-Barrera P, Bouabdallah S and Moreda-Piñeiro A 2019 Aflatoxins screening in non-dairy beverages by Mn-doped ZnS quantum dots—molecularly imprinted polymer fluorescent probe *Talanta* **199** 65–71

[87] Gerdan Z, Saylan Y, Uğur M and Denizli A 2022 Ion-imprinted polymer-on-a-sensor for copper detection *Biosensors* **12** 91

Wang X, Chu Z, Huang Y, Chen G, Zhao X, Zhu Z, Chen C and Lin D 2022 Copper ion imprinted hydrogel photonic crystal sensor film *ACS Appl. Polym. Mater.* **4** 4568–75

Khoddami N and Shemirani F 2016 A new magnetic ion-imprinted polymer as a highly selective sorbent for determination of cobalt in biological and environmental samples *Talanta* **146** 244–52

[88] López F D, Khan S, Picasso G and Sotomayor M D 2021 A novel highly sensitive imprinted polymer-based optical sensor for the detection of Pb (II) in water samples *Environ. Nanotechnol. Monit. Manag.* **16** 100497

[89] Zhao Y, Ma Y, Li H and Wang L 2012 Composite QDs@MIP nanospheres for specific recognition and direct fluorescent quantification of pesticides in aqueous media *Anal. Chem.* **84** 386–95

Lin C I, Joseph A K, Chang C K and Der Lee Y 2004 Molecularly imprinted polymeric film on semiconductor nanoparticles: analyte detection by quantum dot photoluminescence *J. Chromatogr.* A **1027** 259–62

Ren X, Liu H and Chen L 2015 Fluorescent detection of chlorpyrifos using Mn (II)-doped ZnS quantum dots coated with a molecularly imprinted polymer *Microchim. Acta* **182** 193–200

[90] Hart J P and Wring S A 1997 Recent developments in the design and application of screen-printed electrochemical sensors for biomedical, environmental and industrial analyses *TrAC Trends Anal. Chem.* **16** 89–103

[91] Ferrari A G, Carrington P, Rowley-Neale S J and Banks C E 2020 Recent advances in portable heavy metal electrochemical sensing platforms *Environ. Sci. Water Res. Technol.* **6** 2676–90

Hayat A and Marty J L 2014 Disposable screen printed electrochemical sensors: tools for environmental monitoring *Sensors.* **14** 10432–53

Li M, Li Y T, Li D W and Long Y T 2012 Recent developments and applications of screen-printed electrodes in environmental assays—a review *Anal. Chim. Acta* **734** 31–44

Barton J, García M B, Santos D H, Fanjul-Bolado P, Ribotti A, McCaul M, Diamond D and Magni P 2016 Screen-printed electrodes for environmental monitoring of heavy metal ions: a review *Microchim. Acta* **183** 503–17

[92] Wang J and Schultze J W 1996 Analytical electrochemistry *Angew. Chem. Engl. Ed.* **35** 1998

[93] Malon A, Vigassy T, Bakker E and Pretsch E 2006 Potentiometry at trace levels in confined samples: ion-selective electrodes with subfemtomole detection limits *J. Am. Chem. Soc.* **128** 8154–5

[94] You R, Li P, Jing G and Cui T 2019 Ultrasensitive micro ion selective sensor arrays for multiplex heavy metal ions detection *Microsyst. Technol.* **25** 845–9

[95] Vasapollo G, Sole R D, Mergola L, Lazzoi M R, Scardino A, Scorrano S and Mele G 2011 Molecularly imprinted polymers: present and future prospective *Int. J. Mol. Sci.* **12** 5908–45

Cao Y, Feng T, Xu J and Xue C 2019 Recent advances of molecularly imprinted polymer-based sensors in the detection of food safety hazard factors *Biosens. Bioelectron.* **141** 111447

[96] Poma A, Brahmbhatt H, Pendergraff H M, Watts J K and Turner N W 2015 Generation of novel hybrid aptamer—molecularly imprinted polymeric nanoparticles *Adv. Mater.* **27** 750–8

[97] Sullivan M V, Allabush F, Bunka D, Tolley A, Mendes P M, Tucker J H and Turner N W 2021 Hybrid aptamer-molecularly imprinted polymer (AptaMIP) nanoparticles selective for the antibiotic moxifloxacin *Polym. Chem.* **12** 4394–405

[98] Ali G K and Omer K M 2022 Molecular imprinted polymer combined with aptamer (MIP–aptamer) as a hybrid dual recognition element for bio (chemical) sensing applications. Review *Talanta* **236** 122878

IOP Publishing

Molecularly Imprinted Polymers for Environmental Monitoring
Fundamentals and applications
Raju Khan and Ayushi Singhal

Chapter 8

Molecularly imprinted polymers for the extraction and sensing of vitamins

M P Divya, Y S Rajput, Rajan Sharma and K B Divya

The availability of a wide array of monomers and understanding of non-covalent interactions have enabled technology for the synthesis of imprinted polymers for the separation and detection of molecules differing in size, shape, and functional groups. In addition to the use of molecular imprinted polymers (MIPs) as extraction matrices, recent developments in the synthesis of MIPs have focused toward their use in sensors. The classical way of synthesizing MIPs through bulk polymerization results in robust materials with unmatched thermal and chemical stability, but diffusion constraint associated problems restrict their use. The surface imprinting of the polymer on magnetic particles has minimized the diffusion issue. Surface imprinting is now being explored widely for developing chemical sensors by coating the polymer on the electrode surface, quartz crystal microbalance (QCM) sensors and surface plasmon resonance (SPR) sensors. The technique of MIPs is used widely for the separation and sensing of pesticides, antibiotics, drugs, metabolites, vitamins, anti-oxidants, and many more. Vitamins are essential for well-being and their absence in the diet results in some diseases of deficiency. Vitamins differ in solubility and based on this they are categorized into water-soluble and fat-soluble vitamins. Vitamins also vary significantly from each other in their structure. Some vitamins are not soluble in porogenic solvents such as chloroform, acetonitrile, and methanol, and therefore require changes in their solubilization strategy during MIP synthesis. Some vitamins derive their activity from molecules that differ slightly in their structure, synthetic vitamins may differ in chemical structure from natural vitamins, and further some of them exist in forms bound to macromolecules such as proteins. This limits the wider use of MIP technology for the extraction of vitamins from food matrices, however, their use in extracting vitamins from supplements and pharmaceuticals can be achieved successfully. In this chapter, descriptions of the monomers, crosslinkers, and initiators used in bulk or surface imprinting polymerization are provided and application of MIP for separation and sensing of vitamins are described. Successful use of MIPs for the extraction of vitamins bound to proteins will require a pre-step for dissociation. Considerable progress has been made in MIP-based vitamin sensing methods, but sensing in biological samples is still a challenge.

8.1 Overview of molecular imprinting

Molecular imprinted polymers (MIPs) are synthetic polymers with binding sites for target molecules which can be small or large molecules such as proteins and even cells [1–12]. The quest to develop MIPs is derived from known specific interactions in biological processes which include antigen–antibody interactions, ligand–receptor interactions, interactions in signalling pathways, virus–host cell interactions, and enzyme–substrate interactions. These examples have encouraged chemists/biochemists to exploit this knowledge for the preparation of synthetic polymers for recognizing target molecules. This recognition in biological processes involves non-covalent interactions and the cumulative strength of these interactions results in specificity in recognition.

Molecular interactions play a key role in defining the physical properties of molecules and several processes essential for life. The interactions are classified into two broad categories, covalent and non-covalent interactions. Non-covalent interactions include hydrogen bonds, ionic interactions, van der Waal interactions, dipole–dipole interactions, hydrophobic interactions, and pi–pi interactions. Non-covalent interactions depend on the pH, ionic strength, temperature, and polarity of the solvent and changes in these parameters can make the interaction strong or weak. During binding of a target to an MIP reaction conditions which favour interaction are adopted, while during elution of the bound molecule unfavourable conditions for interaction are used.

Intramolecular interactions are confined within a molecule and depend on the chemical nature of the atoms, groups, and their arrangement in space. Intermolecular interactions occur between different molecules which can be the same or different molecules. Intermolecular interactions are widely exploited in MIP synthesis. MIP has imprints in the form of cavities where target molecules can specifically bind and therefore MIPs have an inherent affinity for their target (template) molecules. This recognition or affinity is derived from interactions between atoms and groups present in the cavity of the polymer and atoms and groups present in the target. For improved affinity, it is essential that the atoms and groups present in the target and polymer involve multiple interactions. Synthetic polymers have a rigid structure and restrict flexibility to atoms or groups in space. This eventually will create a hindrance in multiple interactions with the target molecule. On the other hand, molecules in a solution can interact quickly and easily, which facilitates multiple interactions.

MIPs are synthetic polymers and their synthesis requires a template, monomer, crosslinker, initiator, and solvent [2, 13–15]. The hallmark in the synthesis of MIPs lies in the selection of the monomer, crosslinker, and solvent used during polymerization. The monomer should have the capability to undergo multiple interactions with template in the solvent system used during polymerization. A template molecule can be the target molecule, a fragment of it, or a molecule with a size, shape, and functional groups similar to the target molecule [3]. The purpose of the crosslinker molecule is to link different monomer molecules during the polymerization process. Crosslinkers possess at least two groups which are far apart in structure and are capable of making linkages with monomers. At a sufficient concentration of monomer and crosslinker, polymerization yields a solid material which encompasses strength and stability. The polymerization reaction is initiated by the initiator which on decomposition generates free radicals. The free radical reaction is commonly used for the synthesis of MIP.

Monomers, which have an open chain carbon–carbon double bond and can also make hydrogen bonds, ionic bonds, and other non-covalent bonds, are preferred. The most used monomers are acrylic acid, methacrylic acid (MAA), methyl methacrylic acid, diethylene glycol dimethacrylate (DEGDMA), diethylaminoethyl methacrylate (DAM), 2-(dimethylamino)ethylmethacrylate (DMAEMA), ferrocenylmethyl methacrylate (FMMA), bisphenol dimethacrylate (BPADM), 2-hydroxyethyl methacrylate (HEMA), methacrylamide, p-vinylbenzoic acid (VB), itaconic acid, styrene, 4-ethylstyrene, aniline, 4-vinylpyridine (4-VP), 2-vinylpyridine (2-VP), 1-vinylimidazole, acrylamide, acrylamido-2-methyl-1-propane-sulphonic acid, N-isopropylacrylamide, 4-aminobenzoic acid, trans-3-(3-pyridyl)-acrylic acid, pyrrole, m-phenylenediamine, o-phenylenediamine, p-aminothiophenol, and indole-3-acetic acid (figure 8.1). Amongst these, acrylic acid and its derivatives are widely used. Open chain carbon–carbon double bonds participate in the free radical polymerization reaction. The carboxylic, carbonyl, hydroxyl, and amido groups can participate in hydrogen bonds and therefore monomers containing these groups are selected. In the case that the template contains aromatic structures, monomers capable of forming pi–pi interactions are preferred. Also, if the template is anionic or cationic, a monomer enabling ionic interaction is selected.

Commonly used crosslinkers employed in non-covalent molecular imprinting are ethylene glycol dimethacrylate (EDMA or EGDMA), trimethylolpropane trimethacrylate (TRIM), tetramethylene dimethacrylate, pentaerythritol tetraacrylate, divinylbenzene (DVB), N,N'-1-4-phenylenediacrylamine, N,N'-methylenediacrylamide, 3,5-bis(acryloylamido)benzoic acid, 1,3-diisopropenyl benzene, N,O-bisacryloyl-phenylalaninol, 2,6-bisacryloylamidopyridine, and 1,4-diacryloyl piperazine (figure 8.2). The selection of crosslinker depends largely on the monomer used in MIP synthesis.

In addition to the monomer and crosslinker, the synthesis of MIPs requires an initiator for the initiation of the polymerization reaction. Common initiators include azobisisobutyronitrile (AIBN), 2,2'-azobis(2-methylpropionitrile), 2,2'-azobis(2-amidinopropane) dihydrochloride, 4,4'-azobis(4-cyanovaleric) acid, 2,2'-azobis (2,4-dimethylvaleronitrile), benzoylperoxide, and potassium persulfate (figure 8.3). Initiators decompose at a higher temperature and generate free radicals which are used in chain reactions.

Typically, the first step in MIP synthesis requires mixing of the template with the monomer in a porogenic solvent which facilitate non-covalent interactions.

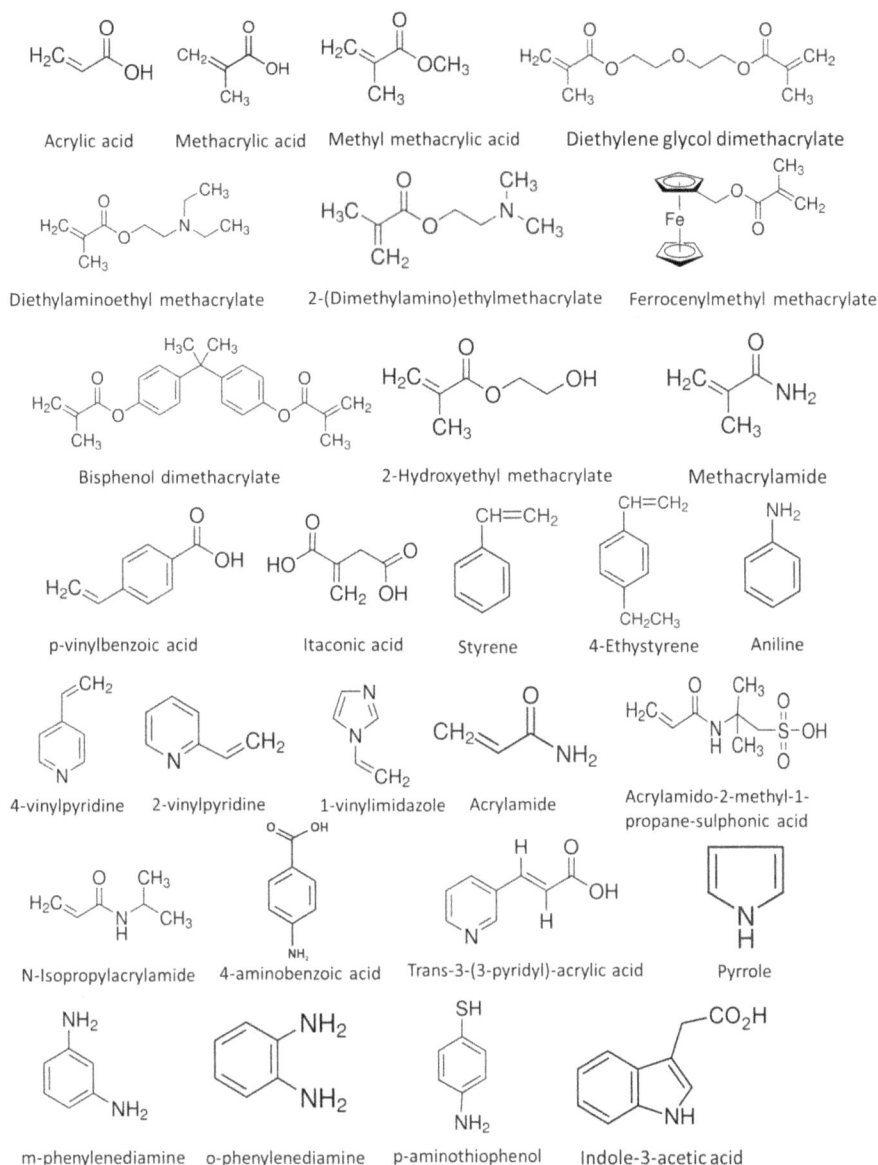

Figure 8.1. The structures of commonly used monomers for MIP synthesis.

A porogenic solvent leaves pores in the polymer when the solvent is removed after the polymerization reaction. The solvent should not interact with the monomer and template. Also, the solvent must not interact with the template–monomer complex. Chloroform, acetonitrile, methanol, acetone, 1,4-dioxane, and tetrahydrofuran are solvents commonly used during MIP synthesis and, essentially, the monomer, crosslinker, and initiator must be soluble in these solvents. A highly hydrophilic

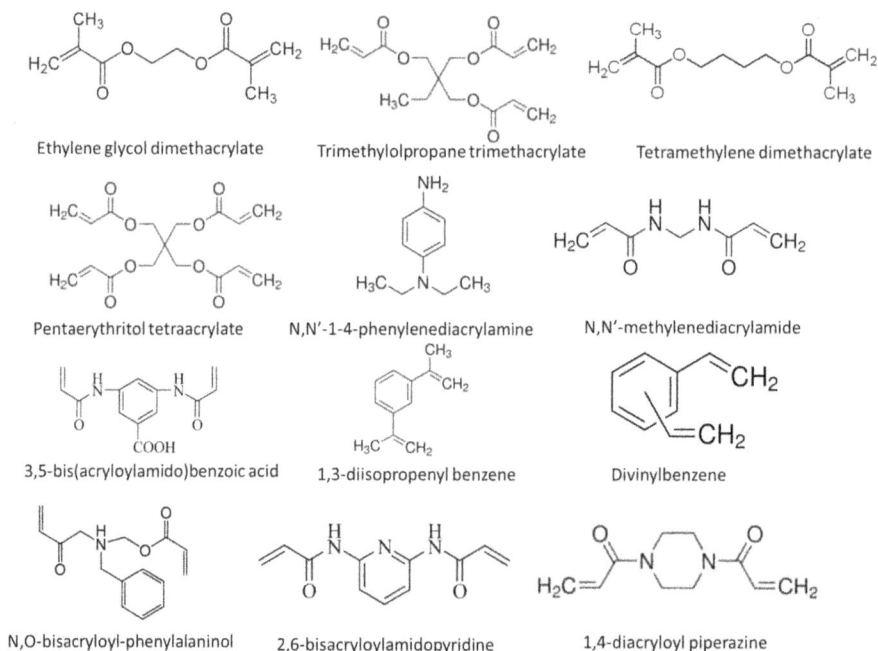

Figure 8.2. The structures of commonly used crosslinkers for the synthesis of MIPs.

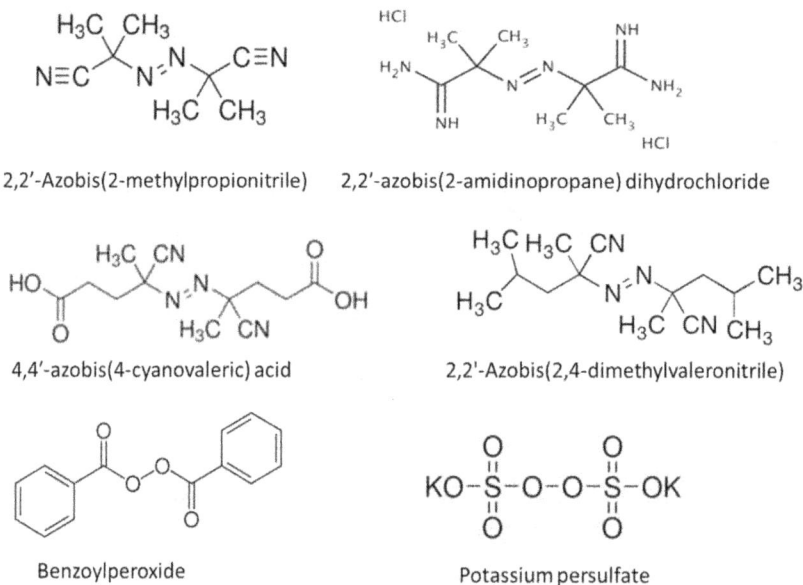

Figure 8.3. The structures of commonly used initiators for the synthesis of MIPs.

template may pose a problem in solubilizing in apolar solvent. After template–monomer complex formation, the crosslinker is added to the reaction mixture. Finally, the polymerization reaction is initiated by the addition of the initiator and the reaction mixture is incubated at 60 °C to 80 °C until the appearance of a solid mass. The bound template is then removed and this leaves cavities into which new incoming targets can bind. This concept is used in MIP synthesis, and the binding of target and its elution are shown in figure 8.4.

In the first step, the template and monomer in the porogenic solvent are pre-mixed, followed by sequential addition of the crosslinker and initiator, and subsequent incubation of the reaction mixture at 60 °C to 80 °C resulting in a solid mass. When the propagation of polymeric chains progresses, the template molecule is surrounded by monomers and consequently trapped in a three-dimensional polymer network. The solid mass is ground into fine particles. After removal of the template the resultant polymer, referred to as an MIP, can be used for the extraction of target molecules from a sample. This method of preparation of the polymer is called bulk polymerization. After binding of the target under appropriate conditions, the bound molecule can be eluted by weakening of the interaction between the polymer and the target molecule. Thus, the successful use of MIPs for the extraction of a target from a sample will require optimization of the conditions in terms of the pH, ionic strength, temperature, and polarity of the solvent for binding as well as for elution of target molecules. MIPs are robust and exhibit stability towards changes in pH, ionic strength, and temperature. The ratios of monomer, template, crosslinker, initiator, and solvent are important for the yield of product as well as the number of binding sites for the target per unit mass of product. Grinding of the bulk is essential for facilitating diffusion of the target during the binding and elution steps. However, grinding results in loss or distortion of binding sites. MIPs have some limitations, which include poor recognition properties in water, long equilibrium binding kinetics, and slow leaching of the template from the polymer matrix [1, 16].

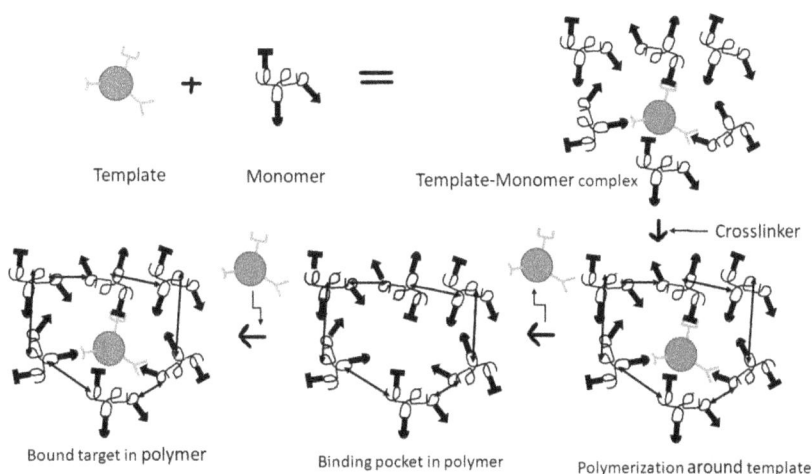

Figure 8.4. Overview of the synthesis of an MIP, template removal, and target binding.

The target and monomer can perhaps interact in more than one arrangement of atoms or groups and each such arrangement will result in an altered cavity with different affinity. This is the basis of heterogeneity in binding pockets. MIPs can act as a solid phase extraction material which can be packed into a column for chromatographic analysis.

The prepared MIPs are evaluated for selectivity and for this purpose the binding of the target to MIPs and non-imprinted polymers (NIPs) is compared. NIPs are prepared by omitting the template during polymerization. The ratio of binding of the target to MIPs and NIPs is referred to as the selectivity. The selectivity of MIPs is also assessed by evaluating the binding of the target molecule *vis-a-vis* a similar molecule. Higher selectivity of MIPs is preferred for the extraction of targets from biological samples.

MIP synthesis involving non-covalent interaction is the most popular as the removal of the template from the polymer is easy. The polymerization process can take place either in an abundance of solvents (emulsion, dispersion, suspension, multistep swelling, and precipitation polymerization) or in the presence of a small amount of solvents that ensures appropriate structure and porosity of the bulk MIPs [17].

8.2 Imprinted polymers for vitamins

Vitamins are essential for the well-being of humans and the absence of vitamins can cause disease of deficiency [18]. Most vitamins are either not produced in the body or are synthesized in limited quantities and must be consumed within the diet to avoid their deficiency. These are broadly classified into (i) water-soluble vitamins and (ii) fat-soluble vitamins. Examples of water-soluble vitamins are thiamine (vitamin B1), riboflavin (vitamin B2), nicotinamide (vitamin B3), pantothenic acid (vitamin B5), pyridoxine (B6), biotin (vitamin B7), folic acid (vitamin B9), cyanocobalamin (vitamin B12), and ascorbic acid (vitamin C). The fat-soluble vitamins are vitamin A, vitamin D, vitamin E, and vitamin K.

8.2.1 Water-soluble vitamins

8.2.1.1 Thiamine (vitamin B1)
Thiamine consists of an aminopyrimidine and thiazolium ring linked through methylene. The thiazole is substituted with methyl and hydroxyethyl groups. It is a cation and is soluble in water but insoluble in apolar solvents. Thiamine pyrophosphate (TPP) is derived from thiamine and acts as a coenzyme for several enzymes involved in the catabolism of carbohydrates and amino acids. It is stable in acidic pH but oxidizes in alkaline pH.

Thiamine is present in whole grains, cereals, meat, and fish and is also added to foods or sold as a supplement. In plant-based foods, thiamine predominantly occurs in its free form, while in animal-based foods it is bound to proteins. Only small amounts are stored in the liver, so a daily intake of thiamine-rich foods is needed. In mammals, deficiency results in Korsakoff's syndrome, optic neuropathy, and a disease called beriberi that affects the peripheral nervous system (polyneuritis) and/ or the cardiovascular system [18].

An imprinted polymer for the determination of thiamine in food samples was prepared by Alizadeh and co-workers in 2018 [19]. These workers applied a unique strategy to address the solubility problem of thiamine in chloroform, a common solvent used for the preparation of MIPs. Thiamine is a cationic agent and thus can make an ion-pair complex with trimethylsilyl-propane sulphonate, an anionic species. The prepared ion-pair complex is then transferred to chloroform for further carrying out the polymerization reaction. A mixture of MAA and VB was used as the monomer while EDMA was employed as the crosslinker. As usual, the initiator used was AIBN. NIP was obtained following the same procedure without the presence of the template molecule. The value of the imprinting factor was 2.97. The maximum theoretical binding capacities of the MIP and NIP materials were obtained as 106.1 and 35.7 μmol g^{-1}, respectively. The extraction efficiency of MIPs for thiamine from brown rice, wheat bran, and soy bean samples was in the range of 40%–46% only [19].

8.2.1.2 Riboflavin (vitamin B2)

Riboflavin contains an aromatic diamino group and is chemically defined as 7,8-dimethyl-10-[(2S,3S,4R)-2,3,4,5-tetrahydroxypentyl]benzo[g]pteridine-2,4-dione. It exists in oxidized as well as reduced forms. The ending 'flavin' is from the Latin word *flavus* which refers to its yellowish colour. In the retina of the eye, in whey, and in urine it exists in its free form while in tissues and cells the vitamin is present as flavin mononucleotide (FMN) and flavin-adenine dinucleotide (FAD). Riboflavin is a precursor of the coenzymes FMN and FAD. It is an essential coenzyme in many oxidation–reduction reactions involved in carbohydrate and protein metabolism. Riboflavin is soluble in water and ethanol but insoluble in chloroform and other organic solvents. The aqueous yellow solution of riboflavin shows a yellowish-green fluorescence with λ_{max} = 565 nm and this property is used for its estimation. The vitamin is heat stable. It is found in milk [20], eggs, malted barley, liver, kidney, heart, and leafy vegetables, but the richest natural source is yeast. Riboflavin-binding protein (~37 kDa) is present in bovine milk and its concentration in mammalian milk is higher than in plasma during lactation. Riboflavin is present in foods mostly (80%–90%) as FAD and FMN. Hydrochloric acid from the stomach readily releases the flavins that are only loosely bound to their proteins. It needs to be consumed regularly as the body does not store much of it. Its deficiency symptoms are cracked lips, sore throat, swelling of the mouth and throat, swollen tongue (glossitis), hair loss, skin rash, anaemia, itchy red eyes, and cataracts in severe cases.

8.2.1.2.1 Constant release of riboflavin

Normally MIPs are prepared for the extraction and sensing of analytes. However, in 2019 Mokhtari and Ghaedi [21] prepared water compatible MIPs for the controlled release of riboflavin as a drug delivery system. The constant release of drugs should cause minimal fluctuations of its concentration in the plasma. Chitosan (a linear chain polysaccharide) which is biocompatible and biodegradable and, in the presence of various functional groups such as $-NH_2$ and $-OH$, capable of hydrogen bond formation, was used for the imprinting of riboflavin. Chitosan loaded riboflavin was prepared by using different ratios of riboflavin and chitosan under an acidic environment at 60 °C. The release of riboflavin was studied under simulated body fluid (SBF) with simulated gastric fluid (SGF) conditions or (1%, w/v) of Na_2SO_4 and NaCl solutions. It was found that 100 mg of bound riboflavin was present in 1 gm of chitosan and about 50% of the bound riboflavin could be released under simulated conditions [21].

8.2.1.2.2 Riboflavin sensor

A surface plasmon resonance (SPR) sensor for the determination of vitamin B2 in infant formula and milk samples has been reported recently [22]. The SPR sensor chip is a metal (mostly gold) film coated on a glass substrate. When polarized light strikes at the metal interface with a medium with a different refractive index at certain angle (the resonance angle), a portion of the light energy couples with the collective oscillations of free electrons (known as surface plasmons) in the metal surface layer. This results in the absorption of light. However, the plasmon is observed when light strikes at the resonance angle. The resonance angle can be obtained by observing a dip in reflection intensity. In an SPR sensor, light is focused onto a metal film through a glass prism and the intensity of the reflected light is detected. Plasmon resonance is dependent on the refractive index of the medium in vicinity of the gold surface. In SPR, the ligand is immobilized on gold film and when the analyte binds to the ligands, it changes the reflection intensity (figure 8.5) [23]. This change in reflection intensity is related to the amount of analyte bound. SPR provides data in real time, meaning that the technique can also be used to measure association and dissociation constants. A thin film of MIPs

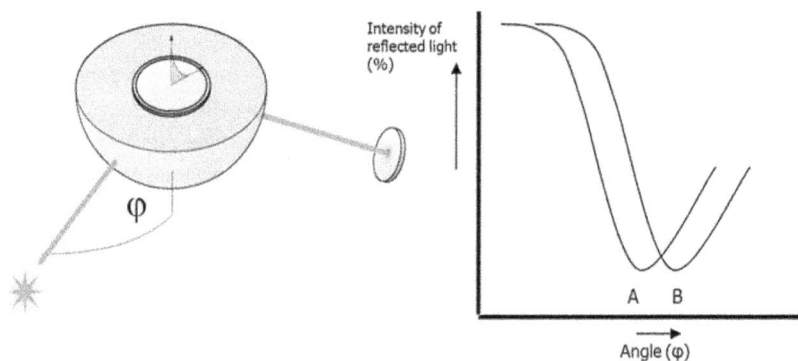

Figure 8.5. The working principle of a surface plasmon resonance sensor. A sensor chip with a gold coating is placed on a hemisphere (or prism). Polarized light shines from the light source (star) on the sensor chip. The reflected light intensity is measured in the detector (disk). At a certain angle of incidence (φ), excitation of surface plasmons occurs, resulting in a dip in the intensity of the reflected light (A). A change of refractive index at the surface of the gold film will cause an angle shift from A to B. (Reproduced with permission from [23]. Copyright 2017 Royal Chemical Society.)

can be coated on gold film and thus MIP-based SPR sensors have the potential to measure analytes. Also, the technique is label-free and the detection limit is of the order of 10 pg ml^{-1} [24].

In 2020 Cimen and Denizil [22] prepared vitamin B2 imprinted film on an allyl mercaptan (CH$_2$CHCH$_2$SH)-modified SPR chip surface. For the preparation of vitamin B2 imprinted polymeric film, 6.1 μl HEMA, 18.8 μl EGDMA, and 210 μl of pre-complex solution of MAA and vitamin B2 in a molar ratio of 10:1 were mixed in a rotator at 200 rpm for 30 min, followed by addition of 4 mg AIBN. Then, 3 μl of this pre-polymerization mixture was dropped onto the modified SPR chip surface. Then, the contents were polymerized by exposing the surface to UV light (100 W, 365 nm) for 20 min. NIPS were prepared in the absence of vitamin B2. The sensor exhibited linearity with an aqueous solution of vitamin B2 in the concentration range of 0.01 ng ml^{-1}–10 ng ml^{-1} and the LOD was 1.6×10^{-4} ng ml^{-1}. These workers also evaluated a sensor for measuring vitamin B2 in infant formula and milk samples, but only after extraction of vitamin B2 from food samples. The extraction step involved the separation of an oil layer by centrifugation, treatment with trypsin and clara-diastase, and passing through a solid phase extraction (SPE) cartridge. The sensor exhibited selectivity over vitamin B1 and vitamin B6.

8.2.1.3 Vitamin B3 (niacin)

Vitamin B3 is a term used for niacin, also known as nicotinic acid (pyridine-3-carboxylic acid) and nicotinamide (pyridine-3-carboxyamide). The coenzyme forms of this vitamin are nicotinamide adenine dinucleotide (NAD) and nicotinamide adenine' dinucleotide phosphate (NADP). Vitamin B3 deficiency results in a condition known as pellagra. The signs and symptoms of pellagra include skin and mouth lesions, anaemia, headaches, dementia, diarrhoea, and tiredness. Other signs of severe niacin deficiency include depression, memory loss, and hallucinations. Also, the consumption of a high amount of niacin may lead to hepatotoxicity [25]. This necessitates the need for the development of the extraction and determination of nicotinamide in food samples.

Vitamin B3 is found in bananas, avocado, mushrooms, green peas, beans, lentils, ginger, sweet pepper, potatoes, peanuts, soy nuts, pumpkin seeds, sunflower seeds, whole grains, soy milk, milk, cheddar cheese, cottage cheese, curd, brown rice, wheat, barley, corn, fish, egg, and chicken. Vitamin B3 is very stable and does not get destroyed easily, hence the food sources can be cooked and consumed.

8.2.1.3.1 Nicotinamide extraction

In 2002 Mistry [26] prepared molecular imprinted polymers for nicotinamide analysis using MAA as the monomer, EGDA as the crosslinker, AIBN as the free radical initiator, and chloroform as the porogenic solvent. The polymers were prepared with a 1:4 ratio between the monomer and crosslinker and showed better selectivity for nicotinamide compared to closely related structures. Nicotinamide-based molecularly imprinted microspheres were also developed by Sole et al in 2010 [27]. These authors synthesized MIP microspheres using MAA as the monomer and EGDA as the crosslinker. The imprinted polymers showed good binding capacity for nicotinamide and selectively. In 2017 Asfaram *et al* [28] synthesized metal organic framework (MOF) based MIPs for the preconcentration of nicotinamide. The prepared polymers showed a recovery rate of 95.85%–101.27% for nicotinamide from urine, milk, and water samples.

8.2.1.4 Vitamin B6 (pyridoxine)

Vitamin B6 is composed of three compounds: pyridoxine, pyridoxal, and pyridoxamine. Pyridoxine contains a methylhydroxyl group ($-CH_3OH$), pyridoxal an aldehyde ($-CHO$), and pyridoxamine an aminomethyl group ($-CH_3NH_2$). All three forms can be activated by phosphorylation. Pyridoxal $5'$ phosphate (PLP) is the active coenzyme form and most common measure of B6 blood levels in the body.

Vitamin B6 is essential for the synthesis of neurotransmitters and proper growth, development, and functioning of the brain, nerves, and skin. Vitamin B6 is used for treating heart problems, alcohol addiction, cataract, glaucoma, diabetic pain, cerebellar syndrome, motion sickness, and kidney disease. Vitamin B6 is present in potato, sweet potato, bitter gourd, bitter melon, carrot, spinach, green peas, avocado, banana, papaya, orange, durian, plum, chickpeas, soybeans, beans, lentils, sunflower seeds, wheat bran, milk, eggs, chicken, and salmon.

8.2.1.4.1 Pyridoxine extraction

A molecularly imprinted polymer for pyridoxine using an ion-pair as template was developed by Alizadeh *et al* [19]. In this method an ion-pair complex was formed between the pyridoxine ion and dodecyl sulphate ion and further polymerization was carried out in chloroform. The imprinted polymer prepared using acrylic acid showed better selectivity for pyridoxine compared to other monomers.

8.2.1.4.2 Pyridoxine sensor

A simple sensor was created for vitamin B_6 detection based on an electropolymerized molecularly imprinted polymer of 3-aminobenzoic acid [29]. The poly (3-aminobenzoic acid) film was electrodeposited by potentiodynamic cycling of potential with and without a vitamin B_6 (template) on a carbon fibre paper electrode. The developed imprinted sensor showed a higher selectivity for pyridoxine among structural analogues and a detection limit of 0.010 μM.

8.2.1.5 Vitamin B7 (biotin)

Biotin is an organic heterobicyclic compound that consists of 2-oxohexahydro-1*H*-thieno[3,4-d]imidazole with a valeric acid substituent attached to the tetrahydrothiophene ring. Biotin is also known as vitamin H or B7 or coenzyme R. Biotin was discovered in the year 1931 by German scientist Paul Gyorgy when he extracted the compound from the liver and named it vitamin H. The letter 'H' represents 'Haar und Haut', which are the German words for 'hair and skin', signifying its relevance to hair and skin health. Biotin deficiency is associated with neurological manifestations, skin rash, hair loss, dry brittle nails, a red rash around the eyes, nose, and mouth, dry eyes, scaly dry skin, conjunctivitis, a sore tongue, nausea, vomiting, and loss of appetite. Avidin (a protein found in egg white) binds tightly with biotin in the intestines and prevents biotin absorption by the body.

Biotin is present in every living cell and occurs mainly bound to proteins or polypeptides. It is abundant in the liver, kidney, pancreas, yeast, and milk. It has a role as a prosthetic group in four biotin-dependent carboxylases in mammals where the biotin moiety is covalently bound to the epsilon amino group of a lysine residue in each of these carboxylases.

Biotin is only synthesized by bacteria, moulds, yeasts, algae, and certain plants. The human body is dependent on dietary sources such as carrot, cauliflower, spinach, broccoli, mushrooms, sweet potato, bananas, avocados, raspberries, almonds, peanuts, walnuts, sunflowers seeds, rice, wheat, oats, milk, cheddar cheese, salmon, tuna, eggs, and meat.

The magnetic MIPs (Mag-MIPs) are a significant improvement over standard MIPs. In Mag-MIPs, the molecular imprinting surface (shell) is coated over supermagnetic nanoparticles (core). It retains magnetic properties, and thus can be easily separated by external magnetic fields. Also, the MIPs present on the surface of Mag-MIPs perform recognition towards template molecules with high efficiency (figure 8.6) [30].

The particle size of Mag-MIPs is far smaller than that of bulk MIPs and therefore they provide more surface area for the binding of the template. Since the recognition sites are located on the surface of the Mag-MIP, the diffusion constraints of template during binding and elution are minimized [31–35]. A Mag-MIP called $Fe_3O_4@MIP$ for the isolation and detection of biotin and biotinylated biomolecules has been developed [35, 36]. Supermagnetic nanoparticles (Fe_3O_4) were treated with tetraethyl orthosilicate to introduce hydroxyl groups on the surface of the magnetic particles, allowing them their subsequent silanization. In the next step, the $Fe_3O_4@SiO_2$ were modified by 3-metacriloxipropiltrimetoxissilano (MPS). Then, the Mag-MIPs ($Fe_3O_4@MIPs$) were prepared by (i) mixing (25 °C, 12 h) 0.2 mmol of biotin and 0.8 mmol of acrylic acid in 30 ml of ethanol, (ii) the addition of 200 mg of $Fe_3O_4@SiO_2$–MPS to the biotin–acrylic acid mixture, followed by shaking (3 h), and (iii) further addition of 4.0 mmol EGDMA and 0.05 mmol of AIBN followed by sonication (5 min). The Mag-MIPs showed a high binding capacity for different

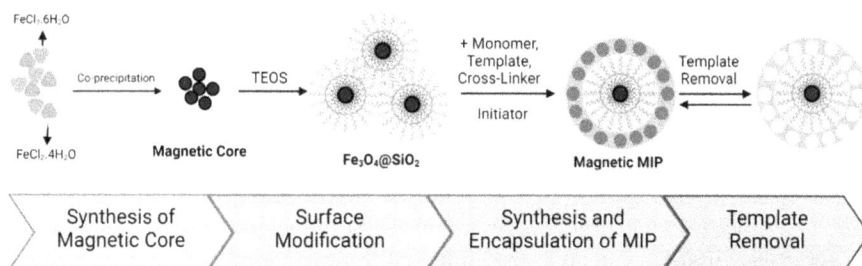

Figure 8.6. The working principle of the Mag-MIPs. (Reproduced from [30]. Copyright 2022 the authors. CC BY 4.0.)

biotinylated molecules and could act as separation tools for biotin and for its estimation [36].

8.2.1.6 Vitamin B9 (folic acid)

Folic acid is a water-soluble heterocyclic molecule belonging to the group B vitamins and comprises three large sub-components, namely (i) a pteridine ring, (ii) *para*-aminobenzoic acid, and (iii) glutamic acid. The colour of folic acid ranges from yellow to orange and it is a flavourless needle-like crystal. Folic acid gets its name from the Latin word *folium* meaning 'leaf', since it is found in many leafy plants. Good sources of folate include dark green leafy vegetables (turnip greens, spinach, lettuce, asparagus, broccoli), whole grains, nuts, beans, peanuts, sunflower seeds, fresh fruits, dairy products, poultry, liver, seafood, eggs, fortified foods, and supplements. Spinach, liver, asparagus, and Brussels sprouts are among the foods with the highest folate levels. Symptoms of folate deficiency are megaloblastic anaemia (fewer red blood cells that are larger in size than normal), weakness, fatigue, irregular heartbeat, shortness of breath, trouble concentrating, headache, irritability, sores on the tongue and inside the mouth, a change in the colour of the skin, hair, or fingernails, hair loss, irritability, headache, and neural tube defects in new-borns. Folate helps to form DNA and RNA and is involved in protein metabolism. It plays a key role in breaking down homocysteine, an amino acid that can exert harmful effects in the body if it is present in high amounts. Folate is also needed to produce healthy red blood cells and is critical during periods of rapid growth, such as during pregnancy and foetal development. It is called a hema-topoietic growth factor for its effects on the production of erythrocytes (anti-anaemic pharmacological function). Synthetic folic acid is commercially used to fortify or enrich the folate content of certain food products and is also utilized in dietary supplements [16, 17, 37].

Folate is the natural form of vitamin B9 in food, while folic acid is a synthetic form. Folate, an essential micronutrient, is a critical cofactor in one-carbon metabolism. Most countries have established recommended intakes of folate through folic acid supplements or fortified foods. External supplementation of folate may occur as folic acid, folinic acid (5-formyl tetrahydrofolic acid or leucovorin) or 5-methyltetrahydrofolate (5-MTHF). Naturally occurring 5-MTHF has important advantages over synthetic folic acid. It is well absorbed even when gastrointestinal pH is altered and its bioavailability is not affected by metabolic defects [38].

8.2.1.6.1 Folic acid extraction

In 2018 Panjan and co-workers [17] prepared MIPs for folic acid for extraction purposes. The preparation of MIP requires solubilization of the template, monomer, and crosslinker in porogenic solvent. The list of the most used solvents for MIP synthesis includes toluene, chloroform, dichloromethane, and acetonitrile. Folic acid is not properly dissolved in acetone, butanol, chloroform, cyclohexane, dichloromethane, dimethylformamide, ethanol, methanol, propanol, isopropanol, heptane, hexane, or toluene. However, folic acid can be solubilized in a mixture of DMSO and acetonitrile in a 5:3 ratio. The above group prepared MIPs using folic acid (10 mg), vinylbenzyl trimethylammonium chloride (VBTMAC, 250 mg), EGDMA (2500 mg), and 4 ml DMSO and acetonitrile mixture. Both MIPs and NIPs were synthesized in glass vials in a similar way except, folic acid was omitted in the NIPs. First, folic acid was dissolved, mixed with VBTMAC (the monomer), and then incubated at room temperature for 10 min for the pre-polymerization process. Then, the EGDMA (the crosslinker) was added, followed by the addition of 50 mg AIBN (the initiator). The vials were sealed with septa, left to cool down on ice, and then purge with nitrogen for 2 min. Polymerization was carried out at 50 °C for 24 h. The prepared polymer was crushed and dispersed in acetone and only sedimented particles were used further. The imprinting factor was dependent on the folic acid concentration and was $\geqslant 37$ at 0.5 M folic acid. Non-specific binding increases with an increase in folic acid concentration. The MIPs were able to extract folic acid from lettuce and cookies samples [17].

8.2.1.6.2 Folic acid sensor

An important step in sensor development is interfacing the MIP films with a transducer or electrode. Bulk polymer cannot make a smooth film and thus these are not used in sensor development. Thin imprinted films of sub-micrometre thickness can be deposited on piezoelectric devices (transducers) using spray- or spin-coating techniques. Alternatively, the electrochemical deposition of polymer on an electrode can be used [16, 37]. These investigators Apodaca and co-workers [16] and Hussain and co-workers [37] have developed sensors for folic acid using quartz crystal wafers as the transducer.

A quartz crystal microbalance (QCM) is an acoustic technology which measures the frequency of sound, typically in the range of MHz (10^6 Hz). The human ear can detect sound up to 20 000 Hz and hence sound in the MHz range goes undetected by the human ear. The core of this technology is the oscillating unit—a thin quartz crystal disk, which has electrodes deposited on each side. When an AC voltage is applied to a quartz crystal, it oscillates at a specific frequency. Also, the change in mass (analyte adsorbed) on the quartz surface (ng cm^{-2}) is related to the change in frequency (Δf) of the oscillating crystal (the Sauerbrey relation). This relationship holds good for elastic thin films adsorbed on QCMs, which do not dissipate any energy during oscillation. The thinner the QCM is, the higher is its resonant frequency and sensitivity [39]. Thinner QCMs are very fragile. The deposition of mass on the surface results in a decrease in resonant frequency. When the change in mass is greater than 2% of the crystal mass, the relationship between the change in frequency and change in mass becomes inaccurate [39].

A QCM device consists of a thin quartz disk with coated electrodes [40]. The quartz crystal plate is cut (AT or BT cut) to a specific orientation with respect to the crystal axes so that the acoustic wave propagates perpendicularly to the crystal surface. An AT cut causes minimum changes in frequency due to variation in temperature. The resonant frequency of the quartz single crystal depends on the angles with respect to the optical axis at which the wafer is cut from the crystal [39]. QCMs, and extended versions, such as quartz crystal microbalance with energy dissipation monitoring (QCM-D), are surface sensitive, real-time technologies that detect mass changes at the sensor surface with nanoscale resolution. In addition to the changes in mass, QCM-D also capture changes in energy loss (ΔD).

Previously, QCM-technology was used to monitor thin-film deposition in the gas phase and vacuum environments. Subsequently, it has also been applied in the liquid phase which requires information about the energy losses in the system, the so-called dissipation. The energy loss is captured with QCM-D. In contrast to standard QCM, which captures one parameter, Δf, as a function of time, QCM-D technology captures two parameters, Δf and ΔD, as a function of time (figure 8.7) [41].

In 2011 Apodaca and co-workers [16] used QCMs coated with an electro-polymerized MIP film of a bis-terthiophene dendron for sensing of folic acid. Polythiophene is a π-conjugated polymer. Like polypyrrole and polyaniline, poly-thiophenes can be both oxidatively or reductively doped in a proper solvent. The presence of sulphur in polythiophenes also enables it to be reduced and thus n-doped. Significantly, polythiophenes can be synthesized using anodic electrodeposition which is a method for directly adsorbing an electropolymerized material on an

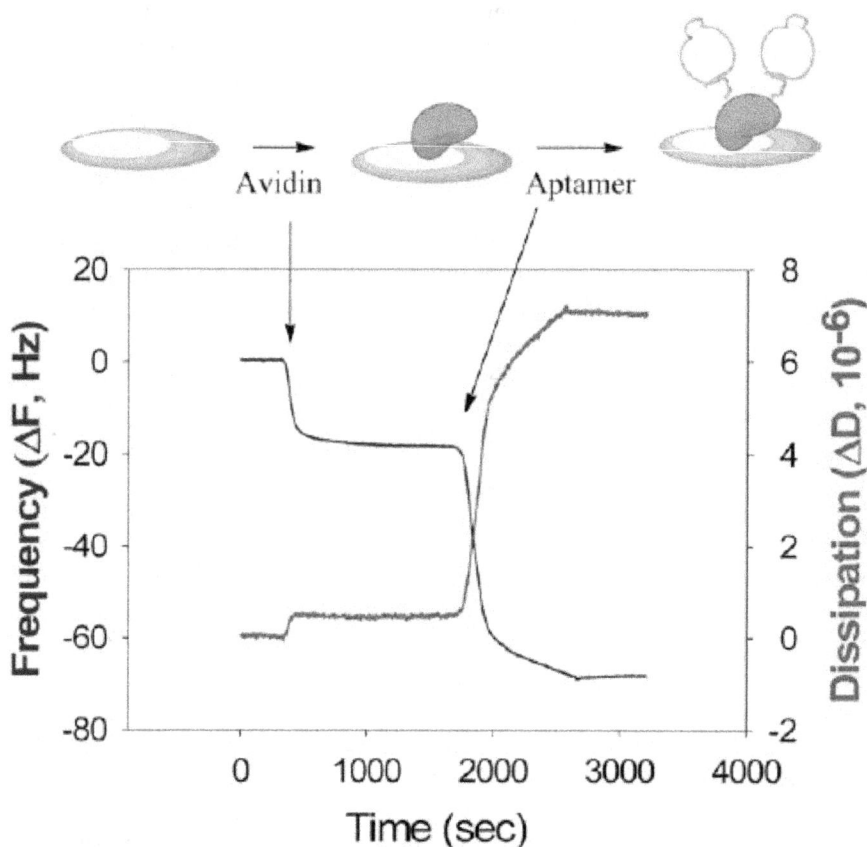

Figure 8.7. The working principle of quartz crystal microbalance dissipation. (Reproduced with permission from [41]. Copyright 2011 The Royal Society of Chemistry.)

electrode interface. Electrodeposition of polymers is accomplished by either anodic or cathodic electropolymerization. The film thickness directly affects the sensor's response time and this can be controlled by the electrochemical parameters in cyclic voltammetry. The developed sensor had an LOD of 15.4 μM (6.8 μg) with a linear range of 0–100 μM. The relative cross-selectivity of the developed QCM sensor for various structurally similar molecules was in the decreasing order of pteroic acid (50%) > caffeine (40%) > theophylline (6%) [16, 39].

In 2013 Hussain and co-workers [37] constructed a folic acid sensor wherein MIP thin films or nanoparticles prepared from methacrylates or acrylate–vinyl pyrrolidone copolymers were spin coated over QCMs. Both yielded sensor characteristics with lower limits of detection of 1–30 ppm, whereas the NIPs did not generate any signals. For methacrylate-based systems, switching from thin films to MIP nanoparticles increases the sensitivity by a factor of three. In contrast to this, in poly vinyl pyrrolidone based materials going from thin films to MIP nanoparticles does not increase sensitivity. Comparatively, vinyl pyrrolidone was a more suitable monomer than methacrylic acid.

In 2020 Cimen and Denizil [22] prepared vitamin B9 imprinted film on an allyl mercaptan (CH_2CHCH_2SH)-modified SPR chip surface. For the preparation of vitamin B9 imprinted polymeric film, 6.1 μl HEMA, 18.8 μl EGDMA, and 210 μl of pre-complex solution of N-vinyl-2-pyrrolidone (VP) and vitamin B9 in molar ratio of 4:1 were mixed in a rotator at 200 rpm for 30 min, followed by addition of 4 mg AIBN. Then 3 μl of this pre-polymerization mix was dropped onto the modified SPR chip surface. Then, the contents were polymerized by exposing the surface to UV light (100 W, 365 nm) for 20 min. NIPs were prepared in the absence of vitamin B9. The sensor exhibited linearity with an aqueous solution of vitamin B9 in the concentration range of 0.1 ng ml^{-1}–8.0 ng ml^{-1} and the LOD was 13.5×10^{-4} ng ml^{-1}. This group also evaluated the developed sensor for measuring vitamin B9 in infant formula and milk samples but only after extraction of vitamin B9 from food sampled. The extraction step involved the separation of the lipid layer by centrifugation, treatment with formic acid, and passing through a solid phase extraction (SPE) cartridge. The sensor exhibited selectivity over vitamin B1 and vitamin B12.

8.2.1.7 Vitamin B12 (cyanocobalamin)

Vitamin B12 is a water-soluble organic structurally complicated red crystalline compound containing cyanide and cobalt. The vitamin is present in dairy products, egg, fish, oysters, meat and poultry, and plant products [42–45]. The vitamin is stored in the liver and its deficiency leads to fatigue, nausea, weakness, weight loss, pernicious anaemia, and impaired brain development [46]. Vitamin B12 contains the mineral cobalt and thus compounds with vitamin B12 activity are collectively called cobalamins. Methylcobalamin and 5-deoxyadenosylcobalamin are the metabolically active forms of vitamin B12. However, two other forms, hydroxycobalamin and cyanocobalamin, become active on their conversion to methylcobalamin or 5-deoxyadenosylcobalamin.

Vitamin B12 plays an essential role in red blood cell formation, cell metabolism, nerve function, and the production of DNA. Vitamin B12 functions as a cofactor for two enzymes, methionine synthase and L-methylmalonyl-CoA mutase. Methionine synthase catalyzes the conversion of homocysteine to the essential amino acid methionine. Methionine is required for the formation of S-adenosylmethionine, a universal methyl donor. L-methylmalonyl-CoA mutase converts L-methylmalonyl-CoA to succinyl-CoA in the metabolism of propionate, a short-chain fatty acid.

Vitamin B12 binds to lysozyme and bovine serum albumin [42, 44]. Vitamin B12 is bound to protein in food and must be released before it is absorbed. The process starts in the mouth when food is mixed with saliva. The freed vitamin B12 then binds with haptocorrin, a cobalamin-binding protein in the saliva. More vitamin B12 is released from its food matrix by the activity of hydrochloric acid and gastric protease in the stomach, where it then binds to haptocorrin. In the duodenum, digestive enzymes free the vitamin B12 from haptocorrin, and this freed vitamin B12 combines with intrinsic factor, a transport and delivery binding protein secreted by the stomach's parietal cells. The resulting complex is absorbed in the distal ileum by receptor-mediated endocytosis. If vitamin B12 is added to fortified foods and dietary supplements, it is already in its free form and therefore does not require the separation step.

8.2.1.7.1 Cyanocobalamin extraction

Vitamin B12 is large molecule and therefore the MIPs prepared for vitamin B12 through bulk polymerization will be prone to diffusion constraint during binding and elution. Thus, in 2018 Li and co-workers [43] attempted to use surface imprinting. Further, this group used an iron magnetic nanoparticle surface for surface imprinting for easy separation of the matrix during the binding and elution steps of the extraction process. Boronic acid can form covalent bonds with cis-diols containing compounds at neutral or alkaline pH while the bond can be broken easily in acidic pH and, further, 2,4-difluoro-3-formyl-phenylboronic acid (DFFPBA) can conjugate to amino-functionalized magnetic nanoparticles. This arrangement has the potential to fix vitamin B12 in one orientation on the surface prior to initialization of the polymerization reaction [43]. 2-anilinoethanol was used in water self-polymerization for preparing imprinting coatings [43]. The prepared imprinted polymer is highly hydrophilic and acquires specificity from imprinting in addition to the affinity of boronic acid for cis-diols. The imprinting factor value was 23 which suggests that the imprinted polymer is highly selective over NIPs. The prepared MIPs exhibited specificity and high binding strength. The prepared MIPs were successfully applied to the analysis of vitamin B12 in human milk. The imprinted polymer quantitatively extracted vitamin B12 from spiked milk sample [43].

8.2.1.7.2 Cyanocobalamin sensor

In 2020 Cimen and Denizil [22] prepared vitamin B12 imprinted film on an allyl mercaptan-modified SPR chip surface. For the preparation of vitamin B12 imprinted polymeric film, 6.1 μl HEMA, 18.8 μl EGDMA, and 210 μl of pre-complex solution of N-methacryloyl-(L)-glutamic acid (MAGA) and vitamin B12 in molar ratio of 5:1 were mixed in a rotator at 200 rpm for 30 min, followed by addition of 4 mg AIBN. Then 3 μl of this pre-polymerization mix was dropped onto the modified SPR chip surface. Then, the contents were polymerized by exposing the surface to UV light (100 W, 365 nm) for 20 min. NIPs were prepared in the absence of vitamin B12. For the removal of vitamins from the polymeric film, the SPR chip was washed with 0.5 M NaCl solution. The sensor exhibited linearity with an aqueous solution of vitamin B12 in the concentration range of 0.01 ng ml^{-1}–1.5 ng ml^{-1} and the LOD was 2.5×10^{-4} ng ml^{-1}. These workers also evaluated the developed sensor for measuring vitamin B12 in infant formula and milk samples but only after extraction of vitamin B12 from the food sample. The extraction step involved separation treatment with β-amylase and passing through a solid phase extraction (SPE) cartridge. The sensor exhibited selectivity over vitamin B1 and vitamin B9.

8.2.1.8 Ascorbic acid (vitamin C)

Ascorbic acid is a water-soluble vitamin. It is a powerful reducing and antioxidant agent that functions in fighting bacterial infections, in detoxifying reactions, and in the formation of collagen in fibrous tissue, teeth, bones, connective tissue, skin, iron absorption, immune response activation, wound healing, as an anticoagulant, and aids in tissue healing. Ascorbic acid is found in citrus fruits, milk [47, 48], and vegetables. It cannot be produced or stored by humans and should be ingested in the diet. It is used in the treatment and prevention of scurvy. In the presence of transition metal ions, heat, light or mildly alkaline conditions, ascorbic acid is oxidized reversibly to dehydroascorbic acid, which still has vitamin activity. Dehydroascorbic acid is further irreversibly oxidized to 2,3-diketogulonic acid, which does not have vitamin activity. [49–52].

Ascorbic acid imprinted polymers are synthesized for dual purposes, both the extraction and/or estimation of ascorbic acid from a sample. In methods where a synthesized polymer is used for the extraction of ascorbic acid, estimation can be achieved by employing additional techniques/methods post-extraction.

8.2.1.8.1 Ascorbic acid extraction

Selective extraction of ascorbic acid from a sample has been achieved using MIPs [49]. The MIPs were prepared in highly diluted solutions of MAA, EGDA, ABIN, and ascorbic acid prepared in anhydrous acetonitrile. As an alternative to the normal initialization of the polymerization reaction, initialization by exposure to UV radiation (350 nm, 20 °C) was also used. After initialization, polymerization required incubation of the content at 65 °C for 24 h. The templates were removed by washing with acetic acid. The MIPs were used for preconcentration of trace amounts of ascorbic acid. The method can be used in real samples, employing a batch process.

8.2.1.8.2 Ascorbic acid sensor

The chemical structure of ascorbic acid determines its physical and chemical properties. It is a weak, water-soluble, unstable organic acid which can be easily oxidized or destroyed by light, aerobic conditions (oxygen), high temperatures, alkali conditions, copper, and heavy metals. Its stability is dependent pH. It is more stable at acidic pH in comparison to neutral or alkaline pH. In milk, ascorbic acid exists in oxidized as well as reduced forms and during the storage of milk ascorbic acid is oxidized. This makes the following sensors of limited use for measuring ascorbic acid in milk.

An ascorbic acid sensor was developed by Prasad and co-workers in 2019 [51] using molecular imprinting technology using melamine and chloranil but avoiding any crosslinker in order to facilitate better mass-transport to the binding sites. The imprinted polymer was drop coated on a hanging mercury drop electrode (HMDE) at +0.4 V (versus Ag/AgCl). The developed sensor showed highly specific binding to the ascorbic acid. This could be used as a tool for the selective analysis of ascorbic acid in aqueous, blood serum, and pharmaceutical samples. The sensor can detect serum ascorbic acid levels as low as 0.26 ng ml^{-1}. The MIP-modified HMDE sensor has shown a quantitative 100% response for ascorbic acid while the interferents, namely uric acid, tyrosine, oxalic acid, and glucose, were non-responsive and other interferents, namely dopamine, urea, and caffeine showed negligible uptake. A limitation in the use of HMDE is the requirement for oxygen free nitrogen gas and the toxicity associated with the use of mercury.

In 2011 Roy and co-workers [52] fabricated molecularly imprinted polyaniline (PANI) film (~100 nm thick) electrochemically onto indium–tin-oxide (ITO) coated glass plate using ascorbic acid (AA) as the template molecule. The presence of ascorbic acid in a PANI matrix was established by Fourier transform infra-red spectroscopy, scanning electron microscopy, cyclic voltammetry, and differential pulse voltammetry (DPV). The studies indicate that ascorbic acid acted as a dopant for PANI. Over-oxidation of the ascorbic acid doped PANI electrode resulted in removal of ascorbic acid leading to the preparation of an ascorbic acid selective molecularly imprinted PANI electrode (AA-MI-PANI/ITO). The molecularly imprinted AA-MI-PANI/ITO electrode could detect ascorbic acid in the range of 0.05–0.4 mM with a detection limit of 0.018 mM. The AA-MI-PANI/ITO electrode can be used eight times and is stable for one week. Amongst glucose (180 MW), uric acid (168 MW), glycine (75 MW), and lactic acid (90 MW), apart from glucose all the other interferants had a negligible effect. The sensor was used for measuring ascorbic acid in vitamin C tablets. The authors have not applied the sensor in real biological samples such as milk or lemon juice, where vitamin C may exist in its reduced as well as oxidized form.

In 2022 Chen et al [50] designed and validated an MIP-based electrochemical sensor for the detection of ascorbic acid. Preparation of the sensor involved three steps, namely (i) synthesis of polyvinylpyrrolidone coated gold nanoparticles (PVP-AuNPs), (ii) a covalent-organic framework (COF) prepared from 1,3,5-tris(p-formylphenyl)benzene (TFPB) and N-boc-1,4-phenylenediamine (NBPDA) in the presence of PVP-AuNPs and a preparation referred to as AuNPs@COFTFPB-NBPDA, and (iii) electrodeposition of AuNPs@COFTFPB-NBPDA, o-phenyl-enediamine, and ascorbic acid on GCEs by cyclic voltammetry at a potential of 0–0.8 V for 20 cycles, in order to obtain MIP/AuNPs@COFTFPB-NBPDA/GCE. The ascorbic acid (the template) is removed by soaking the electrode in water. In the prepared sensor, the signal is generated from the oxidation of ascorbic acid, specificity is provided by MIPs, porosity is introduced by COF, and catalytic efficiency is improved by AuNPs. The non-imprinted sensor was also able to generate a signal. The electrochemical sensor had a linear range of 7.81 μM to 60 mM ascorbic acid and a detection limit as low as 2.57 μM. Dopamine, uric acid, hydrogen peroxide, and glucose did not interfere in the assay. It has been shown that COFTFPB-NBPDA had certain catalytic activity and could catalyze the oxidation of AA. However, after AuNPs@COFTFPB-NBPDA was modified on the electrode surface, the peak current response value was further increased and the peak potential was further negatively shifted. This proved that the AuNPs@COFTFPB-NBPDA has better catalytic activity toward ascorbic acid. The high electrical conductivity of AuNPs@COFTFPB-NBPDA accelerated the electron transfer to ascorbic acid during the oxidation process, and the rich aromatic system in COFTFPB–DHzDS facilitated π–π stacking interactions with ascorbic acid. Furthermore, the large surface area of AuNPs@COFTFPB-NBPDA provided abundant sites for ascorbic acid binding. Presence of AuNPs enables oxidation of ascorbic acid at lower potential.

8.2.2 MIPs against fat-soluble vitamins

8.2.2.1 Vitamin A
Vitamin A is a fat-soluble vitamin and it has a role in cell differentiation, reproduction, the immune system, embryogenesis, and most importantly in vision processes. Deficiency of vitamin A in the diet leads to conditions such as night blindness, xerophthalmia, and ketatinization, etc. Milk is often fortified with the stable form of vitamin A, namely, retinyl acetate or retinyl palmitate. A method for the measurement of the amount of added retinyl acetate in food is required for ensuring that food is fortified with retinyl acetate as per the labels on packed food. An imprinted polymer for vitamin A was developed by Divya *et al* in 2016 [13] using methacrylic acid, ethylene glycol dimethacrylate, and retinyl acetate. The experiment results showed that an aqueous solution of acetonitrile can be used for the extraction of retinyl acetate using the prepared MIPs and bound retinyl acetate can be eluted by enhancing the concentration of acetonitrile. It was concluded that that prepared polymers have imprints of retinyl acetate and the recognition occurred through hydrophobic interactions. The chromatographic profile of the binding and elution of the retinyl acetate to MIPs and NIPs clearly established the presence of cavities specific to retinyl acetate in the imprinted polymers and the abolition of recognition in less polar solvents such as acetonitrile.

8.2.2.2 Vitamin D
Vitamin D is a fat-soluble vitamin. Deficiency of vitamin D in children results in rickets, wherein bones become soft and prone to fracture. Its deficiency in adults increases the risk of osteomalacia, which result in soft or fragile bones. It exists in two forms, namely vitamin D2 (ergocalciferol) and vitamin D3 (cholecalciferol), which differ from each other in their side chains. Vitamin D2 and vitamin D3 have 28 and 27 carbons, respectively. Vitamin D2 has an extra carbon–carbon double

bond at C22–C23 and one methyl group at the C24 position in its side chain. Vitamin D2 and vitamin D3 are synthesized from their respective provitamin forms, namely ergosterol (provitamin D2) and 7-dehydrocholesterol (provitamin D3) on UV irradiation. In the USA, India, and many other countries, dairy milk and plant-derived milk substitutes are fortified with vitamin D using cholecalciferol or ergocalciferol.

MIPs produced using the sol–gel process for solid phase extraction of vitamin D3 have been reported [53]. In the sol–gel method, a metal oxide precursor $(M(OR)_n)$ is dissolved in a low molecular weight solvent medium using a catalyst (acid, base, or ions such as F^-) followed by a hydrolysis (water) and polycondensation step. Sol–gel is a simple method to produce homogeneous and highly porous metal oxide nanosorbents. Sol–gel chemistry can produce imprinted selective cavities with longer lifetimes due to the use of silica-based materials with strong and stable structures. These materials are porous, have more surface area, and are thermally more stable. During template removal a higher temperature can be used and this enables easy removal of the templates [54, 55]. One of the major advances of sol–gel processing is the possibility of synthesizing hybrid materials, combining the properties of inorganic and organic compounds in one material. In the sol–gel material, the template is incorporated in rigid inorganic–organic networks. In 2016 Kia and co-workers [53] synthesized sol–gel MIPs from tetraethyl orthosilicate (an inorganic compound), methacrylic acid (the monomer), ethylene glycol dimethacrylate (the crosslinker), and benzoylperoxide (the initializer) using vitamin D3 as the template. The template was pre-incubated with tetraethyl orthosilicate in the presence of ethanol and HCl. The polymerized mixture containing the monomer, crosslinker, and initializer was prepared separately and the mixture was then added to the pre-incubated template. From the prepared polymer, unreacted monomer and template were extracted by methanol. The prepared polymer exhibited distinct selectivity for vitamin D3 over similar molecules. The MIP-SPE column was prepared by packing a pre-mixed powder of MIP powder and MWCNT and the performance was evaluated by coupling it with the HPLC column. The limit of detection was 1 ng ml^{-1} and the linear range was 1–10 000 ng ml^{-1} [53].

8.2.2.3 Vitamin E

There are eight related compounds which are derivatives of 6-chromanol and have vitamin E activity. Four of these compounds are derivatives of tocopherol and four of tocotrienol. Both tocopherols and tocotrienols have a hydrophobic side chain at the C2 position of the chromanol ring, the main difference being that the side chain is saturated in tocopherols while tocotrienols have three double bonds. Both tocopherols and tocotrienols are further designated as α-, β-, γ-, or δ-tocopherols and tocotrienols based on the number and position of methyl groups on the chromanol ring. Usually, vitamin E activity is expressed in tocopherols equivalent (TE) wherein 1 TE is corresponding to the vitamin E activity of 1 mg α-tocopherol. β-, γ-, or δ-tocopherols and tocotrienols have biological activity equivalent to 0%, 10%, and 33% of the activity of α-tocopherol, respectively.

Being most potent of all the forms of vitamin E, α-tocopherol has been used for the preparation of MIPs by various researchers. MIPs for α-tocopherol have been prepared by the precipitation polymerization process using EDGMA as a cross-linker and acrylamide as functional monomer in acetone solution [56]. AIBN has been used as radical initiator. After the reaction under inert conditions, precipitates of MIPs were obtained because of its high insolubility and higher density in the solution. The results indicated that prepared MIPs were superior to NIPs and were able to bind 38.8 mg g^{-1} α-tocopherol with an imprinting factor of 2.6.

Recently, MIP beads imprinted with α-tocopherol were prepared via a two-step polymerization method, using MAA, EGDMA, and other functional monomers on the surface of polystyrene microspheres [57]. The prepared beads with MIP imprints have a diameter in the range of 1–9 μm with an average of 2.5 μm. The material has an adsorption capacity of 46 mg g^{-1} α-tocopherol and its application in a column for repeated use has been demonstrated.

8.2.2.4 *Vitamin K*
Vitamin K exists naturally in two forms, namely phylloquinone (vitamin K$_1$) and menaquinones (vitamin K$_2$). While phylloquinone exists exclusively in plants, menaquinones are synthesized by bacteria. Apart from these two forms of vitamin K, menadione (vitamin K$_3$) is a synthetic compound with vitamin K activity. Vitamin K$_3$ lacks side the aliphatic chain and is water soluble. Vitamin K has a role in blood clotting and is crucial for the biological synthesis of proteins involved in the clotting process. Vitamin K deficiency is rare but can occur from diminished absorption of fat. Its deficiency results in neurodegenerative diseases [58]. The analytical methods for the estimation of vitamin K in biological samples are costly and the usual techniques used are HPLC and ELISA. An attempt has been made to prepare MIPs for vitamin K3 using the monomer MAA, crosslinker EGDM, and initiator AIBN [59]. The prepared MIPs were used for sensor preparation by their immobilization on aluminium chips which were subsequently stamped using polydimethysiloxane. The prepared sensor was used in a thermal detection platform called thermal wave transport analysis. It has been demonstrated that the prepared MIP sensor could also detect vitamin K$_1$ and a concentration dependent response was established. The limit of detection of the sensor was established at 200 nM and the results agreed with HPLC based methods. The applicability of the developed MIP-based sensor was demonstrated to study vitamin K content in blood serum.

8.3 Future perspective
Some of the vitamins derive their biological activity from similar molecules but these certainly differ in chemical structure. Pharmaceutical preparations exhibiting vitamin A or vitamin D activity are water soluble and these differ structurally from the natural forms of the vitamins. Some vitamins exist in coenzyme form or bound with proteins. Since MIPs possess specificity, MIPs prepared for targeted vitamins may not work effectively when used for their extraction or sensing in food and pharmaceutical preparations. Vitamin C (ascorbic acid) exists in reduced and

oxidized forms but efforts have been focussed only on synthesizing MIPs for reduced ascorbic acid and these preparations are not evaluated for binding to oxidized ascorbic acid. Systematic studies are needed to evaluate the performance of MIPs in aqueous solutions, food, and vitamin supplements.

References

[1] Allender C J 2005 Molecularly imprinted polymers: technology and applications *Adv. Drug Deliv. Rev.* **57** 13–4

[2] Divya M P, Rajput Y S, Sharma R and Singh G 2015 Molecularly imprinted polymer for separation of lactate *J. Anal. Chem.* **70** 1213–7

[3] Fresco-Cala B, Batista A D and Cárdenas S 2020 Molecularly imprinted polymer micro- and nano-particles: a review *Molecules* **25** 4740

[4] Kumar V and Kim K H 2022 Use of molecular imprinted polymers as sensitive/selective luminescent sensing probes for pesticides/herbicides in water and food samples *Environ. Pollut.* **299** 118824

[5] Morsi S M, Abd El-Aziz M E and Mohamed H A 2023 Smart polymers as molecular imprinted polymers for recognition of target molecules *Int. J. Polym. Mater. Polym. Biomater.* **72** 612–35

[6] Singhal A, Parihar A, Kumar N and Khan R 2022 High throughput molecularly imprinted polymers based electrochemical nanosensors for point-of-care diagnostics of COVID-19 *Mater. Lett.* **306** 130898

[7] Singhal A, Ranjan P, Sadique M A, Kumar N, Yadav S, Parihar A and Khan R 2022 Molecularly imprinted polymers-based nanobiosensors for environmental monitoring and analysis *Nanobiosensors for Environmental Monitoring: Fundamentals and Application* (Cham: Springer International) pp 263–78

[8] Singhal A, Sadique M A, Kumar N, Yadav S, Ranjan P, Parihar A, Khan R and Kaushik A K 2022 Multifunctional carbon nanomaterials decorated molecularly imprinted hybrid polymers for efficient electrochemical antibiotics sensing *J. Environ. Chem. Eng.* **10** 107703

[9] Singhal A, Singh A, Shrivastava A and Khan R 2023 Epitope imprinted polymeric materials: application in electrochemical detection of disease biomarkers *J. Mater. Chem.* B **11** 936–54

[10] Singhal A, Yadav S, Sadique M A, Khan R, Kaushik A, Sathish N and Srivastava A K 2022 MXene-modified molecularly imprinted polymers as an artificial bio-recognition platform for efficient electrochemical sensing: progress and perspectives *Phys. Chem. Chem. Phys.* **24** 19164–76

[11] Yang Y and Shen X 2022 Preparation and application of molecularly imprinted polymers for flavonoids: review and perspective *Molecules.* **27** 7355

[12] Yuksel N and Tektas S 2022 Molecularly imprinted polymers: preparation, characterisation, and application in drug delivery systems *J. Microencapsulation* **39** 176–96

[13] Divya M P, Rajput Y S, Sharma R and Nanda D K 2016 Dependence of selectivity of vitamin A imprinted polymer on ratio of water in aqueous solutions of acetonitrile and methanol *Proc. 85th SBC Conf. (Mysore, India, 21–24 November)*

[14] Divya M P, Rajput Y S and Sharma R 2010 Synthesis and application of tetracycline imprinted polymer *Anal. Lett.* **43** 919–28

[15] Lata K, Sharma R, Naik L, Rajput Y S and Mann B 2015 Synthesis and application of cephalexin imprinted polymer for solid phase extraction in milk *Food Chem.* **184** 176–82

[16] Apodaca D C, Pernites R B, Ponnapati R R, Del Mundo F R and Advincula R C 2011 Electropolymerized molecularly imprinted polymer films of a bis-terthiophene dendron: folic acid quartz crystal microbalance sensing *ACS Appl. Mater. Interfaces* **3** 191–203

[17] Panjan P, Monasterio R P, Carrasco-Pancorbo A, Fernandez-Gutierrez A, Sesay A M and Fernandez-Sanchez J F 2018 Development of a folic acid molecularly imprinted polymer and its evaluation as a sorbent for dispersive solid-phase extraction by liquid chromatography coupled to mass spectrometry *J. Chromatogr.* A **1576** 26–33

[18] Combs G F Jr and McClung J P 2016 *The Vitamins: Fundamental Aspects in Nutrition and Health* (New York: Academic)

[19] Alizadeh T, Akhoundian M and Ganjali M R 2018 An innovative method for synthesis of imprinted polymer nanomaterial holding thiamine (vitamin B1) selective sites and its application for thiamine determination in food samples *J. Chromatogr.* B **1084** 166–74

[20] Sharma R and Lal D 1998 Influence of various heat processing treatments on some B-vitamins in buffalo and cow's milks *J. Food Sci. Technol.* **35** 524–6

[21] Mokhtari P and Ghaedi M 2019 Water compatible molecularly imprinted polymer for controlled release of riboflavin as drug delivery system *Eur. Polym. J.* **118** 614–8

[22] Çimen D and Denizli A 2020 Development of rapid, sensitive, and effective plasmonic nanosensor for the detection of vitamins in infact formula and milk samples *Photonic Sens.* **10** 316–32

[23] Schasfoort R B (ed) 2017 *Handbook of Surface Plasmon Resonance* (London: Royal Society of Chemistry)

[24] Nguyen H H, Park J, Kang S and Kim M 2015 Surface plasmon resonance: a versatile technique for biosensor applications *Sensors* **15** 10481–510

[25] Dunbar R L and Gelfand J M 2010 Seeing red: flushing out instigators of niacin-associated skin toxicity *J. Clin. Investig.* **120** 2651–5

[26] Mistry R 2002 Niacinamide analysis using molecularly imprinted polymers *Doctoral Dissertation* University of British Columbia, Vancouver, BC, Canada

[27] Sole R D, Lazzoi M R and Vasapollo G 2010 Synthesis of nicotinamide-based molecularly imprinted microspheres and *in vitro* controlled release studies *Drug Deliv.* **17** 130–7

[28] Asfaram A, Ghaedi M and Dashtian K 2017 Ultrasound assisted combined molecularly imprinted polymer for selective extraction of nicotinamide in human urine and milk samples: spectrophotometric determination and optimization study *Ultrason. Sonochem.* **34** 640–50

[29] Cherian A R, Benny L, Varghese A, John N S and Hegde G 2021 Molecularly imprinted scaffold based on poly (3-aminobenzoic acid) for electrochemical sensing of vitamin B6 *J. Electrochem. Soc.* **68** 077512

[30] Ariani M D, Zuhrotun A, Manesiotis P and Hasanah A N 2022 Magnetic molecularly imprinted polymers: an update on their use in the separation of active compounds from natural products *Polymers* **14** 1389

[31] Aggarwal S, Rajput Y S, Nanda D K, Sharma R and Singh G 2019 Synthesis and characterization of magnetic imprinted polymer for selective extraction of cephalexin from food matrices *Indian J. Chem. Technol.* **26** 431–6

[32] Aggarwal S, Rajput Y S, Sharma R and Pandey A K 2016 Extraction of Cefquinome from Food by Magnetic Molecularly Imprinted Polymer *Pharm. Anal. Acta* **7** 2

[33] Aggarwal S, Rajput Y S, Singh G and Sharma R 2016 Synthesis and characterization of oxytetracycline imprinted magnetic polymer for application in food *Appl. Nanosci.* **6** 209–14

[34] Cheng Y, Nie J, Li J, Liu H, Yan Z and Kuang L 2019 Synthesis and characterization of core–shell magnetic molecularly imprinted polymers for selective recognition and determination of quercetin in apple samples *Food Chem.* **287** 100–6

[35] Uzuriaga-Sánchez R J, Khan S, Wong A, Picasso G, Pividori M I and Sotomayor M D 2016 Magnetically separable polymer (Mag-MIP) for selective analysis of biotin in food samples *Food Chem.* **190** 460–7

[36] Aissa A B, Herrera-Chacon A, Pupin R R, Sotomayor M D and Pividori M I 2017 Magnetic molecularly imprinted polymer for the isolation and detection of biotin and biotinylated biomolecules *Biosens. Bioelectron.* **88** 101–8

[37] Hussain M, Iqbal N and Lieberzeit P A 2013 Acidic and basic polymers for molecularly imprinted folic acid sensors—QCM studies with thin films and nanoparticles *Sensors Actuators* B **176** 1090–5

[38] Scaglione F and Panzavolta G 2014 Folate, folic acid and 5-methyltetrahydrofolate are not the same thing *Xenobiotica* **44** 480–8

[39] Vashist S K and Vashist P 2011 Recent advances in quartz crystal microbalance-based sensors *J. Sens.* **2011** 571405

[40] Akgönüllü S, Özgür E and Denizli A 2022 Recent advances in quartz crystal microbalance biosensors based on the molecular imprinting technique for disease-related biomarkers *Chemosensors* **10** 106

[41] Özalp V C 2011 Acoustic quantification of ATP using a quartz crystal microbalance with dissipation *Analyst* **136** 5046–50

[42] Li D, Yang Y, Cao X, Xu C and Ji B 2012 Investigation on the pH-dependent binding of vitamin B12 and lysozyme by fluorescence and absorbance *J. Mol. Struct.* **1007** 102–12

[43] Li D, Yuan Q, Yang W, Yang M, Li S and Tu T 2018 Efficient vitamin B12-imprinted boronate affinity magnetic nanoparticles for the specific capture of vitamin B12 *Anal. Biochem.* **561–562** 18–26

[44] Li D, Zhang T, Xu C and Ji B 2011 Effect of pH on the interaction of vitamin B12 with bovine serum albumin by spectroscopic approaches *Spectrochim. Acta, Part* A **83** 598–608

[45] Sharma R, Rajput Y S, Dogra G and Tomar S K 2007 Estimation of vitamin B12 by ELISA and its status in milk *Milchwissenschaft* **62** 127–30

[46] Pourié G, Guéant J L and Quadros E V 2022 Behavioral profile of vitamin B12 deficiency: a reflection of impaired brain development, neuronal stress and altered neuroplasticity *Vitam. Horm.* **119** 377–404

[47] Sharma R and Lal D 1999 Effect of various heat processing treatments on vitamin C content in buffalo milk and cow milk *Indian J. Dairy Biosci.* **10** 113–6

[48] Sharma R and Lal D 2005 Fortification of milk with microencapsulated vitamin C and its thermal stability *J. Food Sci. Technol.* **42** 167–73

[49] Baramakeh L 2022 Selective extraction of ascorbic acid by molecular imprinted polymer solid-phase extraction *J. Chem. Lett.* **3** 81–5

[50] Chen Y, Peng X, Song Y and Ma G 2022 An ascorbic acid-imprinted poly(o-phenylenediamine)/AuNPs@COFTFPB-NBPDA for electrochemical sensing ascorbic acid *Chemosensors* **10** 407

[51] Prasad B B, Srivastava S, Tiwari K and Sharma P S 2009 Ascorbic acid sensor based on molecularly imprinted polymer-modified hanging mercury drop electrode *Mater. Sci. Eng.* C **29** 1082–7

[52] Roy A K, Nisha V S, Dhand C and Malhotra B D 2011 Molecularly imprinted polyaniline film for ascorbic acid detection *J. Mol. Recognit.* **24** 700–6

[53] Kia S, Fazilati M, Salavati H and Bohlooli S 2016 Preparation of a novel molecularly imprinted polymer by the sol–gel process for solid phase extraction of vitamin D3 *RSC Adv.* **6** 31906–14

[54] Cummins W, Duggan P and McLoughlin P 2005 A comparative study of the potential of acrylic and sol–gel polymers for molecular imprinting *Anal. Chim. Acta* **542** 52–60

[55] Moein M M, Abdel-Rehim A and Abdel-Rehim M 2019 Recent applications of molecularly imprinted sol-gel methodology in sample preparation *Molecules* **24** 2889

[56] Lu Y, Zhu Y, Zhang Y and Wang K 2019 Synthesizing vitamin E molecularly imprinted polymers via precipitation polymerization *J. Chem. Eng. Data* **64** 1045–50

[57] Zhang Y, Zhu Y, Loo L S, Yin J and Wang K 2021 Synthesizing molecularly imprinted polymer beads for the purification of vitamin E *Particuology* **57** 10–8

[58] Emekli-Alturfan E and Alturfan A A 2023 The emerging relationship between vitamin K and neurodegenerative diseases: a review of current evidence *Mol. Biol. Rep.* **50** 815–28

[59] Eersels K *et al* 2018 A novel biomimetic tool for assessing vitamin K status based on molecularly imprinted polymers *Nutrients* **10** 751

IOP Publishing

Molecularly Imprinted Polymers for Environmental Monitoring
Fundamentals and applications
Raju Khan and Ayushi Singhal

Chapter 9

Molecularly imprinted polymers for the detection of toxic anions

Sadhna Chaturvedi, Yasmin Bano, Shiv Kumar Jayant, Piyush Shukla and Abhinav Shrivastava

Molecular imprinting is a technique to imprint complementary cavities for the template molecules of molecularly imprinted polymers (MIPs) for the detection of target molecules. If they possess a specific molecular recognition capability and a high binding affinity for ionic template molecules, the synthetic products are referred to as ion-imprinted polymers (IIPs). IIPs have all of the advantages of MIPs, including the ability to specifically identify the template ion. IIPs as selective sorbents for template ions have received significant attention in recent years. They have been used in the selective recognition, separation, and enrichment of anions (inorganic and organic) and the removal of them from aqueous and biological media, as described in this chapter.

9.1 Introduction

Molecularly imprinted polymers (MIPs) are high affinity and specific functional molecules with the speciality of complimentary in size, shape, structure, and functionality to an analyte molecule of interest [1, 2]. The credit for developing MIP technology is given to Polyakov [3], but this technology only came to to be well known in the late 1990s when the non-covalent imprinting technique was coming in existence. The non-covalent approach, introduced by Mosbach and coworkers [4], which facilitates advantages such as better options for using multiple flexible monomers and simple detachment of templates from their polymer matrix. However, the utilization of both covalent and non-covalent MIPs has faced the considerable shared problem of the leaching of templates [5]. The problems of developing MIPs and optimizing timings can be reduced via using computer based high through-put screening methods [6]. MIPs are also known as plastic antibodies, as they are the counterparts of natural antibodies, compared to which they have a variety of advantages, including lower cost, facile production that is free from animal sources, thermal and chemical stability, reusability, easy storage, and long shelf-life [7, 8]. Moreover, extensive literature supports the use of MIP-based sensing for the biomimetic recognition of a vast range of targeted molecules, from small ions to larger macromolecules, to whole cell recognition, such as bacteria. Thus, this enormous versatility can be used for numerous applications, including catalysis, purification, solid phase extraction (SPE), sensor development, therapeutic approaches, drug delivery systems, etc [9–13]. One of the most important commercialized applications of MIPs is the selective determination and/or removal of low-level contaminants from compound matrices [14], for example serum, plasma, blood, urine, foodstuffs, aqueous solutions, and environmental samples. MIPs can be utilized to determine the adsorption of these components. This innovation depends on the principle of 'lock and key', as utilized by proteins for substrate recognition. A class of polymerization methods can be utilized to produce MIPs that uses self-assembling monomers around the template particles due to 'functional group template' and 'functional group monomer' cooperation.

The literature also contains records of essential and necessary exploration of MIP innovations, which, with their enormous adsorption limits and reusability, can be utilized creatively for toxicity remediation from current levels to foster a stable and healthy environment. The improvements in synthetic receptors for detecting anionic species have led to a significant field in supramolecular science. Although cationic and neutral designs have been studied for a long time, the utilization of anions in designs is as yet a relatively new area of study [15, 16]. In contrast to cationic and neutral species, anionic particles are larger [17] and thus have a lower charge to radius ratio. Anionic species have various shapes including spherical (F^-, Cl^-, Br^-, I^-), straight (Gracious, N_3^-, CN^-, SCN^-), three-sided (NO_3^-, CO_3^{2-}), tetrahedral (PO_4^{3-}, SO_4^{2-}, VO_4^{3-}), square planar ($[PdCl_4]^{2-}$, $[Pt(CN)_4]^{2-}$), octahedral (PF_6^-, $[Fe(CN)_6]^{4-}$, $[Co(CN)_6]^{3-}$), and complex shapes (DNA, nucleotides), which require a more significant level of planning for engineered receptors. The effect of solvent is a significant element that should be considered for anion recognition. Stable H_2-binding interactions are normally formed between anions and hydroxylic diluents such as water. In addition, anions are usually sensitive to pH values. At low pH, anions will lose their negative charges and become protonated. Consequently, the receptors should to be intended to work in their ideal pH range. The hydrophobicity of anions ought to be likewise be considered as a significant component in designing receptors [18].

According to the Brønsted–Lowry acid theory, molecules or ions have the capability of proton disassociation and can consequently form an anion that falls into the category of acids, a few examples are halide ions, H_2CO_3, HCO^{3-}, HNO_3, H_3PO_4, HPO_4^{2-}, $H_2PO_4^-$, H_2SO_4, HSO_4^-, $HClO_4$, ClO_4^-, $HMnO_4$, MnO_4^-, HCOOH, $HCOO^-$, CH_3COOH, CH_3COO etc. Anions, acids, and molecules containing these groups exist naturally in the environment [19, 20]. Anions are the only molecules which display specific characteristics on dissociation, this phenomenon is the key factor in developing analytical devices for anion moiety sensing [21, 22]. Although the direct sensing of anions has proven problematic, the importance of their analysis cannot be denied. Researchers are continuously trying to find (bio-)chemical and biological tools to resolve the problems related to anion analysis. MIPs are promising potential detection platforms and provide many properties that make MIPs excellent tools for anion analysis [23, 24].

The purpose of this review is to provide notable comparative and quantitative perspectives on the principles, methods, and applications of various MIPs for the identification, adsorption and elimination of anions and anionic group containing molecules and biomolecules (acid targets, antibiotics, pharmaceutical agents, etc). Before going to discuss MIPs applications in the field of anionic detections, we first give a short outline on the different components, technologies and strategies are used in the MIPs development and their integration into sensing mechanisms.

9.2 Molecularly imprinted polymer constituents, technologies, and strategies

MIPs are exceptional cross-linked polymers, which are produced in contact with a template particle [25, 26]. The synthesis steps for preparing MIPs are described in figure 7.1 in chapter 7 [27]. In short, a pre-polymerization monomer complex is produced by combining the monomers and the template particles. Then, the template–monomer complex is polymerized in the presence of a larger number of cross-linkers either under UV light, heat, or different conditions. Finally, the template particles are removed from the completely solid polymer lattice and leave recognition locations that correspond in shape and material properties to the template molecules, and thus an artificially generated three-dimensional polymer network is created.

9.2.1 The different components of MIPs

In classic MIP designs, different components are required, including functional group monomers, template molecules, initiators, cross-linkers, and porogens or solvent. The constituents of MIPs are illustrated in figure 9.1.

Functional monomers. These are small molecules that have functional groups (usually containing one recognition site and one polymerizable site) that bind to the targeted template molecules and substrate to deliver an appropriate pre-polymerized complex. In the process of MIP amalgamation, it is important to choose suitable functional monomers that have a strong interface with the complementary template to form a selective and sensitive donor–acceptor multiplex. These monomers often have functional groups such as double bonds and epoxy bonds (these are easier to cross-link and graft). Functional monomers are divided into four basic categories on the basis of their features and functional groups—those are interact with anions are amino, quaternary ammonium, nitrogen heterocyclic, and carboxyl monomers [28].

The most common amino-compounds are the amines, known as hydrocarbyl derivatives of ammonia. In nature, amines are basic, like ammonia, as they have an unshared electron pair on the atom so have the tendency to combine with protons to gain a plus charge. In this way they interact with anions (minus charged) via

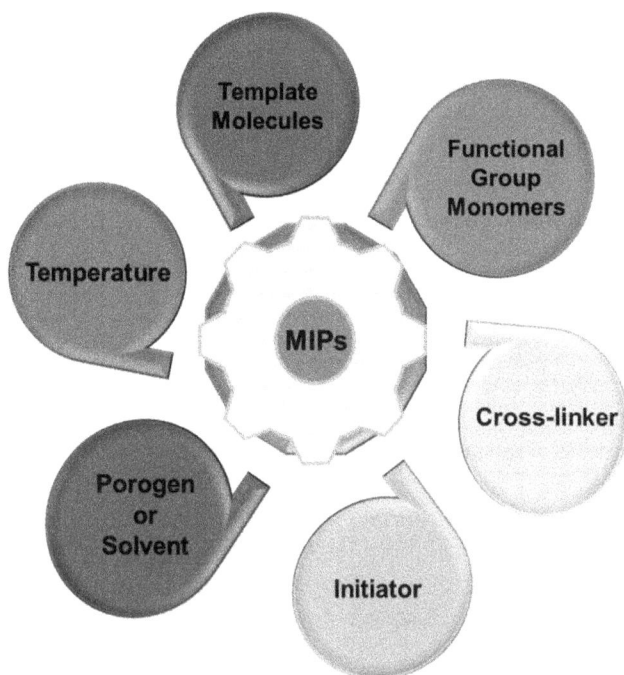

Figure 9.1. Constituents of the molecular imprinting process.

electrostatic interactions. The functional monomer ethylenediamine was bonded to carboxymethyl cellulose through electrostatic pairing, resulting in the fabrication of dichromate $(Cr_2O_7^{2-})$ ion-imprinted polymer (IIP) [29]. Aminosilane, 3-(2-amino-ethylamino) propyltrimethoxysilane, (3-aminopropyl)triethoxysilane, and 3-[2-(2-aminoethylamino)ethylamino]propyltrimethoxysilane are the names of a few mono-mers that are currently used as amino functional monomers that can be cross-linked on hydroxyl-containing materials [30–33]. Quaternary ammonium functional mono-mers are ammonium ions. The quaternary ammonium cation salts are formed after all hydrogens are replaced by hydrocarbon groups, and the properties are similar to ammonium salts. Common nitrogen-containing heterocyclic monomers include pyridines and imidazoles, such as 2-vinylpyridine and 4-vinylpyridine. Due to their heterocyclic structure, the atoms are easily protonated and positively charged to form bonds with anions, hence they can be used as functional monomers in anionic imprinting. In addition, the main carboxyl functional monomer is methacrylic acid (MAA), the carboxyl groups on which can form H_2-bonds with anions [34]. 3-isocyanatopropyltriethoxysilane is an example of a functional monomer used in semi-covalent imprinting. 4-vinyl aniline and 4-vinayl benzene boric acid are commonly used in covalent imprinting, while itaconic acid and trifluoromethyl acrylic acid (both acidic), allylamine and 1-vinylimidazole (both basic), styrene, acylamide, 2-hydroxyethyl methacrylate, and methylmethacrylate are some exam-ples of functional monomers commonly used in non-covalent imprinting.

Template molecules or target molecules. The functioning and characteristics of template molecules allow the communication of the template to the dangling functional groups of the designated functional monomers during MIP synthesis. Easy availability, abundance, low cost, impressive solubility under imprinting condition, and excellent cross-linking capability with the targeted templates are a few properties of the template molecules that should be considered when producing any MIPs. Tyrosine, helical peptides, D-glucose, tetracycline, sulfonamides, 2,4,6-trinitrotoluene (TNT), oestrone, bisphenol, atrazine, benzimidazole fungicides, cinchona alkaloid, bovine serum albumin (BSA), 3,5-cyclic monophosphate (cAMP), adenosine, tobacco mosaic virus, and ions such as Pb(II), Hg(II), $CH_3Hg(I)$, Fe(III), Ni(II), UO_2^{2+}, As(III), and PO_4^{3-} are a few examples of the many template molecules used [18].

Initiators. Initiators are used widely in the process of MIP fabrication as a source of radicals, chiefly in free radical polymerization (FRP), photopolymerization, and electropolymerization. The degree and manner of decomposition of any initiator to radicals are based on the temperature, light, and the nature of chemical. In addition, FRP can be initiated or expediently decomposed photochemically (UV) or thermally, for example, 2,2-azobisisobutyronitrile (AIBN) initiator to polymerize vinyl monomers to poly-methylmethacrylate by producing a stabilized carbon centered radical under certain thermal or photochemical conditions. A few more examples of initiators are potassium persulfate, dimethylacetal of benzyl, benzoylperoxide, 2,2-azobis(2-methylbutyronitrile), and 2,2-azobis(2,4-dimethylbutyronitrile), etc [35].

Cross-linking monomers. Cross-linkers are the fixer molecules used to fix the functional monomers surrounding the template molecules in such a highly rigid manner that the elimination of the template molecule does not affect the arrangement of functional monomers in the process of imprinting. The selection of cross-linker is based on, first, its capability to control the geomorphology of the polymer network, whether it should be macro-porous or a powdered microgel or gel-type, second, its stabilizing property to maintain the imprinted binding entities, and, third, its strength and sustainable quality which provide mechanical strength and durability to the polymer network. Some examples of cross-linkers are divinyl benzene, pentaerythritol triacrylate, pentaerythritol tetraacrylate, N,N-ethylenebismethacrylamide, N,N-tetramethylene bismethacrylamide, etc [18].

Porogens or solvent. In the process of MIP synthesis, porogens not only provide a dispersion medium but also act as pore creating mediators to establish the desired porous structure of MIPs. The following important questions should be asked when choosing a porogen—whether all the components, i.e. functional monomers, template molecules, and cross-linkers, are soluble or not, whether the desired macro-pores are formed or not, and whether the solvent exhibits low polarity or has minimum interference at the time of forming the pre-polymerized complex. A few examples of porogens are acetonitrile, methanol, chloroform, and toluene, which is basically used because they have no or less polarity to achieve the highest imprinting efficiency in the process of non-covalent imprinting [35]. Booker *et al* found room-temperature ionic liquids (RTILs) with excellent properties, and the peripheral vapor pressure of the RTILs reduced the issue of bed shrinking of MIPs and in the polymerization mechanism worked as pore templates. The RTILs encompassing [BMIM][BF$_4$], [BMIM][PF$_6$], [OMIM][PF$_6$], and [HMIM][PF$_6$] showed satisfactory performance for propranolol MIPs [36].

Temperature. Plenty of literature favors the synthesis of MIPs at low temperatures ($\leqslant$60 °C) to produce highly sensitive polymers, rather than polymerizing at higher temperatures. At high temperatures the rapid initiation of polymerization results in low reproducibility of MIPs [37–40]. In 2019 Kempe and Mosbach [39] observed the effect of temperature (0 °C and 60 °C) on the enantioselectivity of L-PheNHPh imprinted polymers in which one was photochemically and other was thermally polymerized at the respective temperatures. The high temperature affected the stability and ultimately reduced the reproducibility of the bulk stationary phase. Consequently, an extended reaction time at low temperatures is generally preferred to generate more reproducible polymers [35]. In 2019 Si *et al* developed thermo-responsive molecularly imprinted hydrogels (T-MIHs) and demonstrate that the T-MIHs are thermosensitive and at a lower temperature (~35 °C) the template molecules were organized with an imprinted cavity and the highest adsorption capacity, while increments in temperature induced shrinking and aggregation at a certain point [40, 41].

9.2.2 Molecular imprinting technologies (MITs)

The concept of 'lock and key', originally utilized to describe enzymes for substrate recognition, serves as the foundation for the development of MIPs and the techniques used in the production process are known as molecular imprinting technologies (MITs). Due to interactions between the 'functional group template' and 'functional group monomer', a variety of polymerization techniques can be used to create an MIP that is initiated by the self-assembly of functional monomers around the molecules of the templates. Based on the final application, different types of MITs are used for MIP synthesis. Free radical polymerization is the most common method of MIP construction. Different techniques are used for MIPs are summarized in figure 9.2.

Bulk polymerization. This is the most widely used traditional method to synthesize MIPs under the category of free radical polymerization. Mechanical force is applied to grind and sieve the rigid bulk material to obtain suitably sized particles with respect to the anticipated application. This is an easy and scalable approach, but major limitations included low yield, slow mass transfer, difficulty eliminating the template from MIP particles, damage to the binding sites reducing recognition ability, and the heterogenetic size of the particles obtained via grinding [42, 43]. An adhesive layer helps to integrate these particles into sensors. Research reports suggest that the functionalization step can be sped up by direct combination of the particles with screen-printing ink [44–46], although this step leads to restricted surface coverage. More sophisticated and complex polymerization techniques have been proposed to overcome the limitations of bulk polymerization, such as precipitation, emulsion, suspension polymerization, and seed polymerization. Using the bulk polymerization method, three monomers—2-

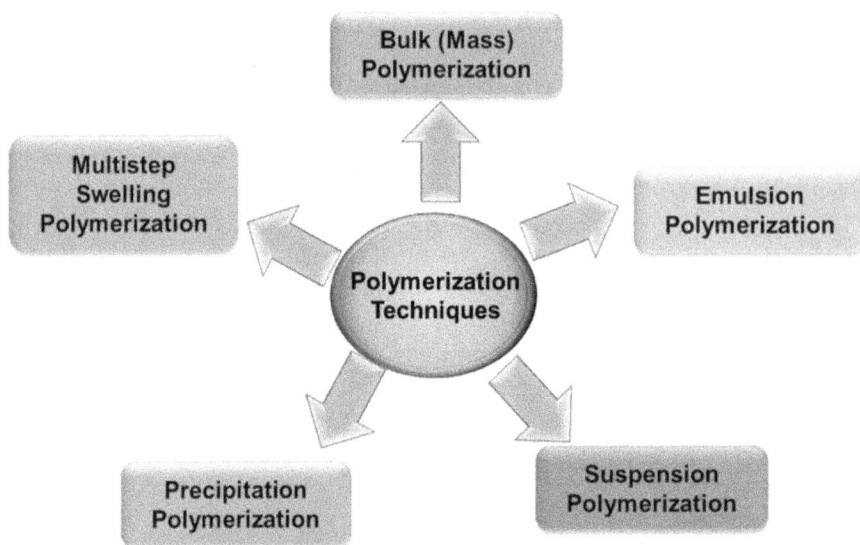

Figure 9.2. Types of polymerization techniques.

methacryloyloxyethyl-trimethyl-ammonium-chloride (DMC), 1-allyl-2-thiourea (AT), and thiourea—were used separately to prepare phosphate ion-imprinted polymers (IIPs). The adsorption capacity for phosphate increased as the size and uniformity of the pores increased and when molar ratio of the monomer species increases to DMC [28].

Precipitation polymerization. This is the most accepted polymerization technique to minimize the limitations of traditional polymerization, as it is a robust technique of MIP fabrication that reduces the size of the particles and forms spherical homogeneous micro- to nanobeads [47]. The method produces polymer microspheres directly with sufficient control of the product shape. Because it does not require stabilizers or other additives, it is currently the most convenient, effective, and thus one of the best methods [48]. The only difference to bulk polymerization is the need for a large amount of template and porogen, which reduces the eco-friendliness of the reaction method [49]. Its advantages of single-step preparation, high-quality, and no need for grinding or sieving processes makes it more attractive [50, 51]. However, it is a relatively expensive technique with a high dilution factor.

Suspension polymerization. This is the easiest one-step polymerization technique to ensure the synthesis of homogeneous spherical polymer beads. It is a method of making imprinted polymers by establishing normal-phase suspension polymerization in the organic phase with water as the dispersion medium. Normal-phase suspension imprinted polymers are quite problematic because random copolymerization of cross-linkers and water-soluble functional monomers is difficult and the transfer of template ions in the continuous aqueous phase leads to a reduced imprinting effect. The opposite of this method, reverse-phase suspension polymerization, suspends a comparatively polar monomer in the form of small droplets in an oil-soluble medium for polymerization [28]. The immense heat transfer capability is a unique advantage making it suitable for industrial platforms [6, 37, 52]. However, binding sites may be reduced as the aqueous medium used works as dispersion mediator. Moreover, the kinetics between the template and functional monomers can be impeded by an aqueous medium [6]. The large sizes of the particles and poor recognition are the limitations of this technique [37].

Emulsion polymerization. This process involves dissolving functional monomers, template ions, and cross-linking agents in organic solvents, then adding water with emulsifiers, stirring and emulsifying, then cross-linking and polymerizing to obtain spherical polymers with relatively uniform particle sizes [53]. Emulsion polymerization is considered a good technique to produce high yield, monodispersed, spherical, and water-soluble polymeric particles 0.1–3 μm in diameter, where the performance of yielded product-by-process material is measured by delicate balance between co-polymer composition, chemical proportions dispersion, molar-mass distribution, polymer architecture, shape and size of the particle distribution and surface composition [54, 55]. Although it is free from the grinding and sieving processes, it is prone to suffering from low imprinting capacity, poor run-to-run reproducibility, and the presence of remnants of surfactants [6, 54, 55].

Different forms of emulsion polymerization techniques have been introduced, such as emulsion core–shell polymerization (has the capability to fine-tune the particle dimensions, ranging from small nanobeads to larger nanoparticles), seeded emulsion polymerization (often used to formulate hybrid colloids that can bind organic as well as inorganic components), semicontinuous emulsion polymerization, and thermocontrolled emulsion polymerization (polymerization can be triggered by controlling the temperature).

Multistep swelling polymerization. In 1994 Hosoya and coworkers introduced this technique of MIP fabrication which was further improved by Haginaka and Sagai in 2009. This technique is used to fabricate homogeneous monodisperse particles of controlled size between 2 and 50 μm diameter [56, 57]. As the size of the monodispersed particles is slightly higher than for those synthesized using precipitation polymerization, it is ideal for high-performance liquid chromatography (HPLC). However, it is limited by being a considerably time-consuming, complex synthesis process and is a less effective product forming technique [49, 50].

In addition to the techniques described above, some other MITs are also described in the literature and used in the synthesis of MIPs. A few are listed here: enzyme-catalyzed polymerization (enzymes are used as biocatalysts in the process of constructing polymers) [58], solution polymerization [59], oxidative polymerization [60], photoactivated polymerization [61], controlled/'living' radical polymerization [62], surface blotting, sol–gel method, etc.

9.2.3 Strategies and approaches

Basically, two strategies are used to produce MIPs—one the covalent approach and the other is the non-covalent approach, based on the interaction found between the template molecules and the monomers' functional groups at the pre-polymerization step. However, further derivative strategies such as as semi-covalent imprinting, metal-mediated imprinting, and host–guest inclusion-based interactions are also being developed and pursued by researchers [37].

Covalent imprinting approach. Reversible covalent binding is used in this method to covalently attach the imprinted molecule to the polymerizable molecule. By synthesizing sugar or amino acid derivatives that contained vinylphenylboronate as a polymerizable function, Wulff and his research team created the first MIPs using this approach, which is also known as the pre-organized approach [63]. The ratios of functional polymers, cross-linkers, and templates differ completely between the two approaches in terms of the polymerization requirements. Reversible covalent interactions have few potential templates, so breaking the bonds between the functional monomers and the templates often requires acid hydrolysis [64].

Non-covalent imprinting approach. Non-covalent imprinting is the most common method for designing MIPs due to its accessibility and ease of use. The specific binding sites are formed by self-assembly between the functional monomers and the chosen templates in the non-covalent approach, which follows a cross-linked copolymerization [65]. During imprinting and rebinding, the imprinting polymer material interacts with the polymerizable solution through non-covalent interactions, such as H_2-bonding and ionic bonding. Because of its large number of compounds, this method has a strong capacity for molecular imprinting [13]. Mosbach and colleagues introduced non-covalent imprinting when they imprinted L-phenylalanineanilide using MAA as the functional monomer.

Semi-covalent imprinting approach. This approach was introduced by Whitcombe and colleague in 1995 and is semi-covalent because both type of bonds, covalent as well as non-covalent bonds, are formed during the synthesis process. In the process of polymerization covalent bonds are formed between the template molecules and functional monomers, while elimination of the template molecules is performed by non-covalent interactions. They used linkers/spacers between the template molecules and functional monomers that were removed at the time of bond rebinding of template molecules. The carbonyl group was used by them for the esterification of cholesterol with 4-vinylphenol, that yielded covalently bound cholesteryl (4-vinyl) phenyl carbonate ester, a template–functional monomer complex. The resulting complex after polymerization with the cross-linker was liberated of cholesterol (the template molecule) and carbonic acid on hydrolysis. Recognition occurred because of H_2-bond formation among the –OH group of cholesterol and formation of phenolic-OH group at the binding sites [66]. For peptide imprinting salicylate (2-hydroxybenzoate) is used as a spacer [67] and for the aromatic heterocyclic based imprinting the dimethyl sillyl groups of sillyl ethers and sillyl esters are reported in the literature [68].

9.3 Molecular imprinting for anion recognition

In the structure of MIPs, the geometry of the recognition cavities in the polymer are dictated by the size and shape of the anionic target atom. The monomers that are appended to the polymer network give the main impetus to pull the anionic analytes into the imprinted cavities. In the following sections, we will discuss anions grouped by similar anionic species. As mentioned before, anions are sensitive to pH values, so the discussing will focus on the environment that the target atom exists in for the anionic configuration [18].

9.3.1 Sensing of inorganic anions

It has been proven that MITs are important and interesting technologies to produce synthetic receptors for inorganic cations as well as anions, however, creating specific binding cavities for the detection of anionic templates is quite difficult because of their tiny size, limited function, or adaptation of shape compared to the oxyanion species. Because of these difficulties only few studies have been done on the molecular detection of inorganic anions using MIPs [69].

Nitrate (NO_3^-)/nitrite (NO_2^-). An electrochemically mediated templating method using a polypyrrole (PPy) polymer sensing method for the detection of nitrate (NO_3^-) was developed by Hutchins and Bachas in 1995. The developed PPy-based nitrite electrode demonstrated the great affinity towards NO_3^- and SCN^- over ions such as Br^-, Cl^-, $H_2PO_4^-$, salicylate, sulfate, and phosphate. The PPy-based nitrite sensors showed a better selectivity coefficient against ClO_4^- and I^- by up to four orders of magnitude when compared to the nitrate-selective sensors available on the market which suffer from interference by lipophilic ions such as ClO_4^- and I^-. The researchers explained that the reason behind this recognition could be dependent on size-exclusion in the PPy-based sensors they offered great selectivity for NO_3^-. Thus, these electrode sensors showed the capability of discriminating anions that are larger than NO_3^-. Anion selective electrodes that have to recruit ionophores that have no anion identification functionalities, apart from positive charges, report just on the basis of the lipophilicity of the anions, where more lipophilic anions responded best. They found that the more lipophilic anions can enter the hydrophobic PPy films very easily because Hofmeister-type selection is superimposed over the NO_3^- imprinting selection. [70]. Further, Lenihan *et al* integrated nitrate PPy polymer film with nano-liter electrochemical vessels to improve the recognition from nano-liter sized samples, where the PPy nano-liter vessels displayed admirable reproducibility along with a low limit of detection (LOD) of about 0.36 ng [71].

The need for an accurate, inexpensive, and easy-to-use technique for the of sensing of nitrate in water samples was fulfilled in 2018 by Alahi *et al*, who developed a continuous real-time monitoring program to monitor the nitrate-nitrogen (N) amount in aqueous models. The IIP based technique allows the development of the low-cost determination of selective elements. Current research is confined to the use of IIPs at the interdigital sensor stage for nitrate-N recognition, particularly in a water medium. The sensing of nitrate-N using electrochemical impedance spectroscopy (EIS) along with IIP coated sensors achieved precise detection of $1-10$ mg l^{-1}. The obtained results of unknown samples were used to validate the sensing ability of the sensing method, while previously obtained results from the sensing method were compared to commercially available sensors and validation was performed using the gold standard UV-spectrophotometer method [72].

Recently, in 2020, Diouf *et al* produced a very easy, rapid, portable, inexpensive, and non-invasive electrochemical sensing method based on IIPs and gold (Au)-nanomaterials, developed for determining the nitrite profile in exhaled breath condensate (EBC). Analysis of human EBC is an interesting research topics for a number of researchers because it can reveal many kinds of inflammatory disease markers. With regard to fabricating the sensor, Na-nitrite was immobilized on the surface of the self-assembled 2-aminothiophenol on a screen printed electrode (SPE) made up of gold (Au-SPE). A polymer of polyvinyl alcohol with established cross-linking with glutaraldehyde was coupled with Au nanoparticles to cover the modified Au-SPE, and finally complied in the fabrication process of the IIP sensor (figure 9.3). The testing of the synthesized sensor was performed by using different standard methods for different parameters, such as electrochemical impedance spectroscopy (EIS), atomic force microscopy (AFM), scanning electron microscopy coupled with energy dispersion x-ray spectroscopy (SEM-EDS), etc. After satisfying all the conditional parameters, the NO_2^- sensing IIP sensor detected 0.5–50 $\mu g\ ml^{-1}$ with a 4 $\mu mol\ l^{-1}$ LOD value (signal-to-noise ratio $S/N = 3$). The results revealed the suitability of this technique in the recognition of NO_2^- ions in EBC and, most importantly, opened a clinical path in the field of respiratory therapeutics [73].

Phosphate (PO_4^{3-}). In terms of sensing of phosphate, no standard or robust sensing alternative has emerged at a commercial level despite several efforts [74]. The reason behind the lack of sensing based alternatives to the colorimetric standards is due to the PO_4^{3-} ions possessing the inherent property of having quite a large oxyanion, which produces difficulty in obtaining an appropriate receptor for phosphate in comparison to anions of the same size and shape such as sulfate [75].

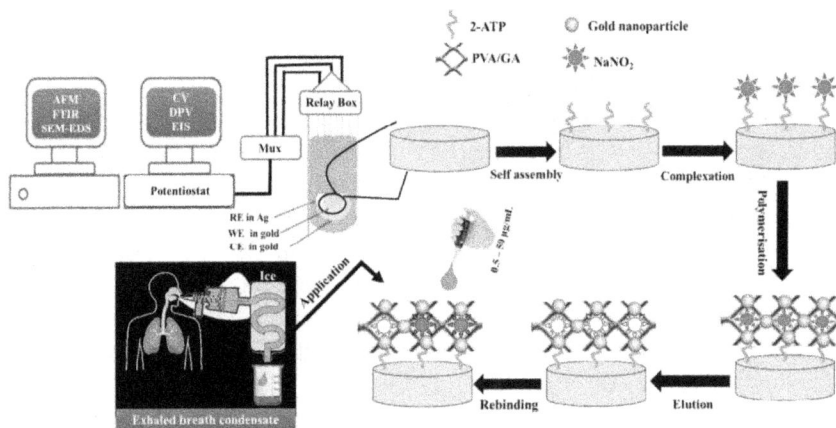

Figure 9.3. The synthesis procedure of the IIP based sensor and determining the nitrite profile in exhaled breath condensate in humans. 2-ATP: 2-aminothiophenol; AFM: atomic force microscopy; CV: cyclic voltammetry; DPV: differential pulse voltammetry; EIS: electrochemical impedance spectroscopy; GA: glutaraldehyde; FTIR: Fourier transform infrared; PVA: polyvinyl alcohol; SEM-EDS: scanning electron microscopy coupled with energy dispersion x-ray spectroscopy; NaNO2: sodium nitrite. (Reproduced with permission from [73]. Copyright 2019 Elsevier.)

Moreover, the charge of oxygen atoms obscures the phosphorus atoms at the center, resulting in a high hydration energy [76]. MIPs may offer a way to minimize these obstacles and represent potential advantages over colorimetric based methods. Warwick *et al* designed an MIP-based sensor for the direct level-free diagnosis of phosphate in water bodies. It was found that the sensor can confirm the presence of phosphate in wastewater with excellent reproducibility and with a good detection range of 0.66–8 mg l^{-1} and an LOD of L0.16 mg l^{-1}. Moreover, the sensor sensitivity was also tested on unfiltered field models, such as the influent of domestic wastewater treatment plants, and the concentrations of phosphate correlated closely with the reference values. Thus, the MIPs provide a new way to detect phosphate either directly in the field or for continuous monitoring in a remote sensing manner [77].

The literature suggests that most of the MIPs involved in the detection of phosphate are based on the imprinting of organo-phosphorylated molecules rather than imprinting of inorganic PO_4^{3-} ions or molecules [78]. Thiourea can also be use as the functional monomer during the preparation of MIPs for the detection of phosphate. Two functional monomers, called AT and N-methyl-N'-(4-vinylphenyl)-thiourea (PT), were used and for the template phenylphosphonic acid (PP) and diphenyl phosphate (DP) were added. This experiment highlighted that the AT made the MIPs display better binding capacity in water samples, while on other hand PT and MAA made the MIPs have zero binding ability to PO_4^{3-} ions under the studied conditions. Despite this, the PP and DP template imprinted polymer reflected a similar binding efficacy to the targeted phosphate. This shows that structural dissimilarity did not affect the ability of the PO_4^{3-} analyte to bind to the imprinted entities. The selectivity of the PP imprinted polymer showed excellent affinity towards PO_4^{3-} ions and great selectivity over the NO^{3-}, SO_4^{2-}, $H_2PO_4^{-}$, Cl^-, and CH_3COO^- ions. The authors of the study fabricated another functional monomer of thiourea N-allyl-N'-methyl-thiourea (AMT) [79, 80]. Luo *et al* [81] constructed an anionic IIP through the incorporation of electrostatic attraction to improve the permanence of the anion and functional monomer in the pre-polymerization phase. The capturing power for phosphate was found to be higher at 35 °C, about 78.88 mg g^{-1}. The experiment concluded that adsorption was carried out through an ion-exchange process (the template ions connected with IIP) where the recognition sites of the IIPs were the N^+ and P–O^- of AMD and $H_2PO_4^-$, respectively, which then moved to the cavity and formed a H_2-bond between N^+–O^- and N–H, which also participated in phosphate adsorption.

In principle, variations in geometric orientations, surface complexing capabilities, and acid–base characteristics have to be kept in mind during the fabrication of selective adsorbents to isolate and adsorb phosphate from the medium even in the presence of other anions. In addition, the geometric configuration-based adsorbents known as MIPs display precise identification motifs. Eventually, MIPs will be tailored not only to specific adsorption but also to the recovery of PO_4^{3-} ions. The metal-dependent adsorbent or metal–adsorbent hybrid technique was incorporated to facilitate surface complexing to obtain selectivity for phosphate via using compressing oxides or hydroxides of aluminum, iron, lanthanum, and zirconium producing porous adsorbents of excellent rigidity [82]. The metal–adsorbent hybrid showed excellent selectivity, affinity, and capability to bind phosphate due to the assembled complexes of phosphate and metal oxides and hydroxides and also because of the synergistic effect of the porous materials. However, uncertain long-term activity, durability, reproducibility, and susceptibility to acidic solutions are the main hurdles to making it powerful tool for anionic detection. Very recently, to help save the environment from eutrophication-associated problems, Chen et al developed an imine-functionalized adsorbent that can efficiently remove and recover PO_4^{3-} ions from effluent in the form of discriminate protonated phosphate, i.e. $H_2PO_4^-$ and HPO_4^{2-}, from a contextual framework of deprotonated anions such as Cl^-, NO_3^-, and SO_4^{2-}, etc, via establishing H_2-bonding [83].

$Halide$ $ions.$ Decades ago, Dong and coworkers produced a method to develop a chloride (Cl^-) ion selective imprinted PPy polymer film electrode via electrochemical polymerization in an aquaeous medium [84]. PPy is one of the most prevalent materials used in sensing mechanism over other polymer molecules [85]. Generally, imprinted PPy film is synthesized through the method of electrochemical polymerization and in the presence of a template molecule it is deposited on the sensing surface. PPy polymer film exhibits a quick response and low interference with a low detection limit (LOD = 3.5×10^{-5} M) as well, which highlights its importance over the other chloride-selective electrodes. The selection of PPy was dependent on two parameters; one was the radius of the ion and the second was the amount of negative charge carried by the anion. Later, a microsensor was also fabricated for the recognition of Cl^- in serum samples by using the same methodology for the development of s microsensor [86].

Kamata et al formulated an idea for the detection of halide ions via utilizing cross-linked polyviologen (CPV) film, which was manufactured using tri-functional monomers combined with three molecules of 4-cyanopyridinium through the reaction of electropolymerization in a solution containing potassium salt along with counter anions, then the prepared CPV film was placed on an indium-coated SnO_2 electrode. The resulting film worked on the basis of size-exclusion theory to select the halide ions. For example if the CPV film was imprinted with Cl^- ions then it became electrochemically inactivated for other halides in the order of KCl, KBr, KI solution and electrochemical activity was found for all the remaining halide solutions if the CPV film was imprinted with KI solution. Moreover, the CPV film presents an equal redox potential in all three solutions, which shows that small-sized halide ions can be caught by larger cavities than those created and exhibit same redox potential as large halide ions [87, 88].

$Thiocyanate$ (SCN^-) $anions.$ Say and coworkers designed a novel imprinting technique for the exclusive detection and absorption determination of SCN^- ions under different laboratory conditions like pH, discrimination power, kinetic and isotherm. The polymers imprinted with SCN^- ions were synchronized using three different polymers: MIP-1, i.e. chitosan-Zn(II)-SCN; MIP-2, i.e. chitosan-Zn(II)-AAPTS-SCN (AAPTS is N-(2-aminoethyl)-3-aminopropyltrimethoxysilan); and MIP-3, i.e. AAPTS-Zn(II)-SCN. MIP-3 displayed the best capability and selectivity for ion loading at a pH of 3 (MIP-3 > MIP-2 > MIP-1). MIPs-3 can be used as selective separation MIPs for SCN⁻ ions from dilute water solutions. Also, the SCN⁻ ion-imprinted particles also displayed higher adsorption capacity towards template molecules over other competitors such as F⁻ and PO_4^{3-} ions [89]. To remove the SCN⁻ from water, Karekar and Divekar used two micro-porous anionic exchangers, Tulsion A-10X and Amberlyst A-21, which have two different resin matrices, an acrylic matrix and a styrene divinyl benzene matrix, respectively. It was seen that the adsorption mechanisms followed the pseudo-second-order kinetic model as well as the Langmuir and Freundlich isothermic properties and showed an adsorption capability of 153.19 mg g^{-1} and 142.87 mg g^{-1} and the removal efficacy was 75% and 90% for the respective anionic exchangers. Although the adsorption mechanism was exothermic, it did not affect the interior chemistry of the resins' structure [90].

Other anions. For the detection of TNT, a surface-enhanced Raman scattering (SERS) based approach was applied by incorporating paminothiophenol activated Ag-nanoparticles with Ag-molybdate nanowires [91]. In another experiment using IIP techniques, IIP-PEI/SiO₂ of chromate (CrO_4^{2-}) ions showed high activity concerning PO_3^{4-} ions [92]. Moreover, a few related studies have shown the recognition of arsenate with the help of using ion imprinting techniques [93].

9.3.2 Sensing of organic anions

The effectiveness of MIT for organic anion recognition, adsorption, and elimination is much higher than for inorganic anions and has taken the attention of researchers towards the use of MIPs in this field. Anionic organic templates demonstrate a very efficient impact on the preparation of synthetic receptors concerning organic anions. In the construction of MIPs, attaching a complex of functional and anionic groups to the imprinted polymer basically drives the recognition of target ions while rebinding ability is derived from the template's morphology and function. In addition, the majority of proteins show a negative charge at neutral pH but form a zwitterion in an acidic pH. As number of reports are available on protein recognition and here we discussed few of them which are specially focused on small-sized organic anions, including carboxylate anions, organophosphate anions, organic sulfonate and sulfate anions, and other anions. These include dyes, antibodies, pharmaceuticals, amino acids, nucleotides, etc [18].

Carboxylate anions. Deprotonation of the carboxyl group/acid gives its conjugate base that produces a carboxylate anion. Carboxylic anions play a central role in living systems as biomolecules such as amino acids, prostanoids, fatty acids, and pharmacophores possess carboxylic acid moieties. Thus carboxylate anion recognition has received a lot of attention, particularly in recent years. A number of studies based on MIPs have been performed and many more are in progress. Most the reports depend on H₂-binding interactions that have powerful recognition applications with carboxylate anions if non-polar organic solvents are present in the detecting media [18]. Willner and his research team [94] are the pioneers in the field who synthesized imprinted ultrathin TiO₂ film with titanium (Ti) butoxide and a covalently bound carboxylated template. Consequently, imprinted TiO₂ was further used for chiral carboxylic acid detection in amino acid derivatives [95, 96], azobenzene [97], and penicillin-G antibiotics [98, 99].

Methotrexate/amethopterin. Gomy and Schmitzer [100] amalgamated photo-induced polymers of decent selectivity using two self-assembled monomers in the form of the template di(ureidoethylenemethacrylate)azobenzene and analog of methotrexate (bis(TBA)-*N*-*Z*-L-glutamate). A photo-inductive response conferred the *cis–trans* isomerization of azobenzene (polymeric backbone) to control the geometry of the receptors' cavities which ultimately governed the binding and unbinding of the substrate molecules. These photo-induced MIPs were effectively applicable in smart drug delivery platforms. Recently, in 2021, Ye *et al* designed a functionalized radical initiator-based MIP with improved molecular imprinting efficacy to mitigate the limitations of conventional MIT (where thermo- or photo-decomposition generated radicals damage the recognition site by causing self-polymerization of the cross-linker monomers). Active radicals hold the template binding sites, and radical polymerization occurs adjacent to the molecular template, prominently improving the imprinting pattern. They used an amidine-functionalized initiator for the specific identification of methotrexate (a cytosolic cancer drug). They selected two glutamic acids (N-terminal protected), *N*-(carbobenzyloxy)-L-glutamic acid and *N*-[(9*H*-fluoren-9-ylmethoxy)carbonyl]-L-glutamic acid, as the template due to the resemblance to methotrexate (pseudo-template), and 2,2'-azobis (2-amidinopropane) dihydrochloride (AAPH) as the initiator molecule. Specific binding of MIPs occurred for a cytosolic drug mediated by two carboxyl group carried by methotrexate and the amidine group of imprinted polymers [101].

Penicillin-G/β-lactams. This is one of the most common antibiotics used in animal husbandry, even when not required. Widespread use with no proper caution has resulted in its presence in milk and milk-products and can cause a serious threat to humankind in the form of bacterial resistance and several side effects. Hence, the simple, fast, selective, and point-of-care detection of such antibiotics is mandatory. Plenty of MIP-based studies are listed for the detection of penicillin-G in different kinds of samples [98, 99, 102]. In this list, Khataee and his team prepared a ratio-metric fluorescent sensor comprising colored carbon dots (CDs) playing the role of dual fluorophores and meso-porous MIPs as the receptor (B/YCDs@mMIP) for the recognition of penicillin-G in milk. The presence of the targeted analyte was detected via changes observed in the color of fluorescence, which turns from yellow to blue on addition of penicillin-G, as the analyte blockage only quenches the yellow fluo-rescence emissive CDs, whereas that of the blue fluorescence emissive CDs remains stable, as shown in figure 9.4. This method offers *in situ* applications not only for penicillin recognition (detection range 1–32 nM and LOD 0.34 nM) but also for its analogs [103]. Furthermore, Denizli *et al* constructed MIPs based on a surface plasmon resonance (SPR) sensor to detect penicillin-G in aqueous and milk samples. Penicillin-G printed poly(2-hydroxyethyl methacrylate-*N*-methacroyl-(L)-cysteine methyl ester)–Au nanoparticles–*N*-methacryloyl-L-phenylalanine methyl ester (MIP-AuNPs) nano-sensors were fabricated. They determined an efficacy for the target molecule of about imprinting efficiency (I.F) = 7.83. They also found that the MIP-AuNP nanosensor was more selective for penicillin-G, which was higher,

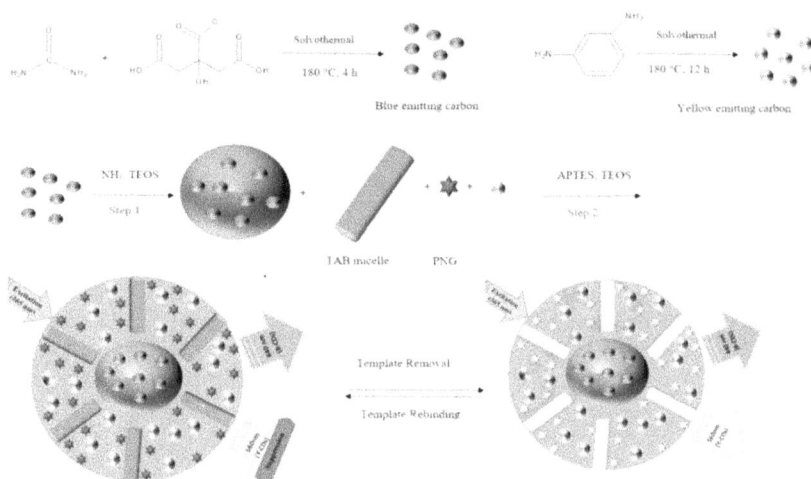

Figure 9.4. The working procedure of the B/YCDs@mMIP sensor for the detection of penicillin. (Reproduced with permission from [103]. Copyright 2020 Elsevier.)

around ~10- and ~17-fold, in contrast to amoxicillin and ampicillin, respectively [104]. In addition, a similar work was also carried out soon after [105], where they used Ag-nanoparticles in place of AuNPs. The linear range of detection was 0.01–10 ng ml^{-1} for the penicillin-G studied.

Acesulfame. Acesulfame is an artificial sweetener with few calories found in various foodstuffs and beverages, and hence consumed in considerable amounts. In the metabolic processes of the human body, a large portion of acesulfame is unaffected, thus making its occurrence more ubiquitous in the aqueous environment [106]. A study done by Scheurer *et al* demonstrated the presence of acesulfame even after performing multi-layer drinking-water treatments, finding that acesulfame is the only compound which can be detected in completely processed drinking-water [107]. Acesulfame is a highly negatively charged, well-known indicator used for the identification of the escape of household wastewater into different water systems. It has been reported in many studies that it could be used as a marker to determine the contamination of wastewater into surface-water as well as ground-water samples, sewage sludge, leachate, and hyporheic and riparian zones [108, 109].

Zarejousheghani and colleagues synthesized MIPs for acesulfame (figure 9.5). This negatively charged acesulfame is extremely soluble in water at environmental pH. The synthesized MIPs were highly useful for the discriminate SPE from inflow and outflow wastewater. They used an innovative phase transfer substance called (vinylbenzyl)trimethylammonium chloride (VBTA). The VBTA was incorporate primarily to enhance the solubility of acesulfame in organic porogen (a phase transfer catalyst) and second it works as a functional monomer molecule during the manufacture of MIPs. They synthesized four different MIPs to augment the extraction capacity of acesulfame and different materials were also assessed using of scanning electron microscopy (SEM), selectivity experiments, and equilibrium rebinding experiments [110]. The most effective MIPs were used in a molecularly

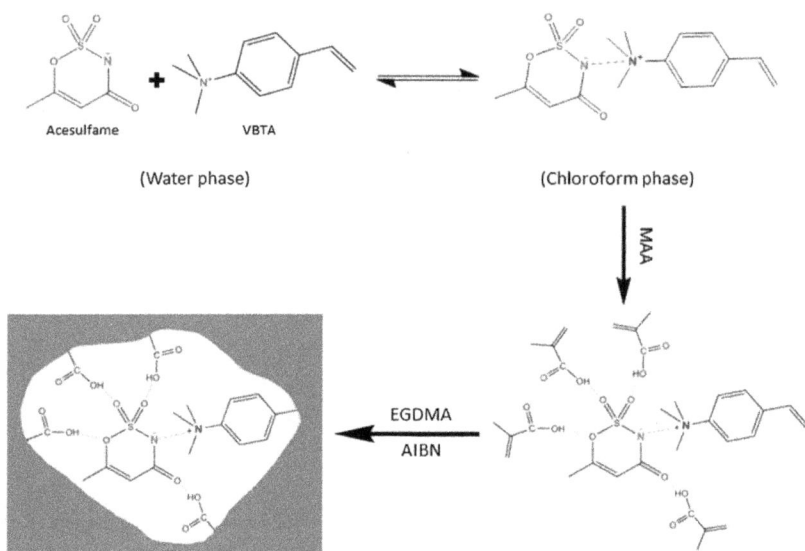

Figure 9.5. The systematic synthesis process of IMP for acesulfame. VBTA: (vinylbenzyl)trimethylammonium chloride; MAA: methacrylic acid; EGDMA: ethylene glycol dimethacrylate; AIBN: 2,2-azobisisobutyronitrile. (Reproduced with permission from [110]. Copyright 2015 Elsevier.)

imprinted-SPE protocol to extract acesulfame from wastewater models. The process of the optimization step (steps of acclimatizing, loading, cleaning, and elution) pre-determined the final SPE performance. As it is expected that imprinted polymer molecules contain NH_4^+ groups (the involved VBTA form an acesulfame–VBTA ion pair) that act as anion-exchangers. In addition, by utilizing HPLC—MS/MS mass spectrometry (MS) analysis, 0.12 g l^{-1} detection and 0.35 g l^{-1} quantification were attained [111].

Gallic acid (GA), $C_7H_6O_5$. GA is a phenolic acid naturally present in fruits, berries, nuts, oak bark, witch hazel, and tea. This bioactive compound exhibits several properties, such as being an antioxidant, cytotoxicity to cancerous cells, and many pharmacological applications such as anti-tumor, anti-inflammatory, anti-microbial, gastrointestinal, neuropsychological, and cardiovascular, etc, and helps in the treatment of internal bleeding, albuminuria, diabetes, and brightening of skin [112–117]. Its extensive usage to solve different problems has made it important to determine, extract, and monitor GA level in samples. Therefore, in 2022 Qin *et al* focused on electrochemical detection from tea samples via developing an MIP-based TiO_2@CNT modified glassy carbon electrode. The sensor efficiently detects 50–700 μM GA with a sensitivity of 0.023 48 μA μM^{-1}, and an LOD of 12 nM, while showing a 97.65% to 99.30% recovery rate from spiked samples [118]. Yang *et al* constructed a low-cost MIP in plasticized polyvinyl chloride (PVC) membrane as a detecting module to identify GA in comestible plants. Bulk polymerization was performed in the presence of trimethylolpropane triacrylate (cross-linker) and AIBN (initiator) in a PVC medium. The imprint polymerized poly-MAA electrochemical

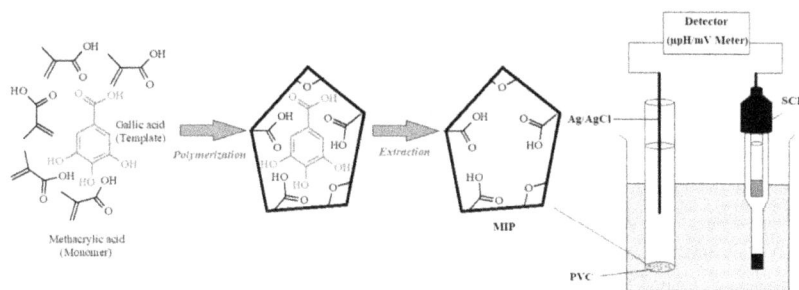

Figure 9.6. The synthesizing procedure and analytical configuration of an MAA-MIP-based electrochemical sensor to measure GA. (Reproduced with permission from [119]. Copyright 2015 Elsevier.)

sensor largely followed the Nernst response in the range of 1×10^{-5} to 3.2×10^{-4} mol l^{-1}, figure 9.6 [119].

Luteolin. In 2021 Tian and colleagues were the first to use GA based MIPs as adsorbents for the selective recognition and capturing of luteolin in herbal medicinal samples [120]. Luteolin, a flavonoid, occurs naturally in herbs such as honeysuckle and feverfew [121] and has important biological and medicinal features, such as inflammatory, cardiovascular, aging, and anti-cancer [122]. GA-MIPs were prepared using luteolin (template), GA (functional monomer), ethylene glycol dimethacrylate (EGDMA) (cross-linker), and 2,2′-azobis(2-methylpropionitrie) (initiator). Under the best optimizing conditions the adsorption capability was observed to be 1.24 mg g^{-1}, which was higher than in many of the reports on the enriching of luteolin, except for one, i.e. the 4-carboxyphenylboronic acid-affinity MIP adsorbent. The range of adsorption was found to be between 0.05 and 100 mg l^{-1} with an LOD of 0.020 mg l^{-1} and a recovery of 93.9%–114.2%, thus they provided a prospective and practical platform to detect luteolin in complex matrices. Bhawani and coworkers synthesized a non-covalent approach-based MIP assay based on a precipitation polymerization technique to detect GA and found about 80% extraction from a spiked urine sample. In the imprinting process, GA, acrylic acid, and ethylene glycol dimethacrylate, participated as the template, functional monomer, and cross-linker, respectively, at a molar composition of 1:4:20. AIBN was added as the initiator and acetonitrile as the solvent in the formation of MIPs and the rebinding efficacy was assessed using a batch binding approach (79.50%) [123].

Phenolics. 2-phenylphenol, a weak organic acid, is in wide use as a fungicide. It is sprayed after the harvesting of fruits and vegetables to protect them from microbial damage. However, it causes adverse effect on the human body when comes into contact and thus its selective identification and extraction is an important step in supporting the health of humans and the environment. Keeping this in mind Bakhtiar *et al* designed an MIP using a non-covalent approach through a precipitation polymerization technique by incorporation of 2-phenylphenol (template), styrene (functional monomer), and divinyl benzene (cross-linker). Brunauer–Emmett–Teller scanning showed the surface area, pore volume, and pore size to be

$131.44 \ m^2 \ g^{-1}$, 5.23 c.c. g^{-1}, and 7.9587 Å, respectively, in comparison to a non-imprinting polymer. The 2-phenylphenol-MIP showed good affinity as an extractant for extraction from spiked blood serum (93%) and natural water modules (88%) [124]. Bi *et al* carried out research under the title 'Separation of phenolic acids from natural plant extracts using molecularly imprinted anion-exchange polymer confined ionic liquids' to detect phenolic acids [125].

In 2022 Xia *et al* developed more stable and reusable magnetic porous cellulose (MPC)–MIPs using the Fe_3O_4 magnetic substance, as illustrated in figure 9.7. Fe_3O_4 worked as an assisted solvent/deep eutectic solvent, while for the functional monomer *N*-isopropylacrylamide was used. Furthermore, the adsorption characteristics of bisphenol-A, bisphenol-F, and bisphenol-AF were also investigated under different temperatures and pH conditions by means of static, dynamic, and selective experiments. They found that the adsorption capability, efficacy of mass transfer, and selectivity of the synthesized MPC–MIPs towards bisphenol-A were 5.9-fold, 4-fold, and 4.4-fold more than those of traditional MIPs. The adsorption characteristics of the MPC–MIPs were also investigated against the target molecules with adsorption isotherm, kinetic, and thermodynamic models and investigating H_2-bonding is the main way of interaction between a target molecule and the MPC–MIPs. MPC–MIP grouping with HPLC or ultra-performance liquid chromatography (LC)–MS can be used for purification and finding of bisphenol-A, bisphenol-F, and bisphenol-AF in milk, canned fruits, and fish samples. The MPC–MIP method proved itself as a very good, sensitive, and selective high accuracy method [126].

Urea. In 2020 Shinde *et al* fabricated oxyanion imprinted polymers integrating neutrally charged urea groups in a hydrophobic scaffold. The synthesis of the polymer was based on H_2-bond derived imprinting of mono- or di-anions of benzoic acid (BA), phenyl sulfonic acid (PSA), and phenylphosphonic acid (PPA), while 3,5-bis-(trifluoromethyl)-phenyl-*N*-4-vinylphenyl urea was used as the functional host monomer. The synthesized polymers present a unique interchangeable anion binder with robust ionic affinity in buffered solutions encompassing both PO_4^{3-} as well as

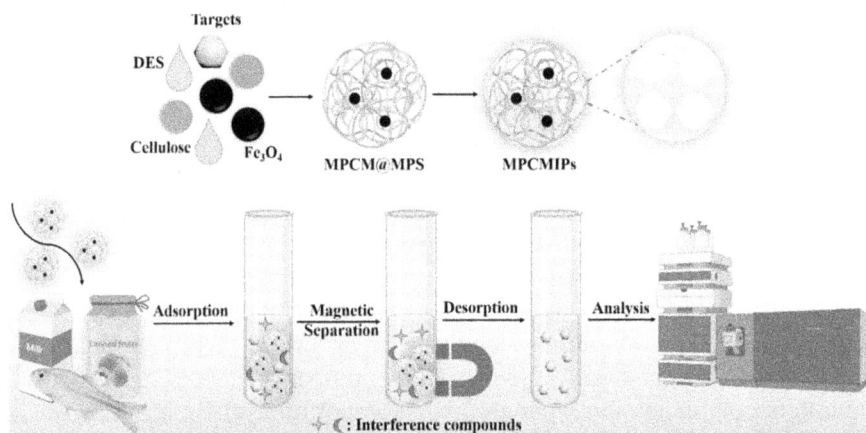

Figure 9.7. The fabrication and analysis process of magnetic porous cellulose (MPC)–MIPs. (Reproduced with permission from [126]. Copyright 2022 Elsevier.)

SO_4^{2-} (strongly hydrated) ions. The external switching between phospho-sulpho anions was driven by small molecule modifiers [69]. Shinde's team provided experimentally proven robust anion binders and the approach will help to develop MIPs for other anions.

Creatinine. In 2021 Cho *et al* prepared hydrogel type MIPs based on precipitation polymerization for the diagnosis of creatinine with the help of acrylic acid, acrylamide, and ethylene glycol dimethacrylate by utilizing a mixture of the solvent of acrylonitrile and toluene in the ratio of 3:1 v/v. The meso-porous nano-sized particles showed a sorption capacity as investigated utilizing 3-amino-1,2,4-triazole, creatinine, creatine, and cytosine species. Langmuir isothermic adsorption and Scatchard analyzed affinity were shown by all species. The pronounced selective binding for creatinine was found to be three-fold, two-fold, and more than two-fold higher in comparison to creatine, cytosine, and aminotriazole, respectively [127].

Nucleotides and nucleic acids. As the blue print of inherited material in all species on the planet, DNA/RNA is vital for species determination and the transportation of genetic information. Reasonable, specific, accurate, selective, and intensive monitoring of DNA/RNA contributes effectively in the disciplines of clinical assistance, pathology, pharmacokinetics, pharmacogenomics, inherited diseases, food-technology, food storage, and protection [128]. Before understanding the affiliation of imprinted polymers with DNA, it is necessary to understand the basic principle that governs the fabrication of MIP receptors. However, partial leaching of templates, a low mass transfer rate, and reduced binding affinity are a few weaknesses of traditional MIPs. Currently, manifold approaches are being applied for the identification and monitoring of nucleic acid analytes, in particular for ssDNA, dsDNA, anti-sense DNA templates, and long DNA molecules [129–131]. Babamiri and coworkers developed an urgently needed assay for the biologically important organism human immunodeficiency virus (HIV). The novel MIPs based on an electrochemiluminescence (ECL) sensing method use a very effective and simple technique which was found to be highly sensitive and selective for the detection of the HIV-1 gene. MIP-ECL used europium sulfide nanocrystals (EsNCs) to produce signals. The basic approach is described in figure 9.8. The hybridization reaction taking place between EsNCs functionalized 5-amino-labeled oligonucleo-tides (capture probes) and oligonucleotides bases of the HIV-1 gene (target molecules). The ECL signal was amplified by $K_2S_2O_8$ (co-reactant). The strong emission signal allowed the HIV-1 gene to be detected in the range of 3.0 fM–0.3 nM with an LOD of 0.3 fM. Pleasing results were obtained when HIV-DNA was detected in actual human serum samples, figure 9.8 [130].

SARS-CoV-2. Recently, for the screening of SARS-CoV-2, MIP-nanoparticles (nanoMIPs) were synthesized by McClements's team. They fabricated nanoMIPs electrografted onto inexpensive SPEs tailored inside 3D-printed measuring cells. In only 2 h the research team could fabricate the nanoMIPs and the scalable process used a short fragment of SARS-CoV-2 proteins only about 10 amino acids long. The heat-transfer based rapid detection greatly validate the patient's samples in short time spam of around 15 min with an excellent efficacy, specificity and sensitivity for the spike protein [132].

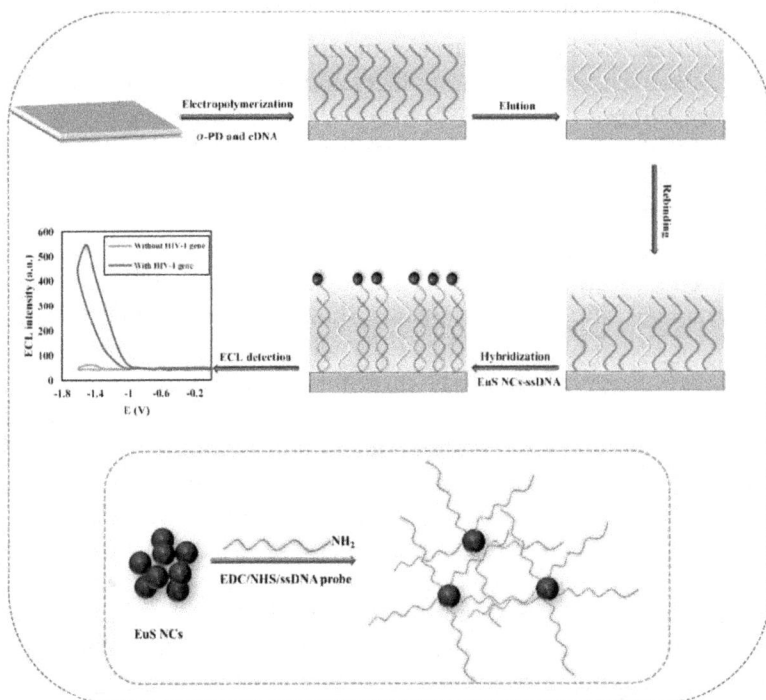

Figure 9.8. The molecularly imprinted MIP-ECL probe for sensing of HIV-1 gene. (Reproduced with permission [130]. Copyright 2018 from Elsevier.)

9.4 Conclusion and outlook

At the current time, MIPs are an emerging technology. MIT is a tremendously promising strategy for the preparation of polymers with tailor-made sub-atomic recognition properties. MIPs are synthesized in a manner which make them complementary in aspects of shape, size, and geometry to the target ions. In order to guide the preparation of anion-imprinted polymers more effectively, the present chapter summarizes the development status of anion-imprinted polymers, and introduces the types of functional monomers that interact with different anions (containing amino groups, quaternary ammonium groups, nitrogen heterocycles, and carboxyl structures). As well as the adsorption properties of anion-imprinted polymers prepared based on the above-mentioned monomers, the preparation methods and synthesis strategies of anion-imprinted polymers and their applications in the field of analytical chemistry, such as electrochemical detection, fluorescence sensing, and SPE technique, were reviewed. The problems existing in the preparation process of the compounds were discussed, and future research directions were prospected.

Although there are not many reports on anion-imprinted polymers, more and more researchers have begun to pay attention to them because anions can also cause water pollution. At present, there are still some shortcomings in the preparation and application of anionic imprinted polymers, such as a small saturated adsorption capacity, poor thermal stability, and small industrial application fields. The point of this chapter is to feature the extraordinary capabilities of the use of MIPs for anion recognition, not only in aqueous media but also in other frameworks. Although they have a tiny size to charge ratio, the asymmetrical geometry of anionic species and the species to species diversity add to the endogenous interference in the development of MIPs, and the irregular dispersion of binding cavities, reduced detection capability, and complex construction optimization impose extra challenges in the construction of MIPs and their successful implementations in water treatment and other areas. In view of these problems, in future research and application work, we should focus on the following. (i) Designing, synthesizing, screening, and selecting new functional monomers, for example allyl mercaptan, mercaptopropionic acid, and other mercapto-containing monomers, and thus extending the range of functional mono-mers. (ii) Co-imprinting with several functional elements in a single micro-structure would optimize the performance of MIPs. (iii) The commercialization of MIPs should be expanded and for this maintenance and storage efficacy should also be a focus. (iv) After a certain cycle MIPs must be desorbed, and at the industrial scale individual maintenance is impossible, so we need to focus on developing stable and reusable MIPs. A new concept for 'self-desorbing MIPs', which can desorb by themselves with the right eluent after a specific cycle of adsorption and desorption, may emerge from this issue. (v) A point of concern is the synthesis of environ-mentally friendly imprinted polymers that are good adsorbers. Achieving this includes examining the consistency of the environment, and better exploration of the binding mechanism and identification of the target. The preparation methods can be explored in the direction from oil phase to water phase synthesis, the development of 'green chemical' imprinted polymers can be pursued, and the search for improved anion transfer kinetics in imprinted polymer methods should be focused. (vi) Moving the molecular imprinted polymers from theoretical studies and laboratories to commercialization that is risk-free for the environment and allows cleaner production of the necessary components is the ultimate objective of this review and parallel researchers.

In general, the research and application of anionic imprinted polymers have great development potential in the future, and they need the attention of experts and scholars in relevant fields. Only in this way can their functions and applications be developed fully. In the future it will be necessary to develop high-performance MIPs, which is the subject of ongoing research and development, with the potential to have positive applications in removing toxic environmental hazards that pose a threat to all the creatures living in them and restoring ecological equilibrium in the surrounding environment.

References

[1] Spivak D A 2005 Optimization, evaluation, and characterization of molecularly imprinted polymers *Adv. Drug Delivery Rev.* **57** 1779–94
[2] Haupt K and Mosbach K 2000 Molecularly imprinted polymers and their use in biomimetic sensors *Chem. Rev.* **100** 2495–504

[3] Polyakov M V 1931 Adsorption properties and structure of silica gel *Zh. Fiz. Khim* **2** 799–805

[4] Vlatakis G, Andersson L I, Müller R and Mosbach K 1993 Drug assay using antibody mimics made by molecular imprinting *Nature* **361** 645–7

[5] Andersson L I 2000 Molecular imprinting for drug bioanalysis: a review on the application of imprinted polymers to solid-phase extraction and binding assay *J. Chromatogr.* B **739** 163–73

[6] Crapnell R D, Hudson A, Foster C W, Eersels K, Grinsven B V, Cleij T J, Banks C E and Peeters M 2019 Recent advances in electrosynthesized molecularly imprinted polymer sensing platforms for bioanalyte detection *Sensors* **19** 1204

[7] Mahony J O, Nolan K, Smyth M R and Mizaikoff B 2005 Molecularly imprinted polymers—potential and challenges in analytical chemistry *Anal. Chim. Acta* **534** 31–9

[8] Svenson J and Nicholls I A 2001 On the thermal and chemical stability of molecularly imprinted polymers *Anal. Chim. Acta* **435** 19–24

[9] Tamayo F G, Turiel E and Martín-Esteban A 2007 Molecularly imprinted polymers for solid-phase extraction and solid-phase microextraction: recent developments and future trends *J. Chromatogr.* A **1152** 32–40

[10] Huang D L, Wang R Z, Liu Y G, Zeng G M, Lai C, Xu P, Lu B A, Xu J J, Wang C and Huang C 2015 Application of molecularly imprinted polymers in wastewater treatment: a review *Environ. Sci. Pollut. Res.* **22** 963–77

[11] Jalink T, Farrand T and Herdes C 2016 Towards EMIC rational design: setting the molecular simulation toolbox for enantiopure molecularly imprinted catalysts *Chem. Cent. J.* **10** 66

[12] Chu L Y, Yamaguchi T and Nakao S I 2002 A molecular-recognition microcapsule for environmental stimuli-responsive controlled release *Adv. Mater.* **14** 386–9

[13] Luliński P 2017 Molecularly imprinted polymers based drug delivery devices: a way to application in modern pharmacotherapy. A review *Mater. Sci. Eng.* C **76** 1344–53

[14] Beltran A, Borrull F, Marcé R M and Cormack P A 2010 Molecularly-imprinted polymers: useful sorbents for selective extractions *TrAC Trends Anal. Chem.* **29** 1363–75

[15] Bianchi G K A, Bowman-James K and Garcia-Espana E 1997 *Supramolecular Chemistry of Anions* (New York: Wiley)

[16] Beer P D and Gale P A 2001 Anion recognition and sensing: the state of the art and future perspectives *Angew. Chem. Int. Ed.* **40** 486–516

[17] Shannon R D 1976 Revised effective ionic radii and systematic studies of interatomic distances in halides and chalcogenides *Acta Crystallogr.* A **32** 751–67

[18] Wu X 2012 Molecular imprinting for anion recognition in aqueous media *Microchim. Acta* **176** 23–47

[19] Hirsch A K, Fischer F R and Diederich F 2007 Phosphate recognition in structural biology *Angew. Chem. Int. Ed.* **46** 338–52

[20] Ward M H, Jones R R, Brender J D, De Kok T M, Weyer P J, Nolan B T, Villanueva C M and Van Breda S G 2018 Drinking water nitrate and human health: an updated review *Int. J. Environ. Res. Public Health* **15** 1557

[21] Li A F, Wang J H, Wang F and Jiang Y B 2010 Anion complexation and sensing using modified urea and thiourea-based receptors *Chem. Soc. Rev.* **39** 3729–45

[22] Yao X 2022 Acid- and anion-targeted fluorescent molecularly imprinted polymers: recent advances, challenges and perspectives *Arabian J. Chem.* **15** 104149

[23] Vaneckova T, Bezdekova J, Han G, Adam V and Vaculovicova M 2020 Application of molecularly imprinted polymers as artificial receptors for imaging *Acta Biomater.* **101** 444–58

[24] Haupt K, Medina Rangel P X and Bui B T 2020 Molecularly imprinted polymers: antibody mimics for bioimaging and therapy *Chem. Rev.* **120** 9554–82

[25] Hosseini M W 2003 Molecular tectonics: from molecular recognition of anions to molecular networks *Coord. Chem. Rev.* **240** 157–66

[26] García-España E, Díaz P, Llinares J M and Bianchi A 2006 Anion coordination chemistry in aqueous solution of polyammonium receptors *Coord. Chem. Rev.* **250** 2952–86

[27] Bano Y 2023 *Molecularly Imprinted Polymers for the Detection of Heavy Metals* (Bristol: Institute of Physics Publishing)

[28] Xu F, Zhang K, Yin F Q, Xu F, Pang Y X and Chen S T 2021 Research progress in preparation and application of anion imprinted polymers *Chin. J. Appl. Chem.* **38** 123–35

[29] Velempini T, Pillay K, Mbianda X Y and Arotiba O A 2017 Epichlorohydrin crosslinked carboxymethyl cellulose-ethylenediamine imprinted polymer for the selective uptake of Cr (VI) *Int. J. Biol. Macromol.* **101** 837–44

[30] Chen L, Liang H and Xing J 2020 Synthesis of multidentate functional monomer for ion imprinting *J. Sep. Sci.* **43** 1356–64

[31] Fan H T, Sun T, Xu H B, Yang Y J, Tang Q and Sun Y 2011 Removal of arsenic (V) from aqueous solutions using 3-[2-(2-aminoethylamino) ethylamino] propyl-trimethoxysilane functionalized silica gel adsorbent *Desalination* **278** 238–43

[32] Xing D Y, Chen Y, Zhu J and Liu T 2020 Fabrication of hydrolytically stable magnetic core–shell aminosilane nanocomposite for the adsorption of PFOS and PFOA *Chemosphere* **251** 126384

[33] Soto R J, Yang L and Schoenfisch M H 2016 Functionalized mesoporous silica via an aminosilane surfactant ion exchange reaction: controlled scaffold design and nitric oxide release *ACS Appl. Mater. Interfaces* **8** 2220–31

[34] Golker K, Olsson G D and Nicholls I A 2017 The influence of a methyl substituent on molecularly imprinted polymer morphology and recognition—acrylic acid versus methacrylic acid *Eur. Polym. J.* **92** 137–49

[35] Yan H and Row K H 2006 Characteristic and synthetic approach of molecularly imprinted polymer *Int. J. Mol. Sci.* **7** 155–78

[36] Booker K, Holdsworth C I, Doherty C M, Hill A J, Bowyer M C and McCluskey A 2014 Ionic liquids as porogens for molecularly imprinted polymers: propranolol, a model study *Org. Biomol. Chem.* **12** 7201–10

[37] Sajini T and Mathew B 2021 A brief overview of molecularly imprinted polymers: highlighting computational design, nano and photo-responsive imprinting *Talanta Open* **4** 100072

[38] O'Shannessy D J, Ekberg B and Mosbach K 1989 Molecular imprinting of amino acid derivatives at low temperature (0 °C) using photolytic homolysis of azobisnitriles *Anal. Biochem.* **177** 144–9

[39] Kempe M and Mosbach K 1991 Binding studies on substrate- and enantio-selective molecularly imprinted polymers *Anal. Lett.* **24** 1137–45

[40] Si Z, Yu P, Dong Y, Lu Y, Tan Z, Yu X, Zhao R and Yan Y 2019 Thermo-responsive molecularly imprinted hydrogels for selective adsorption and controlled release of phenol from aqueous solution *Front. Chem.* **6** 674

[41] Wei H, Cheng C, Chang C, Chen W Q, Cheng S X, Zhang X Z and Zhuo R X 2008 Synthesis and applications of shell cross-linked thermoresponsive hybrid micelles based on poly (N-isopropylacrylamide-co-3-(trimethoxysilyl) propyl methacrylate)-b-poly (methyl methacrylate) *Langmuir* **24** 4564–70

[42] Peeters M *et al* 2014 Thermal detection of histamine with a graphene oxide based molecularly imprinted polymer platform prepared by reversible addition–fragmentation chain transfer polymerization *Sensors Actuators* B **203** 527–35

[43] Zhou T 2016 Molecularly imprinted polymer beads-synthesis, evaluation and applications *Doctoral Thesis* (Lund University) Pure and Applied Biochemistry

[44] Peeters M M, Van Grinsven B, Foster C W, Cleij T J and Banks C E 2016 Introducing thermal wave transport analysis (TWTA): a thermal technique for dopamine detection by screen-printed electrodes functionalized with molecularly imprinted polymer (MIP) particles *Molecules* **21** 552

[45] Casadio S, Lowdon J W, Betlem K, Ueta J T, Foster C W, Cleij T J, Van Grinsven B, Sutcliffe O B, Banks C E and Peeters M 2017 Development of a novel flexible polymer-based biosensor platform for the thermal detection of noradrenaline in aqueous solutions *Chem. Eng. J.* **315** 459–68

[46] Betlem K *et al* 2019 Evaluating the temperature dependence of heat-transfer based detection: a case study with caffeine and molecularly imprinted polymers as synthetic receptors *Chem. Eng. J.* **359** 505–17

[47] Wang J, Cormack P A, Sherrington D C and Khoshdel E 2003 Monodisperse, molecularly imprinted polymer microspheres prepared by precipitation polymerization for affinity separation applications *Angew. Chem.* **115** 5494–6

[48] Pardeshi S and Singh S K 2016 Precipitation polymerization: a versatile tool for preparing molecularly imprinted polymer beads for chromatography applications *RSC Adv.* **6** 23525–36

[49] Chen L, Xu S and Li J 2011 Recent advances in molecular imprinting technology: current status, challenges and highlighted applications *Chem. Soc. Rev.* **40** 2922–42

[50] He C, Long Y, Pan J, Li K and Liu F 2007 Application of molecularly imprinted polymers to solid-phase extraction of analytes from real samples *J. Biochem. Bioph. Methods* **70** 133–50

[51] Yoshimatsu K, Reimhult K, Krozer A, Mosbach K, Sode K and Ye L 2007 Uniform molecularly imprinted microspheres and nanoparticles prepared by precipitation polymerization: the control of particle size suitable for different analytical applications *Anal. Chim. Acta* **584** 112–21

[52] Sarpong K A, Xu W, Huang W and Yang W 2019 The development of molecularly imprinted polymers in the clean-up of water pollutants: a review *Am. J. Anal. Chem.* **10** 202–26

[53] Sun Y, Ren T, Deng Z, Yang Y and Zhong S 2018 Molecularly imprinted polymers fabricated using Janus particle-stabilized Pickering emulsions and charged monomer polymerization *New J. Chem.* **42** 7355–63

[54] Hamzehlou S and Asua J M 2020 On-line monitoring and control of emulsion polymerization reactors *Adv. Chem. Eng.* **56** 31–57

[55] Bandiera M, Balk R and Barandiaran M J 2018 One-pot synthesis of waterborne polymeric dispersions stabilized with alkali-soluble resins *Polymers* **10** 88

[56] Hosoya K, Yoshizako K, Tanaka N, Kimata K, Araki T and Haginaka J 1994 Uniform-size macroporous polymer-based stationary phase for HPLC prepared through molecular imprinting technique *Chem. Lett.* **23** 1437–8

[57] Kloskowski A, Pilarczyk M, Przyjazny A and Namieśnik J 2009 Progress in development of molecularly imprinted polymers as sorbents for sample preparation *Crit. Rev. Anal. Chem.* **39** 43–58

[58] Ochoteco E and Mecerreyes D 2010 Oxireductases in the enzymatic synthesis of water-soluble conducting polymers *Enzymatic Polymerisation* Springer, Berlin pp 1–9

[59] Majumdar S, Goswami B and Mahanta D 2020 Polymer synthesis in water and super-critical water *Green Sustainable Process for Chemical and Environmental Engineering and Science* (Amsterdam: Elsevier) pp 1–29

[60] Luo J, Jiang S, Liu R, Zhang Y and Liu X 2013 Synthesis of water dispersible polyaniline/poly (styrenesulfonic acid) modified graphene composite and its electrochemical properties *Electrochim. Acta* **96** 103–9

[61] Capek I 2016 Photopolymerization of acrylamide in the very low monomer concentration range *Des. Monomers Polym.* **19** 290–6

[62] Jaymand M 2013 Recent progress in chemical modification of polyaniline *Prog. Polym. Sci.* **38** 1287–306

[63] Lasáková M and Jandera P 2009 Molecularly imprinted polymers and their application in solid phase extraction *J. Sep. Sci.* **32** 799–812

[64] Li W and Li S 2006 Molecular imprinting: a versatile tool for separation, sensors and catalysis *Adv. Polym. Sci.* **206** 191–210

[65] Jia M, Zhang Z, Li J, Ma X, Chen L and Yang X 2018 Molecular imprinting technology for microorganism analysis *TrAC Trends Anal. Chem.* **106** 190–201

[66] Whitcombe M J, Rodriguez M E, Villar P and Vulfson E N 1995 A new method for the introduction of recognition site functionality into polymers prepared by molecular imprinting: synthesis and characterization of polymeric receptors for cholesterol *J. Am. Chem. Soc.* **117** 7105–11

[67] Klein J U, Whitcombe M J, Mulholland F and Vulfson E N 1999 Template-mediated synthesis of a polymeric receptor specific to amino acid sequences *Angew. Chem. Int. Ed.* **38** 2057–60

[68] Kirsch N, Alexander C, Davies S and Whitcombe M J 2004 Sacrificial spacer and non-covalent routes toward the molecular imprinting of 'poorly-functionalized' N-heterocycles *Anal. Chim. Acta* **504** 63–71

[69] Shinde S, Incel A, Mansour M, Olsson G D, Nicholls I A, Esen C, Urraca J and Sellergren B 2020 Urea-based imprinted polymer hosts with switchable anion preference *J. Am. Chem. Soc.* **142** 11404–16

[70] Hutchins R S and Bachas L G 1995 Nitrate-selective electrode developed by electrochemically mediated imprinting/doping of polypyrrole *Anal. Chem.* **67** 1654–60

[71] Lenihan J S, Ball J C, Gavalas V G, Lumpp J K, Hines J, Daunert S and Bachas L G 2007 Microfabrication of screen-printed nanoliter vials with embedded surface-modified electrodes *Anal. Bioanal. Chem.* **387** 259–65

[72] Alahi M E, Mukhopadhyay S C and Burkitt L 2018 Imprinted polymer coated impedimetric nitrate sensor for real-time water quality monitoring *Sensors Actuators B* **259** 753–61

[73] Diouf A, El Bari N and Bouchikhi B 2020 A novel electrochemical sensor based on ion imprinted polymer and gold nanomaterials for nitrite ion analysis in exhaled breath condensate *Talanta* **209** 120577

[74] Warwick C, Guerreiro A and Soares A 2013 Sensing and analysis of soluble phosphates in environmental samples: a review *Biosens. Bioelectron.* **41** 1

[75] Katayev E A, Ustynyuk Y A and Sessler J L 2006 Receptors for tetrahedral oxyanions *Coord. Chem. Rev.* **250** 3004–37

[76] Bühlmann P, Pretsch E and Bakker E 1998 Carrier-based ion-selective electrodes and bulk optodes. 2. Ionophores for potentiometric and optical sensors *Chem. Rev.* **98** 1593–688

[77] Warwick C, Guerreiro A, Wood E, Kitson J, Robinson J and Soares A 2014 A molecular imprinted polymer based sensor for measuring phosphate in wastewater samples *Water Sci. Technol.* **69** 48–54

[78] Kugimiya A and Takei H 2008 Selectivity and recovery performance of phosphate-selective molecularly imprinted polymer *Anal. Chim. Acta* **606** 252–6

[79] Kugimiya A and Takei H 2006 Preparation of molecularly imprinted polymers with thiourea group for phosphate *Anal. Chim. Acta* **564** 179–83

[80] Kugimiya A and Takei H 2008 Selective recovery of phosphate from river water using molecularly imprinted polymers *Anal. Lett.* **41** 302–11

[81] Xi Y, Huang M and Luo X 2019 Enhanced phosphate adsorption performance by innovative anion imprinted polymers with dual interaction *Appl. Surf. Sci.* **467** 135–42

[82] Koh K Y, Zhang S and Chen J P 2020 Hydrothermally synthesized lanthanum carbonate nanorod for adsorption of phosphorus: material synthesis and optimization, and demonstration of excellent performance *Chem. Eng. J.* **380** 122153

[83] Chen D, Yu H, Pan M and Pan B 2022 Hydrogen bonding-orientated selectivity of phosphate adsorption by imine-functionalized adsorbent *Chem. Eng. J.* **433** 133690

[84] Dong S, Sun Z and Lu Z 1988 Chloride chemical sensor based on an organic conducting polypyrrole polymer *Analyst* **113** 1525–8

[85] Ramanavičius A, Ramanavičienė A and Malinauskas A 2006 Electrochemical sensors based on conducting polymer—polypyrrole *Electrochim. Acta* **51** 6025–37

[86] Dong S and Che G 1991 An electrochemical microsensor for chloride *Talanta* **38** 111–4

[87] Kamata K, Suzuki T, Kawai T and Iyoda T 1999 Voltammetric anion recognition by a highly cross-linked polyviologen film *J. Electroanal. Chem.* **473** 145–55

[88] Kamata K, Kawai T and Iyoda T 2001 Anion-controlled redox process in a cross-linked polyviologen film toward electrochemical anion recognition *Langmuir* **17** 155–63

[89] Özkütük E B, Özalp E, Ersöz A, Açıkkalp E and Say R 2010 Thiocyanate separation by imprinted polymeric systems *Microchim. Acta* **169** 129–35

[90] Karekar J M and Divekar S V 2019 Removal of thiocyanate from water by using weak base microporous resins of different matrix structure *Desalin. Water Treat.* **167** 176–82

[91] Yang L, Ma L, Chen G, Liu J and Tian Z Q 2010 Ultrasensitive SERS detection of TNT by imprinting molecular recognition using a new type of stable substrate *Chem. Eur. J.* **16** 12683–93

[92] Li Y, Gao B and Du R 2011 Studies on preparation and recognition characteristic of surface-ion imprinting material IIP-PEI/SiO$_2$ of chromate anion *Sep. Sci. Technol.* **46** 1472–81

[93] Gao B, Du J and Zhang Y 2013 Preparation of arsenate anion surface-imprinted material IIP-PDMC/SiO$_2$ and study on its ion recognition property *Ind. Eng. Chem. Res.* **52** 7651–9

[94] Lahav M, Kharitonov A B, Katz O, Kunitake T and Willner I 2001 Tailored chemosensors for chloroaromatic acids using molecular imprinted TiO_2 thin films on ion-sensitive field-effect transistors *Anal. Chem.* **73** 720–3

[95] Lahav M, Kharitonov A B and Willner I 2001 Imprinting of chiral molecular recognition sites in thin TiO_2 films associated with field-effect transistors: novel functionalized devices for chiroselective and chirospecific analyses *Chem. Eur. J.* **7** 3992–7

[96] Lee S W, Ichinose I and Kunitake T 2002 Enantioselective binding of amino acid derivatives onto imprinted TiO_2 ultrathin films *Chem. Lett.* **31** 678–9

[97] Lee S W, Yang D H and Kunitake T 2005 Regioselective imprinting of anthracenecarboxylic acids onto TiO_2 gel ultrathin films: an approach to thin film sensor *Sensors Actuators* B **104** 35–42

[98] Urraca J L, Hall A J, Moreno-Bondi M C and Sellergren B 2006 A stoichiometric molecularly imprinted polymer for the class-selective recognition of antibiotics in aqueous media *Angew. Chem.* **118** 5282–5

[99] Urraca J L, Moreno-Bondi M C, Hall A J and Sellergren B 2007 Direct extraction of penicillin G and derivatives from aqueous samples using a stoichiometrically imprinted polymer *Anal. Chem.* **79** 695–701

[100] Gomy C and Schmitzer A R 2007 Synthesis and photoresponsive properties of a molecularly imprinted polymer *Org. Lett.* **9** 3865–8

[101] Viltres-Portales M, Alberto M D and Ye L 2021 Synthesis of molecularly imprinted polymers using an amidine-functionalized initiator for carboxylic acid recognition *React. Funct. Polym.* **165** 104969

[102] Benito-Peña E, Martins S, Orellana G and Moreno-Bondi M C 2009 Water-compatible molecularly imprinted polymer for the selective recognition of fluoroquinolone antibiotics in biological samples *Anal. Bioanal. Chem.* **393** 235–45

[103] Jalili R, Khataee A, Rashidi M R and Razmjou A 2020 Detection of penicillin G residues in milk based on dual-emission carbon dots and molecularly imprinted polymers *Food Chem.* **314** 126172

[104] Safran V, Göktürk I, Bakhshpour M, Yılmaz F and Denizli A 2021 Development of molecularly imprinted polymer-based optical sensor for the sensitive penicillin G detection in milk *ChemistrySelect* **6** 11865–75

[105] Bakhshpour M, Göktürk I, Bereli N, Yılmaz F and Denizli A 2021 Selective detection of penicillin G antibiotic in milk by molecularly imprinted polymer-based plasmonic SPR sensor *Biomimetics* **6** 72

[106] Buerge I J and Poiger T 2011 Acesulfame: from sugar substitute to wastewater marker: highlights of analytical chemistry in Switzerland *Chimia* **65** 176

[107] Scheurer M, Storck F R, Brauch H J and Lange F T 2010 Performance of conventional multi-barrier drinking water treatment plants for the removal of four artificial sweeteners *Water Res.* **44** 3573–84

[108] Tran N H, Hu J, Li J and Ong S L 2014 Suitability of artificial sweeteners as indicators of raw wastewater contamination in surface water and groundwater *Water Res.* **48** 443–56

[109] Roy J W, Van Stempvoort D R and Bickerton G 2014 Artificial sweeteners as potential tracers of municipal landfill leachate *Environ. Pollut.* **184** 89–93

[110] Zarejousheghani M, Schrader S, Möder M, Lorenz P and Borsdorf H 2015 Ion-exchange molecularly imprinted polymer for the extraction of negatively charged acesulfame from wastewater samples *J. Chromatogr.* A **1411** 23–33

[111] Zarejousheghani M, Fiedler P, Möder M and Borsdorf H 2014 Selective mixed-bed solid phase extraction of atrazine herbicide from environmental water samples using molecularly imprinted polymer *Talanta* **129** 132–8

[112] Zhu W *et al* 2020 Dual-phase $CsPbCl_3$–Cs_4PbCl_6 perovskite films for self-powered, visible–blind UV photodetectors with fast response *ACS Appl. Mater. Interfaces* **29** 32961–9

[113] Karimi-Maleh H, Khataee A, Karimi F, Baghayeri M, Fu L, Rouhi J, Karaman C, Karaman O and Boukherroub R 2022 A green and sensitive guanine-based DNA biosensor for idarubicin anticancer monitoring in biological samples: a simple and fast strategy for control of health quality in chemotherapy procedure confirmed by docking investigation *Chemosphere* **291** 132928

[114] Xie Y, Meng X, Chang Y, Mao D, Yang Y, Xu Y, Wan L and Huang Y 2022 Ameliorating strength-ductility efficiency of graphene nanoplatelet-reinforced aluminum composites via deformation-driven metallurgy *Compos. Sci. Technol.* **219** 109225

[115] Zhang Z, Feng L, Liu H, Wang L, Wang S and Tang Z 2022 Mo^{6+}–P^{5+} co-doped $Li_2ZnTi_3O_8$ anode for Li-storage in a wide temperature range and applications in $LiNi_{0.5}Mn_{1.5}O_4$/$Li_2ZnTi_3O_8$ full cells *Inorg. Chem. Front.* **9** 35–43

[116] Wang Z, Dai L, Yao J, Guo T, Hrynsphan D, Tatsiana S and Chen J 2021 Improvement of *Alcaligenes* sp. TB performance by Fe–Pd/multi-walled carbon nanotubes: enriched denitrification pathways and accelerated electron transport *Bioresour. Technol.* **327** 124785

[117] Liu Y *et al* 2021 Comparative study of photocatalysis and gas sensing of ZnO/Ag nanocomposites synthesized by one- and two-step polymer-network gel processes *J. Alloys Compd.* **868** 158723

[118] Qin F, Hu T, You L, Chen W, Jia D, Hu N and Qi W 2022 Electrochemical detection of gallic acid in green tea using molecularly imprinted polymers on TiO_2@ CNTs nano-composite modified glassy carbon electrode *Int. J. Electrochem. Sci.* **17** 2

[119] Yang T, Zhang Q, Chen T, Wu W, Tang X, Wang G, Feng J and Zhang W 2020 Facile potentiometric sensing of gallic acid in edible plants based on molecularly imprinted polymer *J. Food Sci.* **85** 2622–8

[120] Tong Y, Zhang B, Guo B, Wu W, Jin Y, Geng F and Tian M 2021 Gallic acid-affinity molecularly imprinted polymer adsorbent for capture of *cis*-diol containing luteolin prior to determination by high performance liquid chromatography *J. Chromatogr.* A **1637** 461829

[121] Francisco V, Figueirinha A, Costa G, Liberal J, Ferreira I, Lopes M C, Garcia-Rodriguez C, Cruz M T and Batista M T 2016 The flavone luteolin inhibits liver X receptor activation *J. Nat. Prod.* **79** 1423–8

[122] Jegal K H *et al* 2020 Luteolin prevents liver from tunicamycin-induced endoplasmic reticulum stress via nuclear factor erythroid 2-related factor 2-dependent sestrin 2 induction *Toxicol. Appl. Pharmacol.* **399** 115036

[123] Bhawani S A, Sen T S and Ibrahim M N 2018 Synthesis of molecular imprinting polymers for extraction of gallic acid from urine *Chem. Cent. J.* **12** 19

[124] Bakhtiar S, Bhawani S A and Shafqat S 2019 Synthesis and characterization of molecular imprinting polymer for the removal of 2-phenylphenol from spiked blood serum and river water *Chem. Technol. Agric.* **6** 15

[125] Bi W, Tian M and Row K H 2012 Separation of phenolic acids from natural plant extracts using molecularly imprinted anion-exchange polymer confined ionic liquids *J. Chromatogr.* A **1232** 37–42

[126] Wen Z, Gao D, Lin J, Li S, Zhang K, Xia Z and Wang D 2022 Magnetic porous cellulose surface-imprinted polymers synthetized with assistance of deep eutectic solvent for specific recognition and purification of bisphenols *Int. J. Biol. Macromol.* **216** 374–87

[127] Cho M G, Hyeong S, Park K K and Chough S H 2021 Characterization of hydrogel type molecularly imprinted polymer for creatinine prepared by precipitation polymerization *Polymer* **237** 124348

[128] Liu Y, Liao R, Wang H, Gong H, Chen C, Chen X and Cai C 2018 Accurate and sensitive fluorescence detection of DNA based on G-quadruplex hairpin DNA *Talanta* **176** 422–7

[129] Nawaz N, Bakar N K, Mahmud H N and Jamaludin N S 2021 Molecularly imprinted polymers-based DNA biosensors *Anal. Biochem.* **630** 114328

[130] Babamiri B, Salimi A and Hallaj R 2018 A molecularly imprinted electrochemilumines-cence sensor for ultrasensitive HIV-1 gene detection using EuS nanocrystals as luminophore *Biosens. Bioelectron.* **117** 332–9

[131] Huang L, Wang X, Xie X, Xie W, Li X, Gong X, Long S, Guo H and Liu Z 2018 Synthesis and DNA adsorption of poly (2-vinyl-4,6-diamino-1,3,5-triazine) coated polystyrene microspheres *J. Wuhan Univ. Technol. Mater. Sci. Ed.* **33** 999–1006

[132] McClements J *et al* 2022 Molecularly imprinted polymer nanoparticles enable rapid, reliable, and robust point-of-care thermal detection of SARS-CoV-2 *ACS Sens.* **7** 1122–31

IOP Publishing

Molecularly Imprinted Polymers for Environmental Monitoring
Fundamentals and applications
Raju Khan and Ayushi Singhal

Chapter 10

Molecularly imprinted polymers for the detection of environmental estrogens

Melkamu Biyana Regasa and Tebello Nyokong

Molecularly imprinted polymers (MIPs) as synthetic molecular recognition elements in chemical sensors have become advantageous in detecting emerging contaminants such as environmental estrogens (EEs). In contrast to classical methods such as chromatography and spectroscopy, the MIP-based chemical sensors are essential due to their low cost, user-friendly nature, portability, and ability to provide real-time online monitoring. Thus, MIP-based chemical sensors have become a promising technology in the detection and quantification of ubiquitous EEs. This chapter introduces EEs and provides an overview of MIPs, MIP-based chemical sensor formats, and their applications for detecting EEs using selected publications.

10.1 Introduction

10.1.1 What are environmental estrogens?

Estrogens are sex hormones that play various roles in the body. In addition to estrogens from mammals, there are also other natural and synthetic estrogens in the environment. Natural estrogens are more potent than synthetic ones apart fro a few; however, they do not persist for a long time in the body. On the other hand, synthetic estrogens in the environment are lipophilic and can accumulate within the fat and tissue of humans and other animals. Thus, environmental estrogens (EEs) are endocrine-disrupting chemicals (EDCs) because they affect the endocrine system in mammals [1, 2].

EEs are a group of natural and synthetic ubiquitous chemicals in the environment and function similarly to estrogen when absorbed into a person's system. EEs, as typical endocrine system disruptors, are structurally diverse compounds that can interact with nuclear estrogen receptors and pose significant risks to human and ecological health [3]. EEs can be produced in mammals' bodies or by industry for

doi:10.1088/978-0-7503-4962-8ch10

hormone replacement, pesticides, plastics, electrical transformers, and other applications. There are three groups of natural and synthetic EEs.

The primary sources of EEs are human and animal urine, wastewaters, pharmaceuticals, foods, herbs, spices, fungi, pesticides, preservatives, solvents, plasticizers, and agricultural and industrial companies [4]. These ubiquitous contaminants have health effects, including endocrine disruption, cancer, cardiovascular diseases, reduced abnormal sperm count, and reproductive diseases in mammals. According to a recent report, many environmental samples worldwide were confirmed to contain a significant amount of EEs, and they have received considerable public attention due to their harmful effects on the normal endocrine functions of humans and animals [5]. As a result, it is common to find research reports conducted to develop efficient analytical techniques and platforms for the detection and quantification of EEs.

Various analytical methods have been employed for the monitoring of EEs. These include chromatographic techniques [6], spectroscopy, and electrophoresis [7]. The quantities of estrogenic compounds in environmental matrices are very low, and a clean-up and preconcentration step is usually required in order to minimize interference and improve a method's accuracy and sensitivity. The above-mentioned methods are sensitive enough, but they lack selectivity and also suffer from tedious sample preparation and time-consuming analysis. In addition, other advanced methods such as biochemical sensors have also been used for the detection of EEs. The biochemical sensors mostly rely on biological recognition elements (antibodies, enzymes), which have common limitations, such as being expensive, unstable, difficult to handle, not available for some targets, and incompatible with modern micro-nano fabrication technology. Thus, synthetic molecular recognition elements such as aptamers and molecularly imprinted polymers (MIPs) have become the best alternatives recently [8]. Among the methods described here, MIP-based chemical sensors have attracted considerable research interest as selective sorbent materials and biomimetic recognition elements for EE monitoring applications [9]. This might be due to the advantages of MIPs, including their suitability for real-time analysis, ease-of-use, high stability, reusability, low-cost manufacturing, and thus low purchasing cost and low cost of operation [10]. The application of MIPs can achieve the primary requirements for the sensor recognition element to have specific binding properties. MIP-based sensors can provide the market with simple, fast, and inexpensive methodologies for detecting and quantifying chemical compounds in the environment, including EEs [9, 11]. Therefore, this chapter summarizes the basic principles of MIP preparation, its immobilization techniques, and MIP-based sensor formats reported for detecting EEs in environmental samples.

10.2 Overview of molecularly imprinted polymers

Nature has a wealth of information, such as the molecular recognition process based on host–guest chemistry to sustain life. This molecular recognition is the specific interaction between different molecules (such as antigen–antibody, enzyme–substrate, sugar–lectin, etc) via non-covalent interactions (hydrogen bonding, metal coordination, hydrophobic forces, van der Waals forces, π–π interactions, halogen bonding). After careful observation of this natural phenomenon, researchers were able to design and develop artificial molecular recognition elements (biomimetic recognition elements) to solve the problem of looking for naturally occurring materials. Natural molecular recognition elements are expensive, less stable, and unavailable for some targets, and there is a considerable demand to replace them with synthetic functional materials called biomimetic recognition elements. The synthetic molecular recognition materials include proteinaceous affibodies, immunoglobulins, single-chain, single-domain antibody fragments or aptamers, and MIPs. Biomimicry circumvents the limitations of natural recognition elements by imitating nature and implementing its working mechanisms in artificial systems. Molecular imprinting technology (MIT) is a generic technology that introduces recognition properties into synthetic polymers using appropriate templates. MIPs are tailor-made and highly cross-linked synthetic receptors that recognize and bind targets with a high affinity and selectivity for separating target analytes from complicated matrices. Essential elements for preparing MIPs include template molecules, functional monomers, cross-linking reagents, suitable solvents (porogen), and polymerization temperature. The functional monomer(s) selection depends on the nature of the imprint molecule, called a template or target analyte. The ultimate goal of MIP preparation is to recognize and capture target molecules by generating template-shaped binding cavities in the polymer matrices. During the pre-polymerization and recognition process, the template acts as the imprint molecule and also interacts with functional monomers through covalent, non-covalent, semi-covalent, and metal-ion mediated bonds in solution. The template and functional monomer(s) initially form a stable pre-polymerization complex, which is then polymerized in the presence of a suitable cross-linker after the addition of initiator molecules [12]. After successful polymerization, the imprint molecule is removed from the polymer matrix by generating micro/nano-cavities with complimentary three-dimensional structures compatible with the template in terms of the shape, size, and chemical functionality for selective rebinding. More importantly, the three-dimensional arrangement of binding cavities, which can memorize the imprint molecule structure, drives the template itself, creating the capability to bind the target or similar molecule (figure 10.1).

As molecular recognition elements in biochemical sensors, MIPs can be integrated into different transducers in the form of films or membranes and micro- or nanoparticles. However, due to the larger total surface area and high surface-to-volume ratio, the MIP nanoparticles are more advantageous. When the target molecules interact with the MIP, a 'lock and key model' will be formed, indicating that the binding sites of the imprinted polymer and the template are identical in shape for better recognition. This is the reason why MIPs are called artificial antibodies [13]. MIPs have some advantages over other molecular recognition elements, including their high stability, reusability, ease-to-prepare, high degree of freedom to prepare for any analyte, as well as their capability to recognize targets with high selectivity. Therefore, MIPs have become applicable in many application areas such as chemical separation [14], sorbents for solid-phase extraction [15],

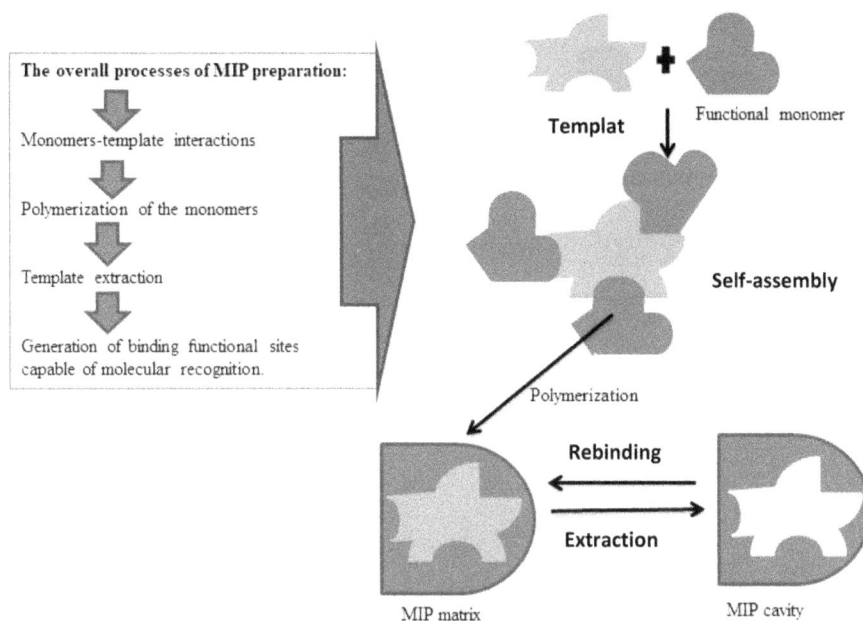

The overall processes of MIP preparation:

↓

Monomers-template interactions

↓

Polymerization of the monomers

↓

Template extraction

↓

Generation of binding functional sites capable of molecular recognition.

Templat

Functional monomer

Self-assembly

Polymerization

Rebinding

Extraction

MIP matrix

MIP cavity

Figure 10.1. Schematic of MIP preparation processes.

the stationary phase in chromatography [16], affinity materials in biochemical sensors [17], chemical catalysis [18], and drug delivery [19].

Most importantly, the combination of MIPs with different substrates has been confirmed as a promising platform for the concurrent separation and determination of trace levels of environmental contaminants. After selecting a suitable transducer technique, the next step is choosing an appropriate method to immobilize the MIP active material on the surface of the transducer. Synthetic receptors, and in particular, MIPs are gaining relevance as selective sorbent materials and biomimetic recognition elements for analyzing emerging contaminants in the environment [20].

10.3 The preparation of MIP-based chemical sensors

A chemical sensor is a device that transforms chemical information, ranging from the concentration of a specific sample component to the analysis of the total composition of a sample, into an analytically helpful signal [21]. A chemical sensor consists of two main elements: the chemical recognition element, called the molecular receptor, and a physicochemical transducer, in addition to signal amplification and readout units (figure 10.2). MIPs were used as recognition elements in chemical sensor development for the first time in 1999 using electrochemical polymerization to modify the surface of a gold electrode using phenylalanine and phenol as the template and functional monomer, respectively [22]. Since then, MIPs have attracted considerable interest from researchers worldwide because of their excellent selectivity, availability for a wide range of targets,

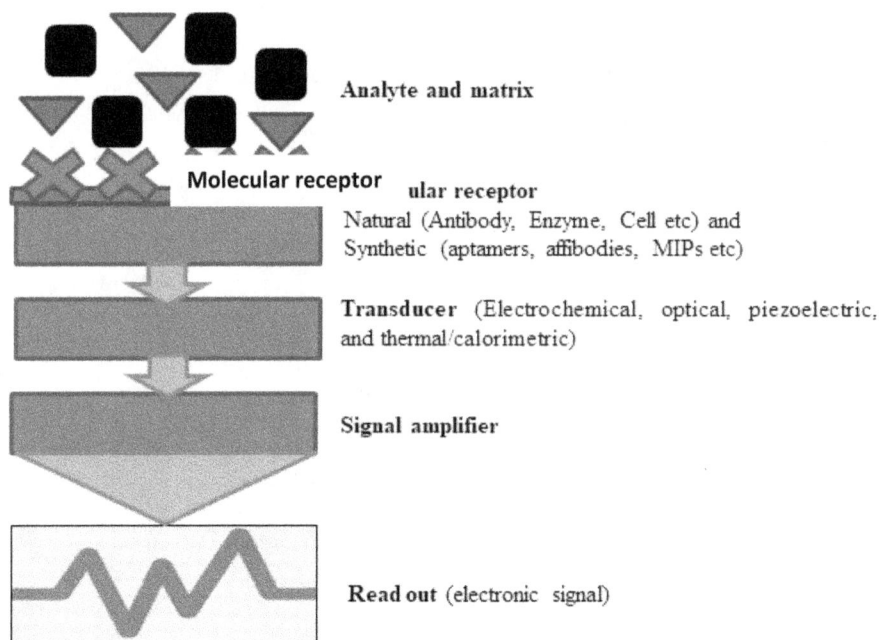

Analyte and matrix

Molecular receptor ular receptor
Natural (Antibody, Enzyme, Cell etc) and
Synthetic (aptamers, affibodies, MIPs etc)

Transducer (Electrochemical, optical, piezoelectric, and thermal/calorimetric)

Signal amplifier

Read out (electronic signal)

Figure 10.2. Schematic diagram of biochemical sensors and their components.

low cost, ease-of-preparation, and higher chemical and physical stability than natural recognition elements [8].

In chemical sensor fabrication, the development and integration of sensing materials with suitable transducers is significant. A proper design and fabrication approach can combine MIPs with different transduction mechanisms to prepare chemical sensors in two general approaches, namely *ex situ* and *in situ* approaches [23], as shown in figure 10.3. At the present time, biochemical sensors have become promising devices for various applications, including clinical diagnostics and environmental monitoring. Thus, suitable biochemical sensor design and fabrication are crucial to facilitate the optimum interaction between the MIP sensing layer and the target to attain extraordinary sensitivity and selectivity for the target. The MIP is prepared separately by different chemical polymerization techniques. Then MIP nanomaterials are coated onto the surface of the substrate by the drop-casting, spin coating, or dip coating methods.

In the second case, the MIP is directly polymerized on the transducer's surface with one-step synthesis using a photochemical, self-assembly, or electrochemical polymerization method [24]. Mimicking the selectivity and sensitivity of biological systems for sensor devices is of increasing interest in biomedical, environmental, and chemical analysis. Synthetic materials with imprinted nano-cavities, acting as highly selective artificial receptors, are a tailor-made solution for obtaining such a sensor. MIP film-based sensors sometimes lack sensitivity, and their combination with different nanomaterials can solve this problem. The introduction of nanomaterials within MIPs can improve the performance of MIP sensors by increasing the surface

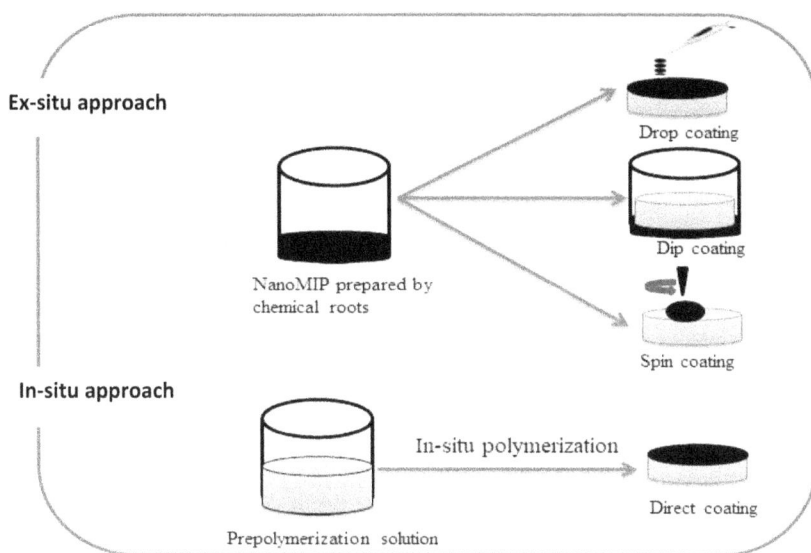

Figure 10.3. MIP sensor preparation approaches.

area-to-volume ratio of the modified surface [25]. Advancing the film thickness can enhance the number of binding sites; however, this slows down the diffusion rate of the analyte to the binding cavities. Commonly used nanomaterials are carbon nanomaterials (carbon nanotubes (CNTs), graphene), metal nanoparticles, magnetic nanoparticles, and other semiconductors. These nanomaterials greatly enhance the electrode surface area, analyte mass transport, sensitivity and selectivity of the sensor, and electrocatalytic properties.

10.4 Types of MIP-based chemical sensors

In sensor technology, the proper integration of MIPs with transducers is the central point in detecting the binding processes using electrochemical, piezoelectric, optical, and thermal mechanisms. Generally, there are many MIP-based sensors according to the transduction mechanisms (electrochemical, optical, mass-sensitive, and thermal) used (figure 10.2).

10.4.1 MIP-based electrochemical sensors

The electrochemical detection methods used to prepare MIP-based sensors are electrochemiluminescence (ECL), electrochemical impedance spectroscopy (EIS), and voltammetry techniques (cyclic voltammetry (CV), linear sweep voltammetry (LSV), differential pulse voltammetry (DPV), square wave voltammetry (SWV), and amperometry). Among these techniques, voltammetry is the most used because of its high sensitivity, simplicity, and fast response [26].

10.4.2 MIP-based optical sensors

Optical sensors convert light rays into electronic signals when the recognition element binds the target molecule selectively. These sensors rely on changes in an optical property such as light absorption, fluorescence, light scattering, refractive index, or reflection as the target rebinds to MIP sensing sites [9]. MIP-based optical sensors can be grouped into UV–visible absorption (UV–vis), fluorescence (FL), phosphorescence, chemiluminescence, surface plasmon resonance (SPR), colorimetry, and surface-enhanced Raman scattering (SERS) techniques. Therefore, the integration of MIPs into an optical transducer leads to the selective adsorption and recognition of different environmental pollutants [27]. MIP-based optical sensors have broad applications in areas such as food safety supervision, environmental pollution monitoring, disease diagnosis, and other fields [28].

10.4.3 MIP-based quartz crystal microbalance sensors

Quartz crystal microbalance (QCM) sensors are a type of transducer made of a piezoelectric material used in chemical sensors based on piezoelectricity—a phenomenon that couples the electrical and the mechanical states of the material. The QCM-based sensor is a sensitive mass balance that measures nanogram to microgram level changes in mass per unit area. It works based on the oscillation of quartz materials at a defined frequency by applying an appropriate voltage, usually via metal electrodes. The addition or removal of a small quantity of analyte mass onto the electrode surface brings changes in the frequency of oscillation. The amount of analyte deposited on the electrode surface monitors this change in frequency in real time to obtain helpful information about molecular interactions or reactions at the electrode surface coated by MIP thin films. Its advantages include rapidity, easy use, high stability, and portability. Generally, QCM sensors are practical and convenient monitoring tools because of their specificity, sensitivity, high accuracy, stability, and reproducibility. These types of sensors are applicable for detecting different target molecules from various environmental samples [29].

10.4.4 MIP-based thermal sensors

Thermal sensors based on MIPs as recognition elements depend on the two thermocouples and a heat source that enables applying the device for real-time measuring. The incorporation of the MIPs onto the electrode surface and mounting it into a thermocouple device leads to a change in thermal resistance at the solid–liquid interface upon binding the target molecules and relating it to the concentration of the target analyte. Researchers developed such sensors to detect various analytes from different matrices [30]. There are many reported studies on the applications of MIP-based sensor platforms to detect EEs. Some examples from the literature are briefly described in the following sections.

10.5 Application of MIP-based sensors in the analysis of EEs

10.5.1 Detection of natural estrogens in the environment

Endogenic hormones or natural estrogens, considered most potent, enter the environment after excretion from humans and animals. Natural hormones such as estrone (E1), estradiol (E2), estriol (E3), and 17-a-ethinylestradiol (EE2) were found in surface water from 94 to 180 pg ml^{-1}, and the drugs E2 and Levonorgestrel from contraceptive pills were found in surface water approximately from 34 to 38 ng l^{-1} [31]. Estrogenic EDCs have received significant public attention due to how they interfere with the normal endocrine functions of mammals. Therefore, monitoring their occurrence in the environment can be carried out by employing biochemical sensors constructed using MIPs as the sensing element. In recent years, tremendous progress has been made in detecting natural estrogens in environmental samples. A list of MIP-based sensors used to detect selected natural estrogens and their limit of detection (LOD) and detection mechanisms are summarized in table 10.1.

Magnetic nanoparticles combined with MIPs (Fe$_3$O$_4$–MIP) were prepared in a core–shell format showing the combination of the best features of both inorganic NPS and MIPs in improving the selectivity and sensitivity. The surface of screen-printed carbon electrodes (SPCEs) modified by a Fe$_3$O$_4$–MIP nanocomposite was reported to bind and detect E2 in a selective way using SWV to obtain 10 pmol detection limits [32]. In another report, MIPs based on a pyrrole metal–organic framework–carbon nanotube–Prussian blue hybrid were used for highly sensitive and selective detection of E2 with a wide linear response range between 10 000 pmol ml^{-1} and 1000.0 pmol ml^{-1} and calculated LOD of 6.19 fmol ml^{-1} using the DPV technique [33].

Table 10.1. Sensing mechanism and detection limit of developed MIP-based sensors for selected natural estrogens.

EE type	Analyte	Sensing mechanism	LOD	Reference
Hormones (natural estrogens)	E1	Electrochemical	27.00 pmol ml^{-1}	[34]
	E2	Electrochemical	10.00 pmol ml^{-1}	[32]
		Electrochemical	6.19 fmol ml^{-1}	[33]
		Electrochemical	2.60 fmol ml^{-1}	[47]
		Optical	250.00 fmol ml^{-1}	[36]
	EE2	Electrochemical	27.00 fmol ml^{-1}	[34]
		Electrochemical	3.30 fmol ml^{-1}	[35]
	E3	Electrochemical	1.18 amol ml^{-1}	[37]
		Electrochemical	27.00 pmol ml^{-1}	[34]
Mycotoxins (natural estrogens)	OTA	Electrochemical	4.10 pmol ml^{-1}	[42]
		Optical	69.00 amol ml^{-1}	[43]
	ZEA	Optical	36.15 fmol ml^{-1}	[38]
		Electrochemical	31.00 amol ml^{-1}	[39]
		Electrochemical	7.85 fmol ml^{-1}	[40]

A very interesting sensor was developed by the combination of a carbon paste electrode with reduced graphene oxide, and MIPs were used to develop an electrochemical sensor for the total determination of estrogenic phenolic compounds (EPCs). The authors reported the linear range (LR) of 0.16–15 μmol ml^{-1} and an LOD of 27 pmol ml^{-1} each, respectively, for the targets E1, E2, E3, and EE2 EPCs [34]. Subsequently, composite nanomaterials were used as a signal amplification strategy for MIP-based electrochemical detection of EE2 [35]. Moreover, optical sensors [36] and electrochemical sensors [37] for the detection of E2 and E3 were reported.

The mycotoxin zearalenone (ZEA) encourages reproductive toxicity due to its strong estrogenic effects. ZEA is one of the most prevalent estrogenic mycotoxins produced by *Fusarium* fungi which acts as an endocrine disruptor and affects the reproductive capacity of animals. An optical sensor based on the electropolymerization of Py in the presence of ZEA on gold chips was developed to attain a wide LR of 0.3–3000 ng ml^{-1} and an LOD of 36.15 fmol ml^{-1} with high selectivity and sensitivity [38]. An electrochemical sensor based on an ionic liquid functionalized boron-doped ordered mesoporous carbon–gold nanoparticle composite was reported to determine ZEA with a very low detection of 31 amol ml^{-1}. The enhanced selectivity and sensitivity are attributed to the multiple molecular interactions between the template and ionic liquid or MIPs through hydrophobic interaction, π–π interaction, and hydrogen bonds, combined with the advanced SWV technique. The sensor was applied for the analysis of real samples such as crop, rice, and beer samples successfully, with percentage recoveries of 96%–109% [39]. The screen-printed gold electrode was modified with imprinted poly(o-phenylenediamine) (MIP/SPGE) and applied to measure ZEA in corn flakes with an LR of 2.5–200 ng ml^{-1} and an LOD of 7.85 fmol ml^{-1} [40].

Ochratoxin A (OTA) is a carcinogenic mycotoxin produced by several species of *Aspergillus* (e.g. *A. ochraceus*) and *Penicillium* (e.g. *P. verrucosum*) fungi [41]. An MIP-based SPR sensor was developed to detect OTA in wheat and wine in the range of 0.123–0.5123 μmol ml^{-1} using imprinted PPy [41]. The composite of MIPPy and multiwalled carbon nanotubes were used to modify the surface of the glassy carbon electrode (GCE) to detect OTA using DPV to maintain an LR of 50–100 nmol ml^{-1} and an LOD of 410 nmol ml^{-1} [42]. Most importantly, label-free SPR MIP sensors were also developed to detect OTA in dried figs with a low detection limit of 69 attomol ml^{-1} (amol ml^{-1}) and successfully applied to real sample analysis [43]. In this work, photopolymerization was employed to synthesize MIPs on SPR chips to achieve an imprinting factor of 2.85 and the Langmuir model for adsorption of OTA.

10.5.2 Detection of synthetic environmental estrogens

In addition to natural hormones (estrogens) from animals and plant origin (phytoestrogens), there are various synthetic EEs categorized as EDCs because of their hormone-disrupting effects. These chemicals include phthalates (PAEs), plasticizers, surfactants, alkylphenols, persistent organic pollutants (POPs), and parabens [44]. Synthetic steroid estrogens include ethinylestradiol (EE2), 17α-ethinylestradiol (EE3), dienestrol (DIS), diethylstilbestrol (DES), estradiol valerate (EV), estropipate (EPP), conjugate esterified estrogen (CEE), and quinestrol (QE). These synthetic estrogens are compounds obtained by chemical synthesis that possess estrogenic activity and are mainly used as an oral contraceptive to treat advanced breast and prostate cancers and treat the deficiencies of amenorrhea and menopause. However, many scientific reports indicate that estrogens in the environment and breast cancer have a causal relationship [44].

DES is a synthetic exogenous estrogen used to treat a pregnant woman to prevent miscarriage and the animal growth promoter and is widely applied in aquaculture and animal husbandry. The residual DES in meat, food, milk, and water is a health threat because of its teratogenicity and carcinogenesis and can cause other serious reproductive and endocrine system diseases. Hence, achieving rapid and accurate analysis of trace amounts of DES in complex matrix environments is important to human health and environmental protection. Thus, Zhao and his co-authors [45] combined MIPs as sensing elements with the ECL transduction mechanism to detect DES at 250 amol ml^{-1} levels, table 10.2. The sensors were applied for the analysis of DES in lake water and milk as well as fish samples with good percentage recoveries. Other sensor formats such as fluorescent [46] and electrochemical [47] were also reported to determine DES in different environmental samples.

For the determination of another synthetic estrogen, EV, a polyurethane composite electrode modified with magnetic nanoparticles (mag; Fe$_3$O$_4$) coated with MIPs was developed. The sensor response for the analyte was evaluated by using SWV to achieve the working range of 1.40–2.10 nmol ml^{-1} with an LOD of 10 pmol ml^{-1} [32]. Generally, based on the literature reports, different MIP-based sensors were developed to detect synthetic steroid estrogens at different levels, as presented in table 10.2.

The current literature reports indicate that there is a need to design and develop MIP-based chemical sensors for the emerging contaminants such as DIS, EE3, EPP, CEE, and QE in the future.

Phenolic environmental estrogens (PEEs) are the most common organic pollutants in water and are structurally stable, highly toxic, and have high mobility [48]. Some examples of PPEs are bisphenol A (BPA), bisphenol B (BPB), bisphenol F (BPF), and bisphenol S (BPS), nonylphenol (NP), octylphenol (OP), diethylstilbestrol (DES), hexestrol (HES), DIS, bisphenol S (BPS), and tetrabromobisphenol A (TBBPA). These PEEs are produced worldwide, commonly used, and thus easily exposed to humans. PEEs can disturb human endocrine function and be detrimental to human health. Therefore, recent research reports on PEEs call for highly selective and sensitive analytical methods relying on the applications of MIPs as receptors [49].

BPA is an industrial chemical produced in large quantities for various applications, including polycarbonate plastics, epoxy resins, food packaging plastic materials, printing inks, flame retardants and surface coatings, tableware, and cookware preparation. It is a weak estrogenic chemical but suspected to be EDC with a wide range of adverse health effects. Thus, several regulatory regulations

Table 10.2. Sensing mechanism and detection limit of developed MIP-based sensors for selected synthetic steroid and phenolic environmental estrogens.

EE type	Analyte	Sensing mechanism	LOD	Reference
Synthetic estrogens	DES	Electrochemical	250.00 amol ml^{-1}	[45]
		Electrochemical	2.60 pmol ml^{-1}	[47]
		Optical	47.70 fmol ml^{-1}	[46]
	EE2	Electrochemical	2.60 pmol ml^{-1}	[47]
	EV	Electrochemical	10.00 pmol ml^{-1}	[32]
PEEs	BPA	Electrochemical	8.83 pmol ml^{-1}	[50]
		Electrochemical	20.00 fmol ml^{-1}	[51]
		Electrochemical	8.00 pmol ml^{-1}	[52]
		Electrochemical	80.00 amol ml^{-1}	[53]
		Mass-sensitive	52.00 pmol ml^{-1}	[54]
		Optical	0.70 fmol ml^{-1}	[55]
		Optical	70.00 pmol ml^{-1}	[56]
	4-NP	Electrochemical	1.00 pmol ml^{-1}	[58]
		Optical	36.00 pmol ml^{-1}	[61]
		Optical	76.00 pmol ml^{-1}	[62]
	4-CP	Electrochemical	300.00 pmol ml^{-1}	[57]
	2,4-DCP	Electrochemical	0.50 pmol ml^{-1}	[59]
		Electrochemical	0.80 pmol ml^{-1}	[60]
		Optical	150.00 pmol ml^{-1}	[63]
	2,4,6-TCP	Optical	35.00 pmol ml^{-1}	[64]
		Optical	83.00 pmol ml^{-1}	[65]
	2,4-DNT	Optical	1.54 nmol ml^{-1}	[69]
	2,4-DNP	Electrochemical	0.40 nmol ml^{-1}	[60]
	2,6-DCP	Optical	67.00 fmol ml^{-1}	[70]
	Toluene	Electrochemical	8.67 nmol ml^{-1}	[66]
		Mass-sensitive	2.50 nmol ml^{-1}	[67]

concerning the use of BPA exist nowadays, requiring fast and sensitive analytical methods [50]. A quick, real, and easy-to-use method for the quantitative determination of BPA is important in the environmental monitoring and health perspectives.

Apparently, a high number of articles is reported based on electrochemical methods using MIPs as recognition elements. Among the different formats of voltammetric techniques, DPV is the most studied one in combination with the MIP sensing layer to detect BPA. Different functional monomers were employed for this purpose. For example, MIPs based on methacrylic acid (MAA) combined with magnetic nanoparticles [50], bifunctional monomers of MAA, and vinylpyridine (4-VPy) with multiwalled carbon nanotubes [51], and Py on laser scribed graphene (LSG) [52], have been reported. A novel electrochemical sensor based on combining MIPs with aptamers obtained a very sensitive sensor to detect atto-level BPA.

In this work, gold nanoparticles were electrodeposited on the surface of a GCE and then treated with a mixture of a thiolated DNA sequence (p-63), with high affinity for free BPA. Pyrrole was then electropolymerized on the surface of the GCE to entrap the BPA@p-63 complex to obtain an LR of 2.19 amol ml^{-1} to 21.9 fmol ml^{-1} BPA concentration range and detection limit of 0.080 fmol ml^{-1} using the EIS technique [53]. The proper immobilization of MIPs onto the surface of QCM chips has been explored for many years for the detection of many analytes based on the mass of substance adsorbed on the coating imprinted polymers. Accordingly, a sensitive MIP film from 4-VP on QCM sensors was fabricated to detect BPA in the wide LR to measure the target down to the nano-level [54]. New fluorescent sensors were constructed from MIP-based on the metal–organic framework of UiO-66-NH$_2$ coated with MIPs using (3-aminopropyl) triethoxysilane and tetraethyl orthosilicate and (3-isocyanatopropyl) triethoxysilane based MIP—carbon dots (CDs) were reported to obtain an LOD of 0.0.7 fmol ml^{-1} [55] and 70 pmol ml^{-1} [56]. Another PEE, 4-chlorophenol (4-CP), was used as a template to design and prepare MIP electrochemical sensors to detect the target in tap and lake water samples with an LOD of 0.3 nmol ml^{-1} [57].

Phenolic compounds such as 4-nitrophenol (4-NP), 2,4-dichlorophenol (2,4-DCP), and 2,4,6-trichlorophenol (2,4,6-TCP) are priority pollutants in water according to the United States Environmental Protection Agency (USEPA). Thus, MIP-based electrochemical sensors were designed and developed for these pollutants to obtain an LOD of 1.0 fmol ml^{-1} on indium thin oxide (ITO), 0.5 fmol ml^{-1}, and 0.8 pmol ml^{-1} on GCE, respectively [58–60]. MIP-based optical sensors for 4-NP detection in water with an LOD of 36 pmol ml^{-1} using MIP-based fluorescence sensor [61], an LOD of 76 pmol ml^{-1}, and MIP-based chemiluminescence sensors [62] were reported. Similarly, 2,4-DCP and 2,4,6-TCP detection in water by MIP-based phosphorescence sensors were also described by different groups [63–65].

Toluene MIP-based electrochemical sensor with an LOD of 8.67 nmol ml^{-1} [66] and a mass-sensitive sensor with an LOD of 2.49 nmol ml^{-1} [67] were also reported. For other PEEs, electrochemical sensors were developed by different groups to detect 2,4-dinitrophenol (2,4-DNP) [68] to reach the detection limit of 400 pmol ml^{-1}. Other PEEs such as 2,4-dinitrotoluene (2,4-DNT) and 2,6-dichlorophenol (2,6-DCP) were detected successfully in water-based optical sensors using MIP materials as recognition elements to achieve an LOD of 1.54 nmol ml^{-1} [69] and 67 fmol ml^{-1} [70].

10.5.3 Persistent organic pollutants

Persistent organic pollutants (POPs) are some of the environmental contaminants of global concern because they are toxic even at minute concentrations. The persistence of POPs is due to their resistance to metabolic degradation and their lipophilic character. They are polyaromatic hydrocarbons (PAH), furans, halogenated aromatic hydrocarbons (HAHs) such as polychlorinated biphenyls (PCBs), and dioxins. These chemicals can cause adverse health effects, including mutagenicity, carcinogenicity, reproductive disorders, immune suppression, and birth defects, and are considered EDCs [71]. Generally, POPs are persistent contaminants in the

Table 10.3. Sensing mechanism and detection limit of MIP-based sensors for selected persistent organic compounds.

EEs type	Group	Analyte	Sensing mechanism	LOD	Reference
POP	PCB	OH-PCB	Chemosensor	888.00 fmol ml^{-1}	[85]
	PAH	Anthracene	Electrochemical	12.00 pmol ml^{-1}	[79]
			Electrochemical	1.30 pmol ml^{-1}	[88]
		1-NA	Optical	180.00 fmol ml^{-1}	[88]
		2-NA	Optical	35.00 fmol ml^{-1}	[88]
		Pyrene	Optical	1.00 pmol ml^{-1}	[87]
			Optical	150.00 fmol ml^{-1}	[74]
			Electrochemical	228.00 pmol ml^{-1}	[75]
			QCM	—	[87]
		Benzoapyrene	Optical	40.00 fmol ml^{-1}	[76]
		Fluoranthene	Optical	170.00 fmol ml^{-1}	[77]
		Phenanthrene	Optical	20.00 pmol ml^{-1}	[78]
	Preservative	BHA	Electrochemical	7.90 pmol ml^{-1}	[81]
			Electrochemical	76.30 pmol ml^{-1}	[82]
			Electrochemical	5.50 pmol ml^{-1}	[83]

environment and bind to sediments and soils. Health effects such as birth defects, developmental delays, and liver changes are caused by high exposure to PCBs [72]. Therefore, the determination of POPs is compulsory using more convenient and robust analytical methods, as summarized in table 10.3.

PAHs are a class of chemicals that occur naturally in coal, crude oil, and gasoline. They also are produced when coal, oil, gas, wood, garbage, and tobacco are burned. PAHs generated from these sources can bind to or form small particles in the air. PAH pollutants have been determined to be highly toxic, mutagenic, carcinogenic, teratogenic, and immunotoxicogenic to various life forms [73]. Dickert and his coworkers have described the synthesis of polyurethane-based imprinted polymers for the analysis of PAHs. Fluorescent sensors using MIPs were developed for pyrene detection in water based on the polymerization of p,p'-diisocyanatodiphenylmethane to achieve an LOD of 30 pg ml^{-1} in aqueous media [74]. Specific optical sensors were prepared from imprinted polymer nanofilm based on its design using hydrogen bonding and π–π interactions of the monomer with the pyrene template. The sensor revealed its potential to determine the target pyrene in different environmental matrices [75]. Furthermore, the development of MIP optical sensors was continued to prepare sensor platforms to detect different PAHs such as benzopyrene [76], fluoranthene [77], and phenanthrene [78] with LODs of 40, 170, and 20 fmol ml^{-1}, respectively.

An MIP-based electrochemical sensor for the detection of trace anthracene in water was reported by Mathieu-Scheers and co-authors [79] to obtain nano-level LOD using the SWV technique. Template extraction was performed using CV in ethanol containing 0.1 M LiClO$_4$, which simplified the elution process. This sensor showed better selectivity for anthracene in the presence of three interferent molecules, namely, isoproturon, benzopyrene, and naphthalene.

Dioxins are a group of polychlorinated compounds released to the environment as by-products of many chemical processes involving chlorine. They are a group of chemically related compounds produced as by-products of industrial processes and incineration processes, including improper municipal waste incineration and burning of trash, and can be released into the air during natural processes such as forest fires and volcanoes. They are persistent environmental pollutants, and every living organism is exposed to dioxins or dioxin-like compounds (DLCs). They are highly toxic and can cause reproductive and developmental problems, damage the immune system, interfere with hormones, and also cause cancer. Nowadays, people are exposed to dioxins primarily (>90%) by eating food, in particular animal products contaminated with these chemicals [80] (2019). Dioxins affect women's reproduction and decline sperm counts in men. Xenoestrogen preservatives such as butylated hydroxyanisole (BHA) and butylated hydroxytoluene (BHT) are known examples.

An electrochemical sensor for ultra-sensitive and specific recognition of BHA was developed. The prepared electrochemical sensor showed a wide LR for BHA in the range of 1.00 pmol ml^{-1} to 10.00 μmol ml^{-1} with a detection limit of 7.9 fmol ml^{-1}. More importantly, the sensor had good reproducibility, stability, and excellent selectivity and was successfully used for the detection of BHA in foods, providing a reliable detection method for sensitive, rapid, and selective detection of BHA in complex food samples [81]. In another research report, different authors developed MIPs layers for the electrochemical detection of BHA with interesting detection capabilities [82, 83].

PCBs are known for their carcinogenic, teratogenic, and mutagenic properties. Thus, their determination in the environment, food, and various products has gained wider attention from scientists [84]. Hydroxy polychlorinated biphenyls (OH-PCBs), a persistent PCB, were imprinted using acrylamide monomer to develop a novel sensor using a timer as readout. A linear relationship was observed between the time for water flow through the column and various concentrations of the OH-PCBs in the range of 2.92–292.04 μmol ml^{-1}. The sensor attained a low detection limit of 0.88 μmol ml^{-1}. The proposed technique has been successfully applied for the detection of 4'-OH-PCB 30 in environmental samples. The detection of different targets could be easily achieved by simply changing the MIPs of various imprints [85].

According to the report by Tiu and co-authors, a pyrene-imprinted polythiophene nanofilm is prepared by electrochemical deposition and used for highly sensitive detection of pyrene and its analogs. This QCM sensor was able to detect pyrene in the range of 0.01–1.0 nmol ml^{-1} pyrene concentrations based on molecular interactions such as π–π stacking and hydrogen bonding. Furthermore, the sensor was claimed to be used for the monitoring of pyrene and other PAHs in aquatic, marine, and air samples [86]. The combination of gold nanoparticles with MIPs was enabled to prepare a novel hybrid plasmatic platform SERS spectroscopy recognition of PAHs by employing the MIPs to trap pyrene PAH close to the gold surface to increase the molecule Raman signal. The sensor was applied for detecting pyrene in two real samples, creek water and seawater, to obtain good detection capability within a broad range of fields such as environmental control, food safety, and biomedicine [87].

Based on the non-covalent imprinting approach, a very promising optical sensor was designed and developed for the simultaneous determination of 1-naphthylamine (1-NA) and 2-naphthylamine (2-NA) in water. In this work, the synergic combination of MIPs and a fluorescence detection technique was explored for analyzing 1-NA and 2-NA separately with detection limits of 180 fmol ml^{-1} and 35 fmol ml^{-1}, respectively, and it also determines 1-NA and 2-NA simultaneously with a detection limit of 0.314 pmol ml^{-1} [88]. Such devices are promising because they are simple, cheap, and fast, thus suitable for environmental pollution monitoring. An MIP-based conductometric sensor based on screen-printed interdigital gold electrodes on a glass substrate coated with molecularly imprinted polyurethane layers was fabricated to detect anthracene PAHs in water. The results prove that modified electrodes are very suitable transducers to manufacture low-cost sensor systems for measuring the change in resistance of anthracene-imprinted layers while exposed to different anthracenes. The sensor showed good selectivity to anthracene molecules and high sensitivity, with a detection limit of 1.3 pmol ml^{-1} found in water which is lower than the WHO's permissible limit [89].

10.5.4 Phthalates

Phthalates (PAEs) are a vast class of chemicals used as plasticizers, lubricants, fragrances and personal-care products. Recently, PAEs have become ubiquitous environmental contaminants in the air, water bodies, soil, and biota [90]. Some forms of PAEs are considered to be EDCs since they act against estrogen and androgen receptors, whose activity depends on the length of the phthalate carbon chain. Due to their adverse health effects, most PAEs are now banned from the market. Some examples of PAEs and their derivatives are di(2-ethylhexyl) phthalate (DEHP), di-n-butyl phthalate (DBP), benzylbutyl phthalate (BzBP), and di-isononyl phthalate (DINP), which have been shown to disrupt reproductive tract development in male rodents in an anti-androgenic manner [91]. Therefore, online monitoring of the occurrence of PAEs resulting from its contamination of soil and water is very important. To address this demand, more advanced techniques such as chemical sensors based on MIPs have become a promising technology. There are many reports on the design and development of MIP-based chemical sensors for the detection of PAEs in environmental samples (table 10.4).

To recognize and detect DEHP, the most commonly used PAE in environmental samples, MIPs were used to design and develop an extended gate field-effect transistor (EGFET) sensor to obtain an LOD of 0.128 nmol ml^{-1} and also successfully applied to water samples with good recovery percentages [92]. A very sensitive and selective surface imprinted MIP using ZnO quantum dot-based fluorescence sensors was prepared for the detection of DEHP in water. In this work, MAA, ethylene glycol dimethacrylate, 2,2′-azobis(2-methylpropionitrile), and ZnO were used as the functional monomers, cross-linker, initiator, and optical

Table 10.4. Sensing mechanism and detection limit of developed MIP-based sensors for selected PAEs.

EE type	Analyte	Sensing mechanism	LOD	Reference
PAEs	DEHP	Optical	1.30 pmol ml^{-1}	[93]
		Optical	23.00 fmol ml^{-1}	[97]
		Electrochemical	128.00 pmol ml^{-1}	[92]
	DINP	Electrochemical	27.00 pmol ml^{-1}	[99]
	DMP	Optical	100.00 fmol ml^{-1}	[96]
		Optical	2.70 fmol ml^{-1}	[97]
	DEP	Electrochemical	110.00 amol ml^{-1}	[98]
	DBP	Optical	40.00 pmol ml^{-1}	[95]
		Electrochemical	4.50 pmol ml^{-1}	[64]

material, respectively. This is best performed in the concentration range of 0.5–40 nmol ml^{-1} and had an LOD of 1.3 pmol ml^{-1} [93].

Bolat and coworkers introduced nanomaterials into MIPs to enhance the signal-to-noise ratio of the electrochemical sensors to detect DBP [94]. The electrochemical sensor showed high sensitivity and selectivity for DBP detection with a nano-level detection limit. In another research work, a label-free MIP-based optical sensor was fabricated using quantum dots [95] to enhance the device's overall performance. The reported sensor response indicates that the prepared materials are promising to fabricate fluorescent MIPs with highly selective recognition ability for the DBP template. Similarly, Yang and his group developed different types of optical sensors to determine dimethyl phthalate (DMP) with an LOD of 0.10 pmol ml^{-1} [96]. A very promising MIP-based SERS sensor was developed to analyze charged phthalates such as dimethyl phthalate (DMP) and di(2-ethylhexyl) phthalate (DEEHP) with detection limits of 2.7 fmol ml^{-1} and 23 fmol ml^{-1}, respectively [97].

Diethyl phthalate (DEP) was detected in Chinese liquor using MIP-based EIS sensors based on the π–π stacking interactions between phthalates and graphene working electrode, which allows direct sampling and analyte preconcentration at 1.08 fmol ml^{-1} with an LR of 9.89 pmol ml^{-1} to 4.99 pmol ml^{-1} [98]. The other type of phthalate, known as di-isononyl phthalate (DINP), was employed as a template to design a sensor based on a non-covalent approach to achieve a detection limit of 27 pM in the range of 50–1000 pmol ml^{-1} [99].

10.6 Conclusion and future perspectives

The MIP-based chemical sensors are versatile for detecting EEs in environmental samples. This chapter described some prominent trends in developing sensors using MIPs as a sensing element. Different EEs considered as EDCs were the most focused molecules. Other formats of MIP sensors, such as electrochemical, optical, and mass-sensitive sensors, are briefly presented with some proof of concepts as the current state-of-the-art. The introduction of various nanomaterials into MIPs has been confirmed to play the role of increasing the surface of the electrode and enhancing the overall performance of the sensor, particularly the sensitivity and selectivity. Many types of nanomaterial-based MIPs sensors were presented as a confirmation of this issue.

For accurate time analysis, portable MIP-based electrochemical and other formats could be proposed for the detection of multiple EEs in the environment and could be commercialized for testing the content of EDCs in many types of environmental samples and consumer products such as drinking water, foodstuffs, etc.

Acknowledgments

This work was supported by the Department of Science and Innovation (DSI) and National Research Foundation (NRF), South Africa, through DSI/NRF South African Research Chairs Initiative for Professor of Medicinal Chemistry and Nanotechnology (UID 62620), DSI/Mintek Nanotechnology Innovation Centre, and Rhodes University, South Africa.

References

[1] Tapiero H, Ba G N and Tew K D 2002 Estrogens and environmental estrogens *Biomed. Pharmacother.* **56** 36–44

[2] Jaffrezic-Renault N, Kou J, Tan D and Guo Z 2020 New trends in the electrochemical detection of endocrine disruptors in complex media *Anal. Bioanal. Chem.* **412** 5913

[3] Paterni I, Granchi C and Minutolo F 2017 Risks and benefits related to alimentary exposure to xenoestrogens *Crit. Rev. Food Sci. Nutr.* **57** 3384–404

[4] Adeel M, Song X, Wang Y, Francis D and Yang Y 2017 Environmental impact of estrogens on human, animal and plant life: a critical review *Environ. Int.* **99** 107–19

[5] Musa A M, Kiely J, Luxton R and Honeychurch K C 2021 Recent progress in screen-printed electrochemical sensors and biosensors for the detection of estrogens *TrAC Trends Anal. Chem.* **139** 116254

[6] Wang D, Xu Z, Liu Y, Liu Y, Li G, Si X, Lin T, Liu H and Liu Z 2020 Molecularly imprinted polymer-based fiber array extraction of eight estrogens from environmental water samples prior to high-performance liquid chromatography analysis *Microchem. J.* **159** 105376

[7] Fonseca A P, Lima D L D and Esteves V I 2011 Degradation by solar radiation of estrogenic hormones monitored by UV–visible spectroscopy and capillary electrophoresis *Water Air Soil Pollut.* **215** 441–7

[8] Gavrila A M, Stoica E B, Iordache T V and Sârbu A 2022 Modern and dedicated methods for producing molecularly imprinted polymer layers in sensing applications *Appl. Sci.* **12** 3080

[9] Kadhem A J, Gentile G J and De Cortalezzi M M F 2021 Molecularly imprinted polymers (MIPs) in sensors for environmental and biomedical applications: a review *Molecules* **26** 6233

[10] Zarejousheghani M, Rahimi P, Borsdorf H, Zimmermann S and Joseph Y 2021 Molecularly imprinted polymer-based sensors for priority pollutants *Sensors* **21** 2406

[11] Cala B F, Batista A D and Cárdenas S 2020 Molecularly imprinted polymer micro- and nano-particles: a review *Molecules* **25** 4740

[12] Patel K D, Kim H W, Knowles J C and Poma A 2020 Molecularly imprinted polymers and electrospinning: manufacturing convergence for next-level applications *Adv. Funct. Mater.* **30** 2001955

[13] Gui R J, Jin H, Guo H J and Wang Z H 2018 Recent advances and future prospects in molecularly imprinted polymers-based electrochemical biosensors *Biosens. Bioelectron.* **100** 56–70

[14] Wu Y, Lu J, Xing W, Ma F, Gao J, Lin X, Yu C and Yan M 2020 Double-layer-based molecularly imprinted membranes for template-dependent recognition and separation: an imitated core-shell-based synergistic integration design *Chem. Eng. J.* **397** 125371

[15] Ye T, Liu A N, Bai L, Yuan M, Cao H, Yu J, Yuan R, Xu X, Yuan H and Xu F 2020 *Microchim. Acta* **187** 412

[16] Liu Z, Zhang X, Cui L, Wang K and Zhan H 2017 Development of a highly sensitive electrochemiluminescence sophoridine sensor using $Ru(bpy)_3^{2+}$ integrated carbon quantum dots—polyvinyl alcohol composite film *Sensors Actuators* B **248** 402–10

[17] Huang K, Wang X, Zhang H, Zeng L, Zhang X, Wang B, Zhou Y and Jing T 2020 Structure-directed screening and analysis of thyroid-disrupting chemicals targeting trans-thyretin based on molecular recognition and chromatographic separation *Environ. Sci. Technol.* **54** 5437–45

[18] Sullivan M V, Dennison S R, Archontis G, Reddy S M and Hayes J M 2019 Toward rational design of selective molecularly imprinted polymers (MIPs) for proteins: computational and experimental studies of acrylamide based polymers for myoglobin *J. Phys. Chem.* B **123** 5432–43

[19] Bai J, Zhang Y, Chen L, Yan H, Zhang C, Liu L and Xu X 2018 Synthesis and characterization of paclitaxel-imprinted microparticles for controlled release of an anticancer drug *Mater. Sci. Eng.* C **92** 338–48

[20] Mostafiz B, Bigdeli S A, Banan K, Afsharara H, Hatamabadi D, Mousavi P, Hussain C M, Keçili R and Bidkorbeh F G 2021 Molecularly imprinted polymer-carbon paste electrode (MIP-CPE)-based sensors for the sensitive detection of organic and inorganic environmental pollutants: a review *Trends Environ. Anal. Chem.* **32** e00144

[21] Hulanicki A, Glab S and Ingman F 1991 Chemical sensors: definitions and classification *Pure Appl. Chem.* **63** 1247–50

[22] Panasyuk T L, Mirsky V M, Piletsky S A and Wolfbeis O S 1999 Electropolymerized molecularly imprinted polymers as receptor layers in capacitive chemical sensors *Anal. Chem.* **71** 4609–13

[23] Uygun Z O, Uygun H D E, Ermiş N and Canbay E 2015 Molecularly imprinted sensors-new sensing technologies *Biosensors: Micro and Nanoscale Applications* (London: InTech) ch 3 pp 86–108

[24] Regasa M B, Soreta T R, Femi O E, Ramamurthy P C and Subbiahraj S 2020 Molecularly imprinted polyaniline molecular receptor–based chemical sensor for the electrochemical determination of melamine *J. Mol. Recognit.* **33** 1–11

[25] Riskin M, Tel-Vered R, Bourenko T, Granot E and Willner I 2008 Imprinting of molecular recognition sites through electropolymerization of functionalized Au nanoparticles: development of an electrochemical TNT sensor based on π-donor−acceptor interactions *J. Am. Chem. Soc.* **130** 9726–33

[26] Elfadil D, Lamaoui A, Pelle D F, Amine A and Compagnone D 2021 Molecularly imprinted polymers combined with electrochemical sensors for food contaminants analysis *Molecules* **26** 4607

[27] Sergeyeva T *et al* 2020 Sensor based on molecularly imprinted polymer membranes and smartphone for detection of fusarium contamination in cereals *Sensors* **20** 4304

[28] Fang L, Jia M, Zhao H, Kang L, Shi L, Zhou L and Kong W 2021 Molecularly imprinted polymer-based optical sensors for pesticides in foods: recent advances and future trends *Trends Food Sci. Technol.* **116** 387–404

[29] Latif U, Qian J, Can S and Dickert F L 2014 Biomimetic receptors for bioanalyte detection by quartz crystal microbalances—from molecules to cells *Sensors* **14** 23419–38

[30] Dilien H *et al* 2017 Label-free detection of small organic molecules by molecularly imprinted polymer functionalized thermocouples: toward *in vivo* applications *ACS Sens.* **2** 583–9

[31] Sornalingam K, McDonagh A and Zhou J L 2016 Photodegradation of estrogenic endocrine disrupting steroidal hormones in aqueous systems: progress and future challenges *Sci. Total Environ.* **550** 209–24

[32] Bergamin B, Pupin R R, Wong A and Sotomayor M D P T 2019 *J. Braz. Chem. Soc.* **30** 2344–54

[33] Duan D, Si X, Ding Y, Li L, Ma G, Zhang L and Jian B 2019 A novel molecularly imprinted electrochemical sensor based on double sensitization by MOF/CNTs and Prussian blue for detection of 17β-estradiol *Bioelectrochemistry* **129** 211–7

[34] Braga G B, Oliveira A E F and Pereira A C 2018 Total determination of estrogenic phenolic compounds in river water using a sensor based on reduced graphene oxide and molecularly imprinted polymer *Electroanalysis* **30** 2176–84

[35] Nodehi M, Baghayeri M, Ansari R and Veisi H 2020 Electrochemical quantification of 17α–ethinylestradiol in biological samples using a $Au/Fe_3O^4@TA/MWNT/GCE$ sensor *Mater. Chem. Phys.* **244** 122687

[36] Xiao L, Zhang Z, Wu C, Han L and Zhang H 2017 Molecularly imprinted polymer grafted paper-based method for the detection of 17β-estradiol *Food Chem.* **221** 82–6

[37] Song H, Wang Y, Zhang L, Tian L, Luo J and Zhao N 2017 An ultrasensitive and selective electrochemical sensor for determination of estrone 3-sulfate sodium salt based on molecularly imprinted polymer modified carbon paste electrode *Anal. Bioanal. Chem.* **409** 6509

[38] Choi S W, Chang H J, Lee N, Kim J H and Chun H S 2009 Detection of mycoestrogen zearalenone by a molecularly imprinted polypyrrole-based surface plasmon resonance (SPR) sensor *J. Agric. Food Chem.* **57** 1113–8

[39] Hu X, Wang C, Zhang M, Zhao F and Zeng B 2020 *Talanta* **217** 1210322

[40] Radi A, Eissa A and Wahdan T 2020 Molecularly imprinted impedimetric sensor for determination of mycotoxin zearalenone *Electroanalysis* **32** 1788–94

[41] Yu J C C and Lai E P C 2005 Interaction of ochratoxin A with molecularly imprinted polypyrrole film on surface plasmon resonance sensor *React. Funct. Polym.* **63** 171–6
Yu J C C and Lai E P C 2010 Molecularly imprinted polymers for ochratoxin A extraction and analysis *Toxins* **2** 1536–53

[42] Pacheco J G, Castro M, Machado S, Barroso M F, Nouws H P A and Delerue-Matos C 2015 Molecularly imprinted electrochemical sensor for ochratoxin A detection in food samples *Sensors Actuators* B **215** 107–12

[43] Akgonullu S, Armutcu C and Denizili A 2021 Molecularly imprinted polymer film based plasmonic sensors for detection of ochratoxin A in dried fig *Polym. Bull.* **79** 4049–67

[44] Rodriguez-Mozaz S, Marco M P, Alda M J L and Barceló D 2004 Biosensors for environmental monitoring of endocrine disruptors: a review article *Anal. Bioanal. Chem.* **378** 588–98

[45] Zhao W R, Kang T F, Xu Y H, Zhang X, Liu H, Ming A J, Lu L P, Cheng S Y and Wei F 2020 Electrochemiluminescence solid-state imprinted sensor based on graphene/CdTe@ZnS quantum dots as luminescent probes for low-cost ultrasensing of diethylstilbestrol *Sensors Actuators* B **306** 127563

[46] Wang Y, Ren S, Jiang H, Peng Y, Bai J, Li Q, Li C, Gao Z and Ning B 2017 A label-free detection of diethylstilbestrol based on molecularly imprinted polymer-coated upconversion nanoparticles obtained by surface grafting *RSC Adv.* **7** 22215–21

[47] Santos A M, Wong A, Prado T M, Fava E L, Filho O F, Sotomayor M D P T and Moraes F C 2021 Voltammetric determination of ethinylestradiol using screen-printed electrode modified with functionalized graphene, graphene quantum dots and magnetic nanoparticles coated with molecularly imprinted polymers *Talanta* **224** 121804

[48] Raza W, Lee J, Raza N, Luo Y, Kim K H and Yang J 2019 Removal of phenolic compounds from industrial waste water based on membrane-based technologies *J. Ind. Eng. Chem.* **71** 1–18

[49] Xie X, Bu Y and Wang S 2016 Molecularly imprinting: a tool of modern chemistry for analysis and monitoring of phenolic environmental estrogens *Rev. Anal. Chem.* **35** 87–97

[50] Messaoud N B, Lahcen A A, Dridi C and Amine A 2018 Ultrasound assisted magnetic imprinted polymer combined sensor based on carbon black and gold nanoparticles for selective and sensitive electrochemical detection of bisphenol A *Sensors Actuators* B **276** 304–12

[51] Anirudhan T S, Athira V S and Sekhar V C 2018 Electrochemical sensing and nano molar level detection of bisphenol-A with molecularly imprinted polymer tailored on multiwalled carbon nanotubes *Polymers* **146** 312–20

[52] Beduk T, Lachen A A, Tashkandi N and Salama K N 2020 One-step electrosynthesized molecularly imprinted polymer on laser scribed graphene bisphenol A sensor *Sensors Actuators* B **314** 128026

[53] Ensafi A A, Amini M and Rezaei B 2018 Molecularly imprinted electrochemical aptasensor for the attomolar detection of bisphenol A *Microchim. Acta* **185** 265

[54] Oh D K, Yang J C, Hong S W and Park J 2020 Molecular imprinting of polymer films on 2D silica inverse opal via thermal graft copolymerization for bisphenol A detection *Sensors Actuators* B **323** 128670

[55] Zeng L *et al* 2021 Simultaneous fluorescence determination of bisphenol A and its halogenated analogs based on a molecularly imprinted paper-based analytical device and a segment detection strategy *Biosens. Bioelectron.* **180** 113106

[56] Zhang J, Wang H, Xu L and Xu Z 2021 A semi-covalent molecularly imprinted fluorescent sensor for highly specific recognition and optosensing of bisphenol A *Anal. Methods* **13** 133–40

[57] Wang B, Okoth O K, Yan K and Zhang J 2016 A highly selective electrochemical sensor for 4-chlorophenol determination based on molecularly imprinted polymer and PDDA-functionalized graphene *Sensors Actuators* B **236** 294–303

[58] Hu Y F, Zhang Z H, Zhang H B, Luo L J and Yao S Z 2012 Sensitive and selective imprinted electrochemical sensor for p-nitrophenol based on ZnO nanoparticles/carbon nanotubes doped chitosan film *Thin Solid Films* **520** 5314–21

[59] Liang Y, Yu L, Yang R, Li X, Qu L and Li J 2017 High sensitive and selective graphene oxide/molecularly imprinted polymer electrochemical sensor for 2,4-dichlorophenol in water *Sensors Actuators* B **240** 1330–5

[60] Liu Y, Liang Y, Yang R, Li J and Qu L 2019 A highly sensitive and selective electrochemical sensor based on polydopamine functionalized graphene and molecularly imprinted polymer for the 2,4-dichlorophenol recognition and detection *Talanta* **195** 691–8

[61] Wei X, Zhou Z, Hao T, Li H, Xu Y, Lu K, Wu Y, Dai J, Pan J and Yan Y 2015 Highly-controllable imprinted polymer nanoshell at the surface of silica nanoparticles based room-temperature phosphorescence probe for detection of 2,4-dichlorophenol *Anal. Chim. Acta* **870** 83–91

[62] Liu J, Chen H, Lin Z and Lin J M 2010 Preparation of surface imprinting polymer capped Mn-doped ZnS quantum dots and their application for chemiluminescence detection of 4-nitrophenol in tap water *Anal. Chem.* **82** 7380–6

[63] Wei X, Yu M, Li C, Gong X, Qin F and Wang Z 2018 Metastable α-AgVO$_3$ microrods as peroxidase mimetics for colorimetric determination of H$_2$O$_2$ *Microchim. Acta* **185** 1–6

[64] Wei X, Zhou Z, Hao T, Li H, Zhu Y, Gao L and Yan Y 2015 A novel molecularly imprinted polymer thin film at surface of ZnO nanorods for selective fluorescence detection of *para*-nitrophenol *RSC Adv.* **5** 44088–95

[65] Lin X, Wu Y, Hao Y, Sun Q, Yan Y and Li C 2018 Sensitive and selective determination of 2,4,6-trichlorophenol using a molecularly imprinted polymer based on zinc oxide quantum dots *Anal. Lett.* **51** 1578–91

[66] Alizadeh T and Rezaloo F 2013 Toluene chemiresistor sensor based on nano-porous toluene-imprinted polymer *Int. J. Environ. Anal. Chem.* **93** 919–34

[67] Iglesias R A, Tsow F, Wang R, Forzani E S and Tao N 2009 Hybrid separation and detection device for analysis of benzene, toluene, ethylbenzene, and xylenes in complex samples *Anal. Chem.* **81** 8930–5

[68] Liu Y, Zhu L, Zhang Y and Tang H 2012 Electrochemical sensing of 2,4-dinitrophenol by using composites of graphene oxide with surface molecular imprinted polymer *Sensors Actuators* B **171–2** 1151–8

[69] Dai J, Dong X and de Cortalezzi M F 2017 Molecularly imprinted polymers labeled with amino-functionalized carbon dots for fluorescent determination of 2,4-dinitrotoluene *Microchim. Acta* **184** 1369–77

[70] Li H, Wang Y, Li Y, Qiao Y, Liu L, Wang Q and Che G 2019 High-sensitive molecularly imprinted sensor with multilayer nanocomposite for 2,6-dichlorophenol detection based on surface-enhanced Raman scattering *Spectrochim. Acta* A **228** 117784

[71] Metcalfe C D, Bayen S, Desrosiers M, Muñoz G, Sauvé S and Yargeau V 2022 An introduction to the sources, fate, occurrence and effects of endocrine disrupting chemicals released into the environment *Environ. Res.* **207** 112658

[72] Balmer B *et al* 2019 Comparison of persistent organic pollutants (POPs) between small cetaceans in coastal and estuarine waters of the northern Gulf of Mexico *Mar. Pollut. Bull.* **145** 239–47

[73] Patel A B, Shaikh S, Jain K R, Desai C and Madamwar D 2020 Polycyclic aromatic hydrocarbons: sources, toxicity, and remediation approaches *Front. Microbiol.* **11** 562813

[74] Dickert F L, Besenbock H and Tortschanoff M 1998 Molecular imprinting through van der Waals interactions: fluorescence detection of PAHs in water *Adv. Mater.* **10** 149–51

[75] Ngwanya O W, Ward M and Baker P G L 2021 Molecularly imprinted polypyrrole sensors for the detection of pyrene in aqueous solutions *Electrocatalysis* **12** 165–75

[76] Traviesa-Alvarez J M, Sánchez-Barragán I, Costa-Fernández J M, Pereiro R and Sanz-Medel A 2007 Room temperature phosphorescence optosensing of benzo[*a*]pyrene in water using halogenated molecularly imprinted polymers *Analyst* **132** 218–23

[77] Sánchez-Barragán I, Costa-Fernández J M, Pereiro R, Sanz-Medel A, Salinas A, Segura A, Fernández-Gutiérrez A, Ballesteros A and González J M 2005 *Anal. Chem.* **77** 7005–11

[78] Li H and Wang L 2013 Highly selective detection of polycyclic aromatic hydrocarbons using multifunctional magnetic–luminescent molecularly imprinted polymers *ACS Appl. Mater. Inter.* **5** 10502–9

[79] Mathieu-Scheers E, Bouden S, Grillot C, Nicolle J, Warmont F, Bertagna V, Cagnon B and Vautrin-Ul C 2019 Trace anthracene electrochemical detection based on electropolymerized-molecularly imprinted polypyrrole modified glassy carbon electrode *J. Electroanal. Chem.* **848** 113253

[80] Tuomisto J 2019 Dioxins and dioxin-like compounds: toxicity in humans and animals, sources, and behaviour in the environment *WikiJ. Med* **6** 8

[81] Han S, Ding Y, Teng F, Yao A and Leng Q 2022 Molecularly imprinted electrochemical sensor based on 3D-flower-like MoS_2 decorated with silver nanoparticles for highly selective detection of butylated hydroxyanisole *Food Chem.* **387** 132899

[82] Cui M, Liu S, Lian W, Li J, Xu W and Huang J 2103 A molecularly-imprinted electrochemical sensor based on a graphene–Prussian blue composite-modified glassy carbon electrode for the detection of butylated hydroxyanisole in foodstuffs *Analyst* **138** 5949–55

[83] Motia S, Bouchikhi B and El Bari N 2021 An electrochemical molecularly imprinted sensor based on chitosan capped with gold nanoparticles and its application for highly sensitive butylated hydroxyanisole analysis in foodstuff products *Talanta* **223** 121689

[84] Cheng J, Wang P and Su X O 2020 Surface-enhanced Raman spectroscopy for polychlorinated biphenyl detection: recent developments and future prospects *TrAC Trends Anal. Chem.* **125** 115836

[85] Lin C, Huang Q, Lu Y, Li Z, Wang P, Qiu B and Lin Z 2022 Detection of hydroxypoly-chlorinated biphenyls using molecularly imprinted polymers as recognition unit and timer as readout *Microchem. J.* **174** 107094

[86] Tiu B D B, Krupadam R J and Advincula R C 2016 Pyrene-imprinted polythiophene sensors for detection of polycyclic aromatic hydrocarbons *Sensors Actuators* B **228** 693–701

[87] Castro-Grijalba A, Montes-García V, Cordero-Ferradas M J, Coronado E, Perez-Juste J and Pastoriza-Santos I 2020 SERS-based molecularly imprinted plasmonic sensor for highly sensitive PAH detection *ACS Sens.* **5** 693–702

[88] Navarro A V, Castillo A S, Sánchez J F F, Carretero A S, Mallavia R and Gutiérrez A F 2009 The development of a MIP-optosensor for the detection of monoamine naphthalenes in drinking water *Biosens. Bioelectron.* **24** 2305–11

[89] Latif U, Ping L and Dickert F L 2018 Conductometric sensor for PAH detection with molecularly imprinted polymer as recognition layer *Sensors* **18** 767

[90] Benjamin S, Pradeep S, Josh M S, Kumar S and Masai E 2015 A monograph on the remediation of hazardous phthalates *J. Hazard. Mater.* **298** 58–72

[91] Luís C, Algarra M, Câmara J S and Perestrelo R 2021 Comprehensive insight from phthalates occurrence: from health outcomes to emerging analytical approaches *Toxics* **9** 157

[92] Venkatesh S, Yeung C, Sun Q, Zhuang J, Li T, Li R K Y and Roy V A L 2018 Selective and sensitive onsite detection of phthalates in common solvents *Sensors Actuators* B **259** 650–7

[93] Wang Y, Zhou Z, Xu W, Luan Y, Lu Y, Yang Y, Liu T, Li S and Yang W 2018 Surface molecularly imprinted polymers based ZnO quantum dots as fluorescence sensors for detection of diethylhexyl phthalate with high sensitivity and selectivity *Polym. Int.* **67** 1003–10

[94] Bolat G, Yaman Y T and Abaci S 2019 Molecularly imprinted electrochemical impedance sensor for sensitive dibutyl phthalate (DBP) determination *Sensors Actuators* B **299** 127000

[95] Zhou Z, Li T, Xu W, Huang W, Wang N and Yang W 2017 Synthesis and characterization of fluorescence molecularly imprinted polymers as sensor for highly sensitive detection of dibutyl phthalate from tap water samples *Sensors Actuators* B **240** 1114–22

[96] Yang Y Y, Li Y T, Li X J, Zhang L, Fodjo E K and Han S 2020 Controllable *in situ* fabrication of portable AuNP/mussel-inspired polydopamine molecularly imprinted SERS substrate for selective enrichment and recognition of phthalate plasticizers *Chem. Eng. J.* **402** 125179

[97] Yang Y Y, Li Y, Zhai W, Li X, Li D, Lin H and Han S 2021 Electrokinetic preseparation and molecularly imprinted trapping for highly selective SERS detection of charged phthalate plasticizers *Anal. Chem.* **93** 946–55

[98] Jiang X, Xie Y, Wan D, Zheng F and Wang J 2020 Enrichment-free rapid detection of phthalates in Chinese liquor with electrochemical impedance spectroscopy *Sensors* **20** 901

[99] Zhao W R, Kang T F, Lu L P and Cheng S Y 2018 Magnetic surface molecularly imprinted poly(3-aminophenylboronic acid) for selective capture and determination of diethylstilbestrol *RSC Adv.* **8** 13129–41

IOP Publishing

Molecularly Imprinted Polymers for Environmental Monitoring
Fundamentals and applications
Raju Khan and Ayushi Singhal

Chapter 11

Molecularly imprinted polymer based detection of pesticides

Priya Chauhan and Annu Pandey

Pesticide residues in foods and other relevant matrices pose a serious hazard to public health around the world owing to misuse and abuse. It has been discovered that controlling food safety depends greatly on very sensitive and selective pesticide detection. One of the most popular methods for locating pesticide residues in food and environmental samples is the use of molecularly imprinted polymers (MIPs). The development of MIPs—synthetic materials that mimic biological receptors—involves the polymerisation of useful monomers in the presence of a target analyte. This chapter highlights the most recent developments in MIP-based electrochemical sensors for pesticide analysis and discusses them critically. The discussion of recent developments for several MIP-based electrochemical sensors for pesticide analysis, chemical synthesis of MIP-based electrochemical sensing, along with the pertinent applications of MIPs, is then coupled to electrochemical analysis after a brief introduction to MIPs and electrochemical sensors has been given.

11.1 Introduction

Pesticides are a broad category of chemicals that are used to protect food, agricultural products, and animal feedstuffs against weeds, pests, illnesses, and insects. Typically, they are divided into insecticides, fungicides, herbicides, rodenticides, molluscicides, nematicides, and defoliants based on the intended target organism [1]. However, a wide variety of commercially available products contain more than 1000 active ingredients that can be applied to the control of pests and disease vectors. With rising use every year, pesticides are considered to have become indispensable in a number of industries, including agriculture, animal husbandry, aquaculture, forestry, and more.

However, because pesticides can be found as residue in a variety of matrices, including water, soil, vegetables, fruits, crops, and even animal-derived products, urine, or blood from humans through the food chain, they have also been widely used for a long time and in excess compared to the recommended doses, which has led to serious social concerns about food safety and the environment [2]. They frequently exhibit severe epithelial, neuro, immune, hepatic, and reproductive toxicity as soon as they enter the bodies of humans or animals, even at low doses, in addition to carcinogenic, mutagenic, and teratogenic effects [3]. More specifically, a small number of lipophilic herbicides can be passed on to nursing newborns, causing risks to the infant's health [4].

As a result, pesticide pollution is currently one of the most concerning global issues. Maximum residue limits (MRLs) in various matrices have been defined by a number of nations and organisations to guarantee the proper application and management of pesticides. Therefore, the creation of consistent methods for determining various pesticides in different matrices remains of great importance in order to meet such strict requirements. Several types of pesticides are in use so as to control different types of pests, such as insecticide, herbicide, fungicide, bactericide, as well as rodenticide. The formulation of pesticides includes complex mixtures of the active ingredient and a huge variety of further substances which are added so as to enhance the efficacy of the product. Pesticides are divided into four classes on the basis of their chemical nature: organophosphorus, organochlorine, carbamates, and pyrethroids. Organophosphorus pesticides and carbamates are extensively utilised due to their high insecticidal activity. Contamination caused by pesticides present significant threats to the environment as well as nontargeted organisms, while in humans it can cause cancer, foetal deformity, sterility, acute intoxications, allergies, and even may cause death [5, 6].

Pesticides are used all around the world to eradicate or control weeds, insects, fungi, and other agricultural pests. Their widespread use has left residues in the environment that are inherently dangerous to human and mammal health. Currently, organophosphorus and organonitrogen pesticides have been extensively substituted by organochloride compounds due to their low cost, comparably quick biodegradation, as well as low environmental toxicity [7]. Currently, food safety as well as quality are known to be prominent aspects in promoting a healthy life [8–10]. The presence of pesticides residues may cause lethal health effects, disturb the ecological balance, and cause pest resistance as well as economic losses [11–14]. Since foreign and domestic experts have lowered the MRLs, detection of pesticides at trace levels now requires highly effective pre-treatment extraction techniques [15]. The most common methods used to detect pesticides are gas chromatography (GC) and GC with a flame photometric detector (GC–FPD), electron capture detector (GC–ECD), mass spectrophotometer (GC–MS), GC–MS/MS, high performance liquid chromatography (HPLC), HPLC-florescent detector (FD), liquid chromatography with fluorescent detector (LC–FLD), and LC–MS/MS [16]. However, due to the complex sample matrix, high cost, labour-intensive sample preparation process, and excessive solvent use, such instruments do not have high sensitivity, high selectivity, or good results. Only a small number of sophisticated instruments, such as ultra-high performance liquid chromatography (UHPLC) and UPLC–tandem mass spectrometry (UPLC–MS/MS), have good sensitivity; however, such instruments are quite expensive or still require difficult sample pre-treatment techniques [17]. Almost 800 pesticides have been known to be currently in use [18–20].

In this context, molecularly imprinted polymers (MIPs) have generated a great deal of interest. MIPs are typically created by co-polymerising and crosslinking functional monomers in the presence of a template. The latter interacts with functional groups to create covalent or noncovalent bonds [21], depending on the context. In order to create sensors with improved performance, such as high specificity, cheap cost, ease of use, and rapid detection, MIPs are known to be ideal coatings for quartz crystal microbalance (QCM) [22]. Modern developments in the field of MIPs have created synthetic materials that resemble biological receptors in function but have fewer stability limitations. Only limited imprinted polymer research has been done up to this point for direct recognition and sensing of non-hydrolysed organophosphates. MIPs, often referred to as artificial recognition materials, are complementary to template molecules in both size and shape and can therefore be used for the specific goal of molecule identification. In comparison to conventional enzymes, antibodies, and receptors, MIPs, novel materials and bionic molecular recognition elements, have unique benefits such as remarkable stability, high selectivity, simplicity of manufacture, stability in organic solvents, and wide compatibility with a variety of scientific domains [23]. Additionally, MIPs offer the benefits of low cost, heat and pressure resistance, storage stability, and suitability for harsh chemical media, in addition to their excellent identification properties. Electrochemical detection techniques offer substantial advantages for pesticide residue detection in food samples over conventional detection methods since they are easy to use, inexpensive, and have a short response time. Molecular imprinting technology (MIT) is a technique for creating artificial MIPs. However, these obtained MIPs have specially designed recognition sites that are complementary to the functional group of the particular chemical or compound known as the template, or the target analyte, in terms of both shape and size. It can only attach to a specific analyte among all the chemical molecules that are closely related to it in the harsh medium [20]. The co-polymerisation process between the template and functional monomer, which is fixed by a crosslinking agent while the template is removed, is recognised to be the typical method for acquiring MIPs [8]. They are excellent sorbents for the detection of pesticide residues with a variety of applications in the separation of analytes in complex matrices coupled with chemical sensors, and they have been used either directly or indirectly with chromatographic procedures [17]. However, due to their high matrix content, real samples containing fruits, vegetables, and biological materials can be challenging to analyse. Biological samples are complex with high matrix interference as they consist mostly of fats and proteins that can often muddle pesticide residue detection. MIPs have been shown to be excellent sorbents for detecting pesticides in such complex real samples [5]. In order to create highly sensitive and selective detection procedures, researchers have proposed combining MIPs with traditional detection methods. Analytical chemistry, biology, and polymeric materials have all used MIPs to selectively attach to templated molecules in manufacturing processes by a 'lock-and-key' mechanism [14].

When molecular imprinting technology was first revealed in 1977, it attracted a lot of interest since it uses a specific target molecule as a template to attach to MIPs in their monomeric form. After the template molecule has been removed and crosslinked, MIPs have specific recognition sites that are identical to the template molecule in terms of size, shape, and functional group [3]. Because of their high selectivity towards the target analyte and their advantages over biological receptors, such as easy and inexpensive preparation, physical and chemical robustness under adverse conditions such as organic solvents, extreme pH values, high temperatures, or high pressures, reusability, stability, and the potential for large-scale production, MIPs have drawn significant attention as recognition elements for the development of sensors. By involving electroactive monomers that can be polymerised by applying an appropriate voltage or performing potential scans in a specific range, electropolymerisation has made it possible to use MIPs in electrochemical sensing. In addition to electropolymerisation, it is also possible to investigate MIP modified electrodes via conventional methods including bulk polymerisation [10]. Recently, screen-printed disposable electrodes have also been combined with MIP technology. Theoretically, MIPs could be made for any target molecule. Molecular imprinting technology, which includes the synthesis of MIPs, is a well-known method for detecting pesticide residues. However, specific template-oriented recognition sites that are created by polymerisation are present in the synthesised imprinted polymers. MIPs have a few restrictions, such as limited application, applicability, and difficult synthesis, which impair their performance as well as efficacy. Due to their extremely high specificity for template molecules, MIPs have a wide range of uses, including purification, separation, isolation, catalysis, and sensing [24]. However, the sensors developed using this method have a strong affinity for template molecules. Additionally, the polymer templates used in MIPs must be simple to synthesise and have improved electrochemical properties in order to be used in the fabrication of electrochemical sensors. Due to their excellent stability, excellent electrochemical characteristics, speed, sensitivity, and relative simplicity of synthesis, conducting polymers have proved particularly advantageous in terms of their application to the field of biosensors. Because of their widespread conjugation and durability across a wide temperature range, polyaniline nanofibers (PANInfs) stand out among the others in that they have excellent electrochemical characteristics and improved electrical conduction [25].

The current chapter discusses the various pesticides that can be detected using MIPs, as well as how to evaluate the application of MIPs to various pesticide groups with a particular emphasis on how they are applied to real samples and how effective various suggested methods are for analysis.

11.2 General concepts of MIPs

MIPs, which have been referred to as 'artificial antibodies' because of their unique characteristics, have captured the interest of researchers in the field of food safety. The typical process for creating an MIP-based electrochemical sensor, which includes both chemical and electrochemical synthesis, is shown in figure 11.1.

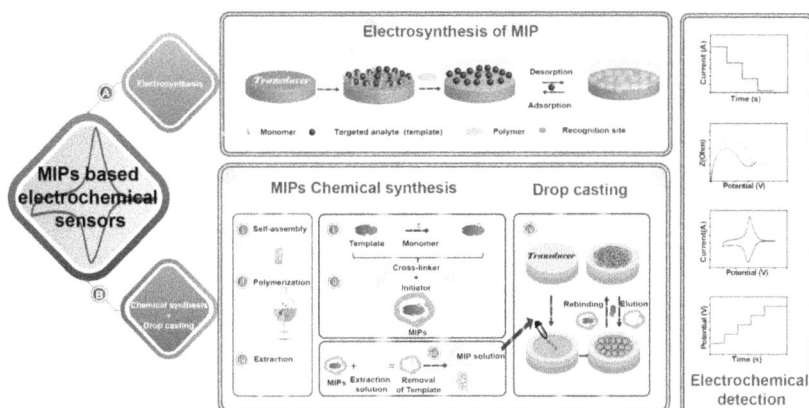

Figure 11.1. Schematic of the preparation of MIP-based electrochemical sensors, including the electrochemical and chemical synthesis. (Reproduced from [17]. Copyright 2021 the authors. CC BY 4.0.)

In the latter, there are primarily four stages: (i) a functional monomer is preassembled with a template molecule in the right solvent to create a complex through a covalent or noncovalent bond or interaction; (ii) a crosslinked polymer is produced after initiator-mediated light or heat polymerisation has taken place; (iii) as the polymer's imprinted molecules (template) are eluted, stable three-dimensional imprinted cavities that match the size and shape of the target molecule are produced; (iv) the MIP solution is dropped onto the electrode's surface. The core of molecular imprinting is the 'lock-key' procedure, which significantly reduces the interference of non-template molecules during the recognition process. This has led to the method also being referred to as 'host–guest' polymerisation or 'template' polymerisation. The types of monomer, crosslinker, solvent, and other polymerisation components are critical and they may also have an impact on the rate of mass transfer, the quantity of useful imprinting cavities, and the degree to which the targets are met [26–28].

11.3 The synthesis procedure for MIPs

MIPs are produced most frequently using synthesis. Functional monomers engage in interactions with the target molecules in solution to build a network of covalent or non-covalently bound complexes. Three forms of interactions—covalent, noncovalent, and semicovalent—can be distinguished according to how the target molecule and the functional monomer in the polymer bond to one another. Reversible covalent bonds between template molecules and functional monomers are typically used in covalent imprinting in order to create stable polymers. The main benefit of such a technique is that binding sites are distributed precisely and uniformly throughout the polymer. The most frequent type of interaction used is noncovalent imprinting, which often relies on the formation of weak binding interactions between the functional monomers and the template in the pre-polymerised mixture, such as hydrophobic or hydrogen bonds, dipole interactions, and ionic interactions.

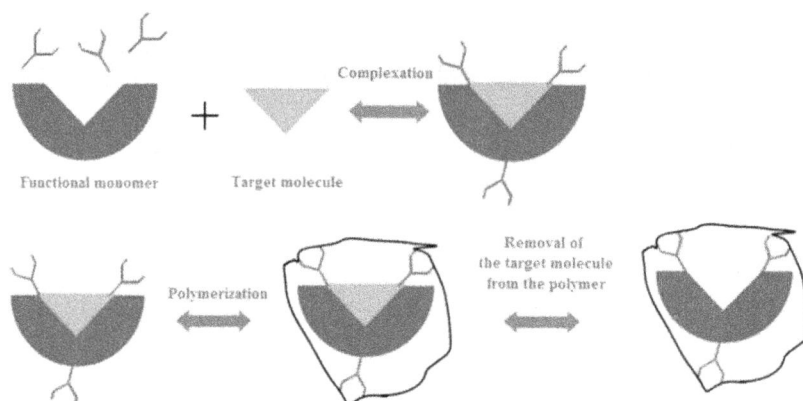

Figure 11.2. MIP preparation steps. (Reproduced with permission from [30]. Copyright 2022 DESIDOC.)

Semicovalent imprinting methods involve both covalent and noncovalent imprinting processes. Although the molecular imprinting process is polymerised in the form of covalent bonds, remarkably the target molecule binds with the monomer using noncovalent interactions [29]. The three fundamental components utilised in the synthesis of MIPs are known as functional monomers, crosslinkers, and initiators. Figure 11.2 shows a few examples of commonly used functional monomers, initiators, and crosslinkers. Functional monomers are crucial for the formation of specific complexes with the templates via covalent or noncovalent interactions, which are necessary for the synthesis of all molecularly imprinted polymers. In fact, the active substituents or functional hydrogen bonds provided by the functional monomers thus affect the affinity of the template to the functional monomer molecules as well as the mechanical stability and porosity of the polymer. In order to immobilise the functional group of the functional monomer onto the imprinted molecule, a crosslinking agent is added while the functional monomer is still reacting appropriately with the template molecule. This complex is formed during the MIP preparation process [30]. Finally, even after the template has been excluded, a rigid, highly crosslinked polymer is created. Figure 11.3 shows a diagram for both the synthesis and identification of MIPs. It is typical that too little crosslinker reduces the polymer's structural stability, which causes functional monomers to shed. Additionally, the number of sites that MIPs recognise will decrease due to an excessive amount of crosslinking agent. Currently, the vast majority of commercial crosslinkers are compatible with molecular imprinting, but some of them, such as ethylene glycol dimethacrylate (EGDMA), have the capacity to interact with the templates. In recent times, initiation techniques have been divided into three categories: thermal, optical, and electrical. It has been determined that the involvement of an initiator is necessary to ensure that the molecularly imprinted polymer proceeds normally and to shorten the reaction cycle. The material properties, however, might potentially be impacted by this. Polymers with a target molecule recognition layer have been made using the molecular imprinting technique. Additionally, the monomers that have been crosslinked forming a polymer around

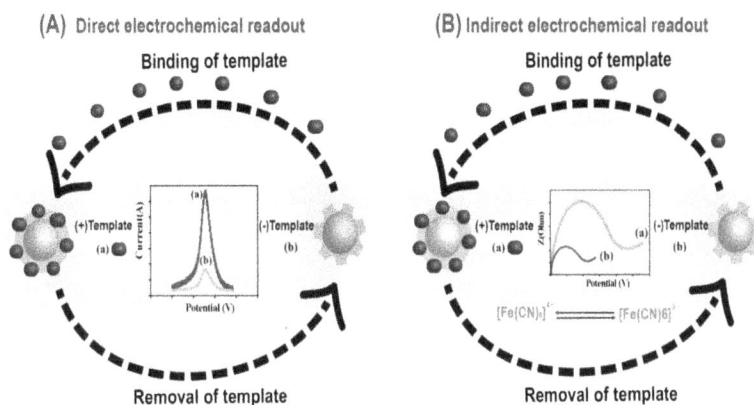

Figure 11.3. A graphical diagram of generic (A) direct and (B) indirect electrochemical measurements combined with MIPs. (Reproduced from [17]. Copyright 2021 the authors. CC BY 4.0.)

the target molecule. So, following the polymerisation and the target molecule is eliminated [31].

11.4 MIPs as recognition elements for electrochemical sensors

MIPs are synthetic materials specialised binding cavities serve as recognition sites for a particular target molecule, simulating natural receptors. Typically, a pre-polymerisation complex is formed by combining the analyte, or template molecule, with a functional monomer in an inert porogenic solvent to create MIPs [32]. In order to create a three-dimensional polymer network, this complex polymerises in the presence of a crosslinking agent that fixes the monomer close to the template. The template has been removed from the polymeric matrix at the end of the polymerisation process, leaving behind particular cavities whose size, shape, and functional groups are complementary to the template molecule. The result is the creation of a molecular memory within the polymer, where the target molecule has been discovered to be suitable for rebinding with an extremely high specificity [33]. In this approach, the MIP develops the ability to detect the target molecule only when other, closely related molecules are present, a process that is very reminiscent of the 'lock-and-key' mechanism of enzymes [34]. The creation of an MIP, signal augmentation techniques, and detection processes are depicted graphically in figure 11.2. Additionally, a non-imprinted polymer (NIP), which is organised under comparable circumstances but without the enclosure of the template molecule, is used for comparison to the MIP's recognition characteristics. Finding an MIP with the best selectivity for the target molecule is, however, noticeably the more difficult task [35].

Electrochemical MIP-based sensors change their electrical signal, or current or potential, in a manner that has been determined to be proportional to the concentration of the target analyte and is therefore recorded. Voltammetry and electrochemical impedance spectrometry (EIS), the two main approaches, have been used to identify the individual recognition events. The voltammetric methods include square wave voltammetry (SWV), differential pulse voltammetry (DPV), cyclic voltammetry (CV), and linear sweep voltammetry (LSV). In such procedures, the analyte's electroactivity influences the detection procedure [36]. For electroactive molecules, on the other hand, the current is directly measured, and the preferred signal outputs are the faradaic currents that come about as a result of the analytes' oxidation or reduction upon binding to the MIP. As a result, it is possible to connect the measured current and analyte concentration under controlled conditions. Redox probes such as $[Fe(CN)_6]^{3-/4-}$ may provide the signal once non-electroactive targets are included. This is also known as the 'gate-controlled' mechanism; it has been discovered that the current response of the redox probe is inversely proportional to the concentration of the analyte, whereas higher concentrations lead to fewer channels for the redox probe's diffusion to the electrode's surface, and consequently to a reduction in the signal. Though EIS is hailed as a potent technique for investigating electron transfer as well as diffusion processes that take place at the electrode/electrolyte interface, [37]. The usual Nyquist plot found in EIS contains two distinct regions: one at high frequencies where a semicircle component relates to the electron transfer resistance, and the other at low frequencies where a linear section has been used to represent the diffusion limited process. Variations in the electrode's surface properties occur during the MIP's assembly process, yet distinct Nyquist charts have been seen. Semi-circles with greater diameters have been seen as a result of the charge transfer resistance on the electrode surface typically increasing as a result of the template rebinding in the polymeric film's cavities [38].

11.5 The development of MIP-based electrochemical sensors

The first electrochemical sensor based on MIPs was developed by the Mosbach group at the start of the 1990s. The first thin molecular imprint utilising a polymer membrane was reported by Hedborg and co-authors within three years. In this instance, the membrane was made of aniline and had L-phenylalanine imprinting sites. The MIP was applied to a field-effect capacitor as a film. Two different approaches have been used to combine MIPs and electrochemical sensors, including the production of MIPs and their deposition on the electrode surface, as well as the electropolymerisation of MIPs directly onto the electrode surface. The latter is possible through drop-casting, coating, or magnetic capture onto magnetoactuated electrodes, taking advantage of MIPs created on magnetic particles [39].

11.6 Electrochemical techniques combined with MIP sensors

Electrochemical sensors rely on changes at the electrode surface, namely changes in current and voltage. The voltage (V) in potentiometry, the current (A) in voltammetry and amperometry, the resistance (Ω) in impedance, and the siemens (S) in the conductance technique can all be specified in accordance with the analytical signal output. Voltammetry, amperometry, and impedance spectroscopy, in that order, are used as detection methods in conjunction with the MIP sensor at a rates of 82%, 8%, and 7%, respectively [38]. Due to its high sensitivity, simplicity, and quick reaction, voltammetry is widely used. The target molecule's electrochemical characteristics are typically used to inform the selection of the measurement method. The current generated by amperometry or voltammetry can be used to quantify electroactive targets. Changes in the conductivity and/or porosity of the MIP film could result from non-electroactive targets. Through the use of an external redox probe and CV, conductimetry, or EIS, such an change can be monitored and controlled indirectly. Due to their quick charge transfer for various modified and/or unmodified transducers, redox probes such as ferri/ferrocyanide, hexaammine ruthenium chloride, or ferrocene have been most frequently used in combination with MIPs for indirect detection [26]. The porosity of the MIP film and the absorptivity of the redox probes to the electroactive sites decrease proportionately when a non-electroactive target analyte rebinds to it at various concentrations; this could consequently lead to a decrease in the current intensity or an increase in the sensor impedance [18]. As a result, this method could be used to indirectly measure the analyte. Additionally, a particular interaction may cause morphological changes in the polymer itself. These changes in the redox probe's diffusion rate may be detected as variations in the faradic current. Figure 11.2 illustrates the general idea of MIPs in combination with direct and indirect transduction methods. The target analyte must be reintroduced before the electrodes are washed to remove non-specific binding, followed by the electrochemical measurement. This is how measurements using MIP-based electrochemical sensors typically take place [40].

11.6.1 Voltammetric detection

The group of electroanalytical techniques known as voltammetry operate at a controlled variable potential. The current is thus measured as a function of the applied potential. The resultant faradic current is found to be proportional to the analyte's concentration for any analyte as long as it can be reduced or oxidised electrochemically. Voltammetry has a number of benefits, including the ability to estimate kinetic parameters and assess the processes of chemical and/or electrochemical reactions, as well as selectivity and sensitivity over a wide concentration range for a variety of analytes. DPV, CV, SWV, and LSV are the voltammetric techniques that are frequently used. Because of its ease of use, high sensitivity, and ability to reduce noise brought on by capacitive currents, the DPV approach has become one of the most used voltammetric techniques in the field of MIP-based sensors [41].

11.6.2 Amperometric detection

Amperometric procedures often operate in a stirred solution or in a flow system and the mass transport of electroactive species must only be regulated by the convection approach. Through the use of amperometry, an analyte can be detected by measuring the current at a fixed applied potential and relating the measured current to the analyte concentration. The signal in amperometric devices depends on the mass transfer rate of electrochemically active analytes into the MIP layer in MIP coated electrodes. Amperometric sensors based on MIPs are known to be easy to use and can be fully incorporated into microfluidic systems for continuous and real-time analysis [42].

11.6.3 Electrochemical impedance spectroscopic detection

The EIS method is quick, easy, and inexpensive. The term 'impedance' refers to a physical variable that investigates how an electrical circuit behaves when an alternating current is applied between the electrodes. In such a system, the current flow's response to the application of a small sinusoidal potential has been measured, allowing for the detection of changes in frequency (f) from the applied potential over a broad frequency range. EIS may examine specific processes involved in the conductivity, resistivity, or capacitance of the electrochemical system in addition to assessing the intrinsic material properties. Such systems are incredibly helpful for the characterisation, analysis, and transduction of materials as well as biosensors [43]. As the electrode surface is gradually modified or passivated, the charge transfer resistance (R_{ct}) values change. The increase in R_{ct} values shows that the electron transfer at the sensor surface evolves into a very complex process [44].

11.6.4 Potentiometric detection

In potentiometry, the amount of an ion in the solution can be determined or quantified using the change in potential that has been measured under flow or batch circumstances. The potentiometric sensor works when there is almost no current flow. As a result, it also detects the potential difference between the working electrode and a reference electrode. Membranes made from MIPs covering particular neutral carriers or ion exchangers are used to announce selectivity. Field-effect transistors that are chemically sensitive may also be used in these devices. Such semiconductor devices respond to the charge or surface electric gradient at the gate electrode. Additionally, the analyte's binding to the MIP at the silicon chip's surface shifts the surface potential, which affects the current and makes it possible to monitor the reaction rate [38].

11.7 Recent advances of MIP-based electrochemical sensors for pesticide detection

11.7.1 Pesticides

Due to their importance in agricultural production, pesticide sales have increased over the past 50 years on a global scale. The growth of harmful animals, insects, invasive plants, weeds, and fungi is both hampered and prevented by them. However, frequent application causes them to accumulate in soils and they may be carried into the aquatic environment by surface runoff. The health of ecosystems and biodiversity are put at risk because of their chemical characteristics and persistence, and their destructive activity is not limited to pests. Herbicides, insecticides, and fungicides are the three most commonly utilised types of pesticides. Pesticides have been quantified using a variety of techniques [35]. Although HPLC

Table 11.1. Detection of pesticides.

S. No.	Target molecule	Electrode	LOD (mol l^{-1})	Real sample	Reference
1.	Malathion (MAL)	MIP–Au–SPE	2×10^{-16}	Olive fruits and oils	[45]
2.	Methyl parathion (MP)	MIP–ILGr/GCE	6×10^{-9}	Cabbage and apple	[46]
3.	Clortoluron	MIP–NiO–GCE	2×10^{-9}	Water sample	[47]
4.	Clortoluron	MIP–NiHCF–GCE	9×10^{-10}	Water sample	[48]
5.	Carbofuran	MIP–rGO–AuNPs–GCE	2×10^{-8}	—	[49]
6.	Phoxim	MIP–Gr–GCE	2×10^{-8}	Cucumber	[50]
7.	Profenofos (PFF)	3D-CNT–MIP–GCE	2×10^{-9}	Vegetable sample	[51]
8.	Imidacloprid (IDP)	GN–VBA–MIP–GCE	1×10^{-7}	Brown rice	[52]
9.	Dinotefuran	—	1.7×10^{-6}	Cucumber and soil sample	[53]
10.	Cyanazine (CZ)	MIP–DFT–CPE	3×10^{-9}	—	[54]
11.	Hexazinone	MIP–CPE	—	River water	[55]

coupled to mass spectrometry is the standard procedure, other sensoristic approaches have been tried that make use of biosensors, nanomaterials-based sensors, and electrophoretic chips. The evaluation of food safety relies heavily on pesticide sensing with a high degree of selectivity and sensitivity [29]. The use of MIPs in conjunction with sensors for the detection of pesticides may help to overcome the high selectivity-related restrictions of conventional analytical procedures, providing a significant possibility to also build practical, affordable, and quick detection techniques. Highly significant factors that affect both the analytical performance and the polymerisation of MIPs, such as the target analyte, monomer, extraction solution, electrochemical technique, linear range, limit of detection (LOD), and application in a real sample, have been reported. Different studies of MIP-based detection of pesticides are discussed below (table 11.1).

11.7.2 Insecticides

Organophosphates persist for days or weeks in the aquatic environment, and studies have revealed that fish and crustaceans may also accumulate them [55]. The majority of electrochemical MIP sensors have been developed for these insecticides. A study using a molecular self-assembly approach for electropolymerising p-aminothiophenol (PATP) on the surface of gold nanoparticle (AuNP) modified glassy carbon electrode (GCE) through the formation of Au–S bonds was published by Xie *et al* in 2010 [56]. An LOD of nM and a linear range between 5.0×10^2 nM and 1.0×10^4 nM were attained. As a result, this sensor showed strong interclass selectivity. The researchers also tested its use with spiked tap water. Through straightforward bulk polymerisation of MAA, an improved LOD of approximately 4.08 nM was attained [57]. The MIPs that resulted from the binding of Chlorpyrifos (CPF) were discovered to be suspended on the surface of the GCE. This coating was subsequently examined by DPV utilising the redox probe $[Fe(CN)_6]^{3-/4-}$. This inexpensive sensor offered good selectivity, and river water samples were used to assess its applicability. Compared to previously published work, an electrochemical MIP sensor based on carbon nitride nanotubes (C_3N_4 NTs) decorated with graphene quantum dots (GQDs) [58] provided a much lower LOD in the range of 2.0×10^{-3} nM. These nanoparticles have drawn a lot of interest, and this was one of the first works to be published that combined MIP with their extraordinary capabilities for use in wastewater samples. Following the hydrothermal treatment used to develop the C_3N_4 NTs@GQDs composite, a suspension of this nanohybrid material was applied to the GCE's surface. Then, using Py as a functional monomer to produce electrostatic interactions and hydrogen bonds with CPF, the electropolymerisation was carried out. SWV was used to perform a direct evaluation of the developed MIP sensor's performance. It was confirmed that adding a limited bandgap to graphene enhanced its gapless nature and increased the electrode surface's conductivity, suggesting a potential use for GQDs in the development of electrochemical sensors. The other use for nanomaterials is the production of electrochemical MIP sensors for the detection of methylparathion (MP) via bulk and electropolymerisation. In 2014 Wu et al [59] constructed a sensor utilising carbon nanotube (CNT)-decorated AuNPs. Electrodeposition of functionalised AuNPs on a multi-walled CNT (MWCNT)/GCE surface was followed by immersion of the electrode in a PATP-containing solution. Considering that MP is electroactive, LSV may be able to directly monitor it by recognising it through the MIPs. The developed sensor allowed for a low LOD of about 0.30 nM and was successfully used to determine MP in spiked distilled water, tap water, apple, and cucumber samples with recoveries ranging from 95% to 106%. A different quantitative approach for MIPs was developed using nitrogen-doped graphene sheets (N-GS), and it involved electro-polymerising Ph to produce MIPs [60]. Despite the fact that N-GS has been acknowledged as a superior sensing material for molecular imprinting applications, the fabricated MIP sensor shown worse analytical performance compared to those previously mentioned [61]. A few advantages of using carbon paste-based electrodes (CPE) include facile surface modification and extremely low ohmic resistance [61, 62]. These benefits allowed for the development of MIP sensors for parathion [63] and diazinon [64], where the MIPs were produced via bulk polymerisation with MAA serving as the functional monomer. After SWV successfully detected these 558 organophosphorus pesticides, the authors were able to confirm that optimising the CPE's composition is necessary to increase the sensor's sensitivity. To attain the highest performance, adding nanomaterials to CPEs looks to be essential. It has been observed that research interest in metal organic frameworks (MOFs), connected by the self-assembly of transition metal ions/clusters and organic ligands, has rapidly increased in recent years for the purpose of detection. In order to identify phosalone (PAS), Xu et al [65] recently developed a new disposable carbon paste microelectrode (CPME) MIP sensor on a zirconium (Zr) based MOF (UiO-66). This was done on the basis of the MOF's beneficial properties, such as its highly ordered structure and exposed sites. 3-aminopropyltriethoxysilane (APTES), the functional monomer, was used to combine

UiO-66 with Pt nanoparticles to create the MIP imprinted with PAS. The resulting sensor showed an LOD of about 0.078 nM, and its viability was subsequently evaluated in soil samples and lake water. When compared to organophosphate insecticides, pyrethroid insecticides such as cyclomethrin (CYP) have a much less toxic and highly effective insecticidal action. However, this kind of pesticide is widely used all over the world. An MIP-based sensor for detecting CYP in samples of spiked wastewater has recently been reported [66]. The sensing phase of the sensor, however, was made on the surface of core–shell type nanoparticles (Fe@AuNPs) by polymerising Ph, which joined two-dimensional hexagonal boron nitride (2D-hBN) nanosheets. Additionally, the highly sensitive characteristics of the developed nanocomposite allowed for a greater ability to bind CYP specifically, and an astounding LOD was as a result obtained (3.0×10^{-5} nM). The MIPs produced with two or more 594 functional monomers performed better according to recent studies. A dual-monomer MIP was created by Li et al in 2019 [67] for the study of CYP. Specific steps were required to prepare the MIP sensor in order to create a hybrid material by mixing activated carbon (AC) with zinc oxide that had been co-doped with nitrogen (Ag–N@ZnO). AC was selected as a sensitising material on the GCE surface due to its low cost and ease of availability in order to develop an expanded sensing surface and, consequently, amplify the signal. However, the authors found that using only AC prevents the electrode's surface from forming a rigid layer. With the help of the inclusion of ZnO for greater immobilisation, the drawback was overcome. ZnO may be mixed with other materials without much difficulty, and doping it could improve its electrical conductivity. The electrochemical performance was evaluated by CV utilising $[Fe(CN)_6]^{3-/4-}$ as a redox probe and, as a result, the sensor was used with soil samples in addition to water samples. Even though this dual-monomer MIP introduced a novel approach to improve binding and affinity with the target analytes, the LOD was still higher than that of the MIP sensor created by Atar and Yola in 2018 [66], which only used one functional monomer. Combining MIPs and electrochemiluminescence (ECL) may improve the assay's sensitivity and selectivity. The application of ECL have been found to possess an exponential growth in electrochemical sensors, using ECL as a detection method for MIP sensors with real water sample applications is still uncommon. This combination uses the sensor developed by Xu et al in 2020 [68] to recognise the pesticide cyfluthrin (CYF) as an example. They created an MIP platform using QDs as the luminophore while H_2O_2 served as a co-reactant. In the ECL procedure, the electrode produces radical species as the luminophore and the co-reactant are both oxidised or reduced. Their redox products undergo an electrochemical reaction, which causes the radiation to be released. Luminol is regarded as a traditional ECL luminophore due to its high emission quantum yield. However, because of the restrictions during its ECL procedure, there has been a increase in the demand for other replacements. In the work suggested by Xu et al in 2020, MWCNTs were used to advance the electrocatalytic activity and to lessen surface fouling on the GCE, and the combination of an MIP with ECL based on QDs confirmed high selectivity, good stability, and controllability. The aim remains to simplify the electrode preparation process and to increase the

Table 11.2. Detection of insecticides.

S. No.	Target molecule	Electrode	LOD (nM)	Real sample	Reference
1.	Chlorpyrifos (PATP)	GCE–AuNPs	3.3×10^2	Tap water	[56]
2.	Chlorpyrifos (MAA)	Bulk polymerisation, GCE	4.1 nM	River water	[57]
3.	Chlorpyrifos (Py)	Electropolymerisation, GCE, C_3N_4 NTs, GQDs	2.0×10^{-3}	Industrial wastewater	[58]
4.	Methylparathion (PATP)	Electropolymerisation, GCE, AuNPs, CNTs	0.30	Distilled water and tap water	[59]
5.	Methylparathion (Ph)	Electropolymerisation, SPAuE, N-GS	38	River water	[60]
6.	Parathion (MAA)	Bulk polymerisation, CPE	0.5	Tap, river, and lake water	[63]
7.	Diazinon (MAA)	Bulk polymerisation, CPE	0.79	Well water	[64]
8.	Phosalone (APTES)	Sol–gel method, CPME, Pt–UiO-66	0.078	Lake water and soil	[66]
9.	Cypermethrin (Ph)	Electropolymerisation, GCE, Fe@AuNPs, 2D-hBN	3.0×10^{-5}	Wastewater	[67]
10.	Cypermethrin (DA and RC)	Electropolymerisation, GCE, Ag–N@ZnO, CHAC	6.7×10^{-5}	Tap water and soil	[68]
11.	Cyfluthrin (APTES)	Sol–gel method, QDs, Nafion–MWCNTs	0.12	Seawater	[70]

electrocatalytic activity. The detection of several insecticides has been discussed and are listed in table 11.2.

11.7.3 Herbicides

Over the past 40 years, glyphosate (Gly) has been among the herbicides most frequently used by farmers. Although it has not been proven, Gly has been linked to human cancer, which has sparked an animated discussion about its potential harm [69]. It is important to detect trace amounts of Gly in drinking water because of its persistence in seawater [70,71]. In 2017 Zhang *et al* [72] described an electrochemical MIP sensor employed for Gly detection in water samples. This work involved performing a straightforward electropolymerisation technique on an Au electrode using Py as a functional monomer. The sensor offered a virtuous binding kinetics to Gly and possessed good stability, selectivity, as well as sensitivity, and an LOD of 1.60 nM was achieved. However, the $[Fe(CN)_6]^{3-/4-}$ redox probe and DPV were used to generate an analytical signal. When compared to the MIP sensor developed by Prasad *et al* in 2014 [73], where *N*-methacryloyl-L-cysteine (MAC) molecules were used as the monomer during the bulk polymerisation process, this MIP sensor demonstrated greater sensitivity. However, the development of an electrochemical sensor based on MIP–MOF films made on Au surfaces by the electropolymerisation of PATP functionalised with AuNPs yielded the highest sensitivity for the detection of GLY in water samples [74]. Thus, a low LOD of 4.7×10^{-6} nM was achieved, and tap water was used to test the sensor's sensing capabilities. Additionally, other MIP sensors that have been published to identify the herbicides used in water samples were based on modifications of GCEs (AL-Ammari *et al* 2019), pencil graphite electrode (PGE) [75], or CPEs [76,78]. Additionally, a recently developed carbon fibre paper (CFP) used as a working electrode was also reported [80]. Several nanomaterials were combined in this type of sensor to increase selectivity and precision. The presence of GO was observed to significantly reduce the LOD for 2,4-dichlorophenol (2,4-DCP), a typical chlorophenol that is widely used in the production of herbicides and insecticides [79]. In addition, compared to Zhang *et al* in 2013 [80], in 2017 Liang *et al* proposed a relatively straightforward method for the fabrication of a 2,4-DCP-MIP electrochemical sensor. They did this by employing the same functional monomer (MAA). In 2013 Zhang *et al* used MAA as a functional monomer in addition to a co-monomer called chlorohemin to insert chemically active sites into the MIPs and a mixture of chitosan (Ch) and Nafion to immobilise the MIPs and hence increase conductivity. This assembly may have prevented mass from moving to the electrode's surface, which would have improved the LOD. Due to its porosity and potential for a large number of reaction sites, CFP was chosen as the working electrode for the identification of 2,4-DCP. The LOD of about 0.07 nM was improved by this invention and the use of PEDOT as a conductive polymer, as well as by the stability and reproducibility. Eriochrome blue-black B (EBB) was the functional doping ion present when the MIPs were electro-polymerised on top of a PGE. EBB is a type of anionic complexing agent that was used to keep Py electroneutral during the reduction process and, thus, to preserve its electroactivity. Then, in 2015, Wong *et al* [80] created a CPE modified with MIPs and MWCNTs to enhance the electrochemical signal of diuron. The bulk polymerisation method was used to synthesise the MIPs. The numerous interactions between chloridazon (CLZ) and two functional monomers, 2-vinylpiridine (2-VP) and MAA, were used to develop the MIP sensor.

11.7.4 Fungicides and biocides

The number of papers based on the study of fungicides is far lower than that of insecticides and herbicides. Table 11.3 provides an example of electrochemical MIP sensors for identifying fungicides and biocides.

Table 11.3. MIPs for the detection of fungicides and biocides.

S. No.	Target molecule	Electrode	LOD (nM)	Real sample	Reference
1.	Dicloran	MIP–CNT–CPE	0.48	Tap and river water	[81]
2.	Tributyltin	MIP–Fe_3O_4–GCE	5.4×10^{-3}	—	[82]

11.8 Challenges and limitations of MIP application in pesticide detection

Despite the latest advanced methods as outlined in this chapter, MIPs have been found to face certain challenges in selective trace level pesticide extraction, including as follows:

1. The optimisation of polymerisation methods is required for several templates due to the morphological nature of the template as well as selective extinction. MIPs possess single-analyte tailor-made selectivity, and generally have high efficiency in comparison to multi-pesticide detection, and to avoid the complexity, few works have been done on multi-pesticide analysis. However, further research must be done on this technique to obtain high selectivity in a single complex matrix for multi-pesticide detection.

2. Commercial MIP manufacturing is still restricted to a few chosen pesticides, and it has to be improved for additional/novel pesticides with optimisation to unique real samples.

3. Large-scale production is necessary for the commercialisation of MIPs because its application to real sample analysis is at a very low level and it is not expedient for large sample sizes.

4. To create quick detection methods, it is necessary to decrease matrix interference in complex samples and further optimise the analyte's extraction time for a variety of real samples.

5. Because MIPs are less effective in water, it can be difficult to detect pesticides in aqueous media. However, for some analytes, this problem has been solved by using hydrophilic functional monomers. Still, it is crucial to assess novel functional monomers and novel MIPs with universal applicability to different polar and non-polar matrices.

6. In order to ensure food safety, efforts are necessary to build standardised MIPs techniques with effective applications in conjunction with analytical tools for both single-pesticide and multi-pesticide analysis in actual samples at a commercial level.

11.9 Conclusion and future perspectives

The present chapter critically assesses the most recent advancements made in MIP-based electrochemical sensors for pesticide monitoring. The most recent developments for various MIP-based electrochemical sensors are presented after a brief introduction to MIPs and electrochemical sensors, emphasising the most noteworthy developments. The sheer number of papers in this field demonstrates the intense interest that scientists from all over the world have in it. This could be attributed to the requirement for the development of quick, sensitive screening tests that also demonstrate the necessary selectivity. Electroanalytical tools combined with MIPs represent a promising answer in this approach. However, it can be challenging to compare the effectiveness of each analytical strategy because the target and the type of polymerisation have a significant impact on the selectivity and sensitivity of the methods. However, the method is viable for the implementation of pesticide screening assays due to the very low detection limits accessible, the good reproducibility, as well as the good recovery reported for a few targets. Here, we have summarised several current MIP applications for pesticide detection exploiting its efficacy for the extraction of a specific analyte. Thus, the application of MIPs in pesticide residue detection is found to be an outstanding advancement, particularly for pre-treatment clean-up as well as trace level detection in complex samples. In the future, MIP applications to pesticide detection could result in extremely sophisticated research methods, commercial methods with great selectivity and sensitivity, and quick analytical methods.

References

[1] Jiang L, Hassan M M, Ali S, Li H, Sheng R and Chen Q 2021 Evolving trends in SERS-based techniques for food quality and safety: a review *Trends Food Sci. Technol.* **112** 225–40

[2] Cao Y, Feng T, Xu J and Xue C 2019 Recent advances of molecularly imprinted polymer-based sensors in the detection of food safety hazard factors *Biosens. Bioelectron.* **141** 111447

[3] Rojas D, Della Pelle F, Del Carlo M, Fratini E, Escarpa A and Compagnone D 2019 Nanohybrid carbon black-molybdenum disulfide transducers for preconcentration-free voltammetric detection of the olive oil o-diphenols hydroxytyrosol and oleuropein *Microchim. Acta.* **186** 1–9

[4] Della Pelle F, Rojas D, Scroccarello A, Del Carlo M, Ferraro G, Di Mattia C, Martuscelli M, Escarpa A and Compagnone D 2019 High-performance carbon black/molybdenum disulfide nanohybrid sensor for cocoa catechins determination using an extraction-free approach *Sensors Actuators B* **296** 126651

[5] Della Pelle F, Rojas D, Scroccarello A, Del Carlo M, Ferraro G, Di Mattia C, Martuscelli M, Escarpa A and Compagnone D 2020 Class-selective voltammetric determination of hydroxycinnamic acids structural analogs using a WS2/catechin-capped AuNPs/carbon black–based nanocomposite sensor *Microchim. Acta.* **187** 1–13

[6] Scroccarello A, Della Pelle F, Ferraro G, Fratini E, Tempera F, Dainese E and Compagnone D 2021 Plasmonic active film integrating gold/silver nanostructures for H2O2 readout *Talanta* **222** 121682

[7] Rojas D, Della Pelle F, Del Carlo M, Compagnone D and Escarpa A 2020 Group VI transition metal dichalcogenides as antifouling transducers for electrochemical oxidation of catechol-containing structures *Electrochem. Commun.* **115** 106718

[8] Dong C, Shi H, Han Y, Yang Y, Wang R and Men J 2021 Molecularly Imprinted Polymers by the. Surface Imprinting Technique *Eur. Polym. J.* **145** 110231

[9] Li W, Zhang X, Li T, Ji Y and Li R 2021 Molecularly Imprinted Polymer Enhanced Biomimetic Paper Based Analytical Devices: A Review *Anal. Chim. Acta.* **1148** 238196

[10] Beluomini M A, da Silva J, de Sá A C, Buffon E, Pereira T C and Stradiotto N R J 2019 Electrochemical sensors based on molecularly imprinted polymer on nanostructured carbon materials: a review *Electroanal. Chem.* **840** 343–66

[11] Wulff G 2013 Fourty years of molecular imprinting in synthetic polymers: origin, features and perspectives *Microchim. Acta.* **180** 1359–70

[12] Martín-Esteban A 2021 Membrane Protected Molecularly Imprinted Polymers: Towards Selectivity Improvement of Liquid-Phase Microextraction TrAC Trends *Anal. Chem.* **138** 116236

[13] Liu H, Jin P, Zhu F, Nie L and Qiu H 2021 A Review on the Use of Ionic Liquids in Preparation Molecularl y Imprinted Polymers for Application in Solid Phase Extraction 11-19 TrAC Trends *Anal. Chem.* **134** 116132

[14] Boysen R I 2019 Advances in the development of molecularly imprinted polymers for the separation and analysis of proteins with liquid chromatography *J. Sep. Sci.* **42** 51–71

[15] Cai Y, He X, Cui P L, Liu J, Bin Li Z, Jia B J, Zhang T, Wang J P and Yuan W Z 2019 Preparation of a chemiluminescence sensor for multi-detection of benzimidazoles in meat based on molecularly imprinted polymer *Food Chem.* **280** 103–9

[16] He S, Zhang L, Bai S, Yang H, Cui Z, Zhang X and Li Y 2021 Advances of Molecularly Imprinted Polymers (MIPs) and the Application in Drug Delivery *Eur. Polym. J.* **143** 110179

[17] Elfadil D, Lamaoui A, Pelle F D, Amine A and Compagnone D 2021 Molecularly imprinted polymers combined with electrochemical sensors for food contaminants *Molecules* **26** 4607

[18] Lahcen A A, Arduini F, Lista F and Amine A 2018 Label-free electrochemical sensor based on spore-imprinted polymer for *Bacillus cereus* spore detection *Sensors Actuators* B **276** 114–20

[19] Vasilescu A, Nunes G, Hayat A, Latif U and Marty J-L 2016 Electrochemical affinity biosensors based on disposable screen-printed electrodes for detection of food allergens *Sensors* **16** 1863

[20] Ansari S 2017 Application of magnetic molecularly imprinted polymer as a versatile and highly selective tool in food and environmental analysis: recent developments and trends *TrAC Trends Anal. Chem.* **90** 89–106

[21] Yáñez-Sedeño P, Campuzano S and Pingarrón J M 2017 Electrochemical sensors based on magnetic molecularly imprinted polymers: a review *Anal. Chim. Acta.* **960** 1–17

[22] Wang P, Sun X, Su X and Wang T 2016 Advancements of molecularly imprinted polymers in the food safety field *Analyst* **141** 3540–53

[23] Cowen T, Karim K and Piletsky S 2016 Computational approaches in the design of synthetic receptors—a review *Anal. Chim. Acta.* **936** 62–74

[24] Yao J, Li X and Qin W 2008 Computational design and synthesis of molecular imprinted polymers with high selectivity for removal of aniline from contaminated water *Anal. Chim. Acta.* **610** 282–8

[25] Li Y, Li X, Li Y, Dong C, Jin P and Qi J 2009 Selective recognition of veterinary drugs residues by artificial antibodies designed using a computational approach *Biomaterials* **30** 3205–11

[26] Marc M, Kupka T, Wieczorek P P and Namiesnik J 2018 Computational modeling of molecularly imprinted polymers as a green approach to the development of novel analytical sorbents *TrAC Trends Anal. Chem.* **98** 64–78

[27] Andersson L I, Miyabayashi A, O'Shannessy D J and Mosbach K 1990 Enantiomeric resolution of amino acid derivatives on molecularly imprinted polymers as monitored by potentiometric measurements *J. Chromatogr.* A **516** 323–31

[28] Hedborg E, Winquist F, Lundström I, Andersson L I and Mosbach K 1993 Some studies of molecularly-imprinted polymer membranes in combination with field-effect devices *Sensors Actuators* A **37** 796–9

[29] Lahcen A A and Amine A 2019 Recent advances in electrochemical sensors based on molecularly imprinted polymers and nanomaterials *Electroanalysis* **31** 188–201
Zamora-Gálvez A, Lahcen A A, Mercante L, Morales-Narváez E, Amine A and Merkoçi A 2016 *Anal. Chem.* **88** 3578–84

[30] Lahcen A A, Baleg A A, Baker P, Iwuoha E and Amine A 2017 Synthesis and electro-chemical characterization of nanostructured magnetic molecularly imprinted polymers for 17-β-Estradiol determination *Sensors Actuators* B **241** 698–705

[31] Turner N W, Holdsworth C I, Donne S W, McCluskey A and Bowyer M C 2010 Microwave induced MIP synthesis: comparative analysis of thermal and microwave induced polymer-isation of caffeine imprinted polymers *New J. Chem.* **34** 686–92

[32] Lamaoui A, Palacios-Santander J M, Amine A and Cubillana-Aguilera L 2021 Fast microwave-assisted synthesis of magnetic molecularly imprinted polymer for sulfamethox-azole *Talanta* **232** 122430

[33] Lamaoui A, Lahcen A A, Guzmán J J G, Palacios-Santander J M, Aguilera L C and Amine A 2019 Study of solvent effect on the synthesis of magnetic molecularly imprinted polymers based on ultrasound probe: application for sulfonamide detection *Ultrason. Sonochem.* **58** 104670

[34] Breton F, Euzet P, Piletsky S A, Giardi M T and Rouillon R 2006 Integration of photosynthetic biosensor with molecularly imprinted polymer-based solid phase extraction cartridge *Anal. Chim. Acta.* **569** 50–7

[35] Messaoud N B, Lahcen A A, Dridi C and Amine A 2018 Ultrasound assisted magnetic imprinted polymer combined sensor based on carbon black and gold nanoparticles for selective and sensitive electrochemical detection of bisphenol A *Sensors Actuators* B **276** 304–12

[36] Ahmad O S, Bedwell T S, Esen C, Cruz A G and Piletsky S A 2019 Molecularly imprinted polymers in electrochemical and optical sensors *Trends Biotechnol.* **37** 294–309

[37] Ashley J, Shahbazi M-A, Kant K, Chidambara V A, Wolff A, Bang D D and Sun Y 2017 Molecularly imprinted polymers for sample preparation and biosensing in food analysis: progress and perspectives *Biosens. Bioelectron.* **91** 606–15

[38] Ghanam A, Mohammadi H, Amine A, Haddour N and Buret F 2021 *Reference Module in Biomedical Sciences* (Amsterdam: Elsevier)

[39] Wang J 2002 Electrochemical detection for microscale analytical systems: a review *Talanta* **56** 223–31

[40] Lisdat F and Schäfer D 2008 The use of electrochemical impedance spectroscopy for biosensing *Anal. Bioanal. Chem.* **391** 1555–67

[41] Sergeyeva T, Piletsky S, Panasyuk T L, El'Skaya A V, Brovko A A, Slinchenko E A and Sergeeva L M 1999 Conductimetric sensor for atrazine detection based on molecularly imprinted polymer membranes *Analyst* **124** 331–4

[42] Ding J and Qin W 2020 Recent advances in potentiometric biosensors *TrAC Trends Anal. Chem.* **124** 115803

[43] Bratov A, Abramova N and Ipatov 2010 Recent trends in potentiometric sensor arrays—a review *Anal. Chim. Acta.* **678** 149–59

[44] Aghoutane Y, Diouf A, Österlund L, Bouchikhi B and El Bari N 2020 Development of a molecularly imprinted polymer electrochemical sensor and its application for sensitive detection and determination of malathion in olive fruits and oils *Bioelectrochemistry* **132** 107404

[45] Zhao L, Zhao F and Zeng B 2013 Electrochemical determination of methyl parathion using a molecularly imprinted polymer–ionic liquid–graphene composite film coated electrode *Sensors Actuators* B **176** 818–24

[46] Li X, Zhang L, Wei X and Li J 2013 A sensitive and renewable chlortoluron molecularly imprinted polymer sensor based on the gate-controlled catalytic electrooxidation of H_2O_2 on magnetic nano-NiO *Electroanalysis* **25** 1286–93

[47] Zhang L, Li J and Zeng Y 2015 Molecularly imprinted magnetic nanoparticles for determination of the herbicide chlorotoluron by gate-controlled electro-catalytic oxidation of hydrazine *Microchim. Acta.* **182** 249–55

[48] Tan X, Hu Q, Wu J, Li X, Li P, Yu H, Li X and Lei F 2015 Electrochemical sensor based on molecularly imprinted polymer reduced graphene oxide and gold nanoparticles modified electrode for detection of carbofuran *Sensors Actuators* B **220** 216–21

[49] Tan X, Wu J, Hu Q, Li X, Li P, Yu H, Li X and Lei F 2015 An electrochemical sensor for the determination of phoxim based on a graphene modified electrode and molecularly imprinted polymer *Anal. Methods* **7** 4786–92

[50] Amatatongchai M, Sroysee W, Sodkrathok P, Kesangam N, Chairam S and Jarujamrus P 2019 Novel three-dimensional molecularly imprinted polymer-coated carbon nanotubes (3D-CNTs@MIP) for selective detection of profenofos in food *Anal. Chim. Acta.* **1076** 64–72

[51] Zhang M, Zhao H, Xie T, Yang X, Dong A, Zhang H, Wang J and Wang Z 2017 Molecularly imprinted polymer on graphene surface for selective and sensitive electrochemical sensing imidacloprid *Sensors Actuators* B **252** 991–1002

[52] Abdel-Ghany M, Hussein L A and El Azab N F 2017 Novel potentiometric sensors for the determination of the dinotefuran insecticide residue levels in cucumber and soil samples *Talanta* **164** 518–28

[53] Gholivand M B, Torkashvand M and Malekzadeh G 2012 Fabrication of an electrochemical sensor based on computationally designed molecularly imprinted polymers for determination of cyanazine in food samples *Anal. Chim. Acta.* **713** 36–44

[54] Toro M J U, Marestoni L D and Sotomayor M D P T 2015 A new biomimetic sensor based on molecularly imprinted polymers for highly sensitive and selective determination of hexazinone herbicide *Sensors Actuators* B **208** 299–306

[55] Carvalho F P 2017 Pesticides, environment, and food safety *Food Energy Secur.* **6** 48–60

[56] Xie C, Gao S, Guo Q and Xu K 2010 Electrochemical sensor for 2,4-dichlorophenoxy acetic acid using molecularly imprinted polypyrrole membrane as recognition element *Microchim. Acta.* **169** 145–52

[57] Xu W, Wang Q, Huang W and Yang W 2017 Construction of a novel electrochemical sensor based on molecularly imprinted polymers for the selective determination of chlorpyrifos in real samples *J. Sep. Sci.* **40** 4839–46

[58] Yola M L and Atar N 2017 A highly efficient nanomaterial with molecular imprinting polymer: carbon nitride nanotubes decorated with graphene quantum dots for sensitive electrochemical determination of chlorpyrifos *J. Electrochem. Soc.* **164** B223–9

[59] Wu B, Hou L, Du M, Zhang T, Xue Z and Lu X 2014 A molecularly imprinted electrochemical enzymeless sensor based on functionalized gold nanoparticle decorated carbon nanotubes for methyl-parathion detection *RSC Adv.* **4** 53701–10

[60] Xue X, Wei Q, Wu D, Li H, Zhang Y, Feng R and Du B 2014 Determination of methyl parathion by a molecularly imprinted sensor based on nitrogen doped graphene sheets *Electrochim. Acta.* **116** 366–71

[61] Toro M J U, Marestoni L D and Del Pilar Taboada Sotomayor M 2015 A new biomimetic sensor based on molecularly imprinted polymers for highly sensitive and selective determination of hexazinone herbicide *Sensors Actuators* B **208** 299–306

[62] Vytřas K, Švancara I and Metelka R 2009 Carbon paste electrodes in electroanalytical chemistry *J. Serbian Chem. Soc.* **74** 1021–33

[63] Alizadeh T 2009 High selective parathion voltammetric sensor development by using an acrylic based molecularly imprinted polymer-carbon paste electrode *Electroanalysis* **21** 1490–8

[64] Motaharian A, Motaharian F, Abnous K, Hosseini M R M and Hassanzadeh-Khayyat M 2016 Molecularly imprinted polymer nanoparticles-based electrochemical sensor for determination of diazinon pesticide in well water and apple fruit samples *Anal. Bioanal. Chem.* **408** 6769–79

[65] Xu L, Li J, Zhang J, Sun J, Gan T and Liu Y 2020 A Disposable molecularly imprinted electrochemical sensor for the ultra trace detection of the organophosphorou s insecticide phosalone employing monodisperse Pt –doped UiO-66 for signal amplification *Analyst* **145** 3245–56

[66] Atar N and Yola M L 2018 Core–shell nanoparticles/two-dimensional (2D) hexagonal boron nitride nanosheets with molecularly imprinted polymer for electrochemical sensing of cypermethrin *J. Electrochem. Soc.* **165** H255–62

[67] Li Y, Zhang L, Dang Y, Chen Z, Zhang R, Li Y and Ye B C 2019 A robust electrochemical sensing of molecularly imprinted polymer prepared by using bifunctional monomer and its application in detection of cypermethrin *Biosens. Bioelectron.* **127** 207–14

[68] Xu J, Zhang R, Liu C, Sun A, Chen J, Zhang Z and Shi X 2020 Highly selective electrochemiluminescence sensor based on molecularly imprinted-quantum dots for the sensitive detection of cyfluthrin *Sensors* **20** 884

[69] Li H, Wang Y, Zha H, Dai P and Xie C 2019 Reagentless electrochemiluminescence sensor for triazophos based on molecular imprinting electropolymerized poly(luminol-*p*-aminothiophenol) composite-modified gold electrode *Arab. J. Sci. Eng.* **44** 145–52

[70] Silva V, Montanarella L, Jones A, Fernández Ugalde O, Mol H G J, Ritsema C J and Geissen V 2018 Distribution of glyphosate and aminomethylphosphonic acid (AMPA) in agricultural topsoils of the European Union *Sci. Total Environ.* **621** 1352–9

[71] Mercurio P, Flores F, Mueller J F, Carter S and Negri A P 2014 Glyphosate persistence in seawater *Mar. Pollut. Bull.* **85** 385–90

[72] Zhang C *et al* 2017 A highly selective electrochemical sensor based on molecularly imprinted polypyrrole-modified gold electrode for the determination of glyphosate in cucumber and tap water *Anal. Bioanal. Chem.* **409** 7133–44

[73] Prasad B B, Jauhari D and Tiwari M P 2014 Doubly imprinted polymer nanofilm-modified electrochemical sensor for ultra-trace simultaneous analysis of glyphosate and glufosinate *Biosens. Bioelectron.* **59** 81–8

[74] AL-Ammari R H, Ganash A A and Salam M A 2019 Electrochemical molecularly imprinted polymer based on zinc oxide/graphene/ poly (o- phenylenediamine) for 4-cholorophenol detection *Synth. Met.* **254** 141–52

[75] Sayyahmanesh M, Asgari S, Meibodi A S E and Ahooyi T M 2016 Voltammetric determination of paraquat using graphite pencil electrode modified with doped polypyrrole arXiv:1604.07853

[76] Ghorbani A, Ganjali M R, Ojani R and Raoof J 2020 Detection of chloridazon in aqueous matrices using a nano-sized chloridazon-imprinted polymer-based voltammetric sensor *Int. J. Electrochem. Sci.* **15** 2913–22

[77] Maria G C A, Akshaya K B, Rison S, Varghese A and George L 2020 Molecularly imprinted PEDOT on carbon fiber paper electrode for the electrochemical determination of 2,4-dichlorophenol *Synth. Met.* **261** 116309

[78] Liang Y, Yu L, Yang R, Li X, Qu L and Li J 2017 High sensitive and selective graphene oxide/molecularly imprinted polymer electrochemical sensor for 2,4-dichlorophenol in water *Sensors Actuators* B **240** 1330–5

[79] Zhang J, Lei J, Ju H and Wang C 2013 Electrochemical sensor based on chlorohemin modified molecularly imprinted microgel for determination of 2,4-dichlorophenol *Anal. Chim. Acta.* **786** 16–21

[80] Wong A, Foguel M V, Khan S, Oliveira F M, De Tarley C R T and Sotomayor M D P T 2015 Development of an electrochemical sensor modified with MWCNT-COOH and MIP for detection of diuron *Electrochim. Acta.* **182** 122–30

[81] Khadem M, Faridbod F, Norouzi P, Foroushani A R, Ganjali M R and Shahtaheri S J 2016 Biomimetic electrochemical sensor based on molecularly imprinted polymer for dicloran pesticide determination in biological and environmental samples *J. Iran. Chem. Soc.* **13** 2077–84

[82] Zamora-Gálvez A, Mayorga-Matinez C C, Parolo C, Pons J and Merkoçi A 2017 Magnetic nanoparticle-molecular imprinted polymer: a new impedimetric sensor for tributylin detection *Electrochem. Commun.* **82** 6–11

IOP Publishing

Molecularly Imprinted Polymers for Environmental Monitoring
Fundamentals and applications
Raju Khan and Ayushi Singhal

Chapter 12

Molecularly imprinted polymers for the detection of pharmaceutical residues

Antony Nitin Raja, Ayushi Singhal, Priya Chauhan, Raju Khan and Hitesh Malviya

Our planet Earth is facing numerous issues and many of them are interconnected in some or another way. Typically, pollution and population growth are the root causes of all issues. More medical care is needed to treat the ageing and expanding human population, which directly or indirectly drives worldwide pharmaceutical production and consumption. Pharmaceutical research has grown significantly over the years in an effort to provide safe and healthy lives, but at the same time many studies show alarming signs that these pharmaceutical medications are now endangering the environment. These pharmaceutical residues are like a slow poison—an emerging threat to biotics. Molecularly imprinted polymer (MIP) technology has been around for more than five decades, but it has started gaining more attention as an emerging technique with numerous potential uses in the detection of pharmaceuticals. Molecular imprinting is a potent tool which can customize artificial receptors. It has brought about a revolution in the field of sensor development. Although there are many existing techniques which can detect pharmaceutical residues they have certain drawbacks. The current chapter focuses on discussing the integration of MIPs with sensing equipment, the current advancements and techniques in molecular imprinting procedures, their applications, in particularly in the detection of drug molecules as residues in the environment, as well as future perspectives on this topic.

12.1 Introduction

Molecularly imprinted polymers (MIPs) are synthetic matrices that resemble biological receptor systems. They may bind a target molecule with comparable affinity and specificity to antibodies as well as enzymes. This distinguishing characteristic of MIPs is added during the synthesis process, during which functional as well as crosslinking monomers are co-polymerized in the presence of a target analyte which

doi:10.1088/978-0-7503-4962-8ch12

is the imprinting molecule. Since Wulff and Sarhan first introduced the concept in 1972 [1], molecularly imprinted materials have been the subject of significant research. They have attracted scientific attention, inspired by their favourable characteristics: simplicity, robustness, stability, ease of preparation, as well as high affinity and selectivity towards the target molecule. The binding processes that take place when natural molecular species such as antibodies and biological receptors interact have fascinated generations of scientists. Consequently, many methods have been employed over time to imitate these interactions [2]. The development of strong molecular recognition materials that can imitate these natural processes is now possible through the use of molecular imprinting technology.

The applications of molecularly imprinted materials have expanded to cover the areas such as the separation sciences and purification such as extraction and chromatographic sorbents, chemical sensors, catalysis, drug delivery, and biological antibodies along with receptors systems as well [3]. MIPs are known to be among the most desirable materials among the polymeric materials produced in the field of bioanalytical and pharmaceutical applications. Because of the high selectivity MIPs possess toward pharmaceutical compounds, there are several companies that have followed the route to commercialize MIP sorbents.

The most commercially viable use of MIP is for purification, notably in analytical chemistry; other applications still require further research. The electrochemical sensing of active pharmaceutical entities in pharmaceutical formulations as well as in biological matrices has evolved greatly with the use of various electrochemical signal transduction techniques that use MIPs as a recognition component [4]. Molecular imprinting technology is being upgraded by using nanomaterials such as graphene, carbon nanotubes, and nanoparticles as well as several electrode modifiers so as to enhance the performance of the sensor. This has given scientists the opportunity to identify and rectify problems related to environmental health. Numerous factors are responsible for the deterioration of environmental health. Slowly but steadily, pharmaceutical use is increasing in day-to-day activity [5] and pharmaceutical residues are an emerging factor in the deterioration of environmental health, known as an emerging pollutant. These are the unmetabolized drug molecules found in surface and ground waters in low concentrations but which have a magnified impact. Rachel Carson—the American marine biologist, writer, and conservationist—first raised concerns about emerging pollutants in 1962. Emerging pollutants have been defined as all synthetic or naturally occurring chemicals which are not encompassed in the routine monitoring programme, while they possess the potential to enter the environment and may cause known or suspected adverse ecological as well as human health effects. Pharmaceutical residue as an emerging contaminant is receiving serious attention from researchers and scientists. Pharmaceutical residue is a serious problem because of its negative effects on ecosystems [6]. Moreover, the presence of pharmaceutical residues in drinking water, even at levels below the permissible concentration, poses a serious threat to human health because they are persistent pollutants, extremely poisonous, and poorly biodegradable substances. It was found that groundwater, drinking water, surface water, and wastewater all contained substantial amounts of pharmaceutical residues [7]. Several antibiotics, for example tetracycline, ciprofloxacin, amoxicillin, and metronidazole, are among the most widely used medicines because of their powerful capacity to treat bacterial infections and other ailments [8]. As a result, a significant volume of these drugs accumulate in wastewater, potentially endangering ecosystems by causing an increase in drug resistance genes. The reason behind the conversion of life-saving drugs to life threatening drugs is the incomplete metabolization of drugs in humans, plants, and animals. The route of entry for these incompletely metabolized pharmaceutical drugs into the environment is through a variety of excretion channels, such as wastewater discharge, agricultural land runoff, and human excreta. The different routes taken by pharmaceutical contaminants are shown in figure 12.1 [9, 10].

These leftover pharmaceutical drugs and their metabolites have the potential to cause deadly diseases such as cancer in humans, cause allergic reactions in animals, and enhance the chance of bacterial resistance in vulnerable aquatic creatures. Further they can also have detrimental impacts, including deteriorating ecological health or water, producing aberrant growth in aquatic creatures, and upsetting ecosystems [11]. The administration of medications plays a crucial role in both animal and human lives. These drugs are created to have specific impacts on the host organisms. These medications are often disposed off carelessly when they are no longer needed or have become spent and expired, particularly down sinks as well as bathrooms, where they enter leachates, landfills, and sludge, and while in wastewater

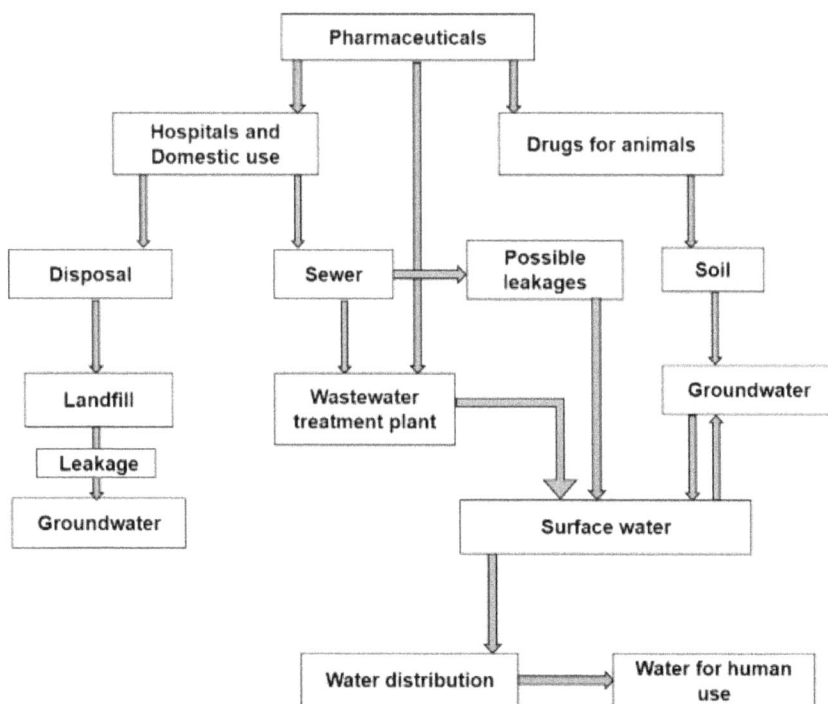

Figure 12.1. Routes of pharmaceutical contaminants. (Reproduced with permission from [10]. Copyright 2022 Elsevier.)

treatment facilities they end up polluting aquatic ecosystems and receiving water bodies as well [12]. The presence of pharmaceutical formulations and their concentration over time in ecosystems is found to have an impact on microbial populations by promoting the growth of drug resistant pathogens and microorganisms [13]. Several analytical techniques, including potentiometry, voltammetry, atomic absorption and emission methods, polarography, high-performance liquid chromatography, gas chromatography, spectrophotometry, spectrofluorimetric analysis, and electrochemical detection, including various nanocomposites used to modify glassy carbon electrodes, have been discussed for the estimation of pharmaceutical residues in different matrices [14]. Additionally, numerous potentiometric sensors using several ion-exchange materials as well as ion pairs as electroactive materials have been reported for the electrochemical assessment of drug residues [15]. Analytical chemistry researchers have been working on developing alternatives to conventional analytical techniques in a variety of fields recently, including environmental pollution and health monitoring. Such techniques often require sophisticated as well as expensive instrumentation, a high consumption of organic solvents, lengthy analysis times, specialized analysts, and, occasionally, costly and time-consuming sample preparations as well [16].

MIPs are known to be helpful for several analytical techniques, including separation, solid-phase extraction, etc, and are commonly utilized as mock antibodies in pharmaceutical and biochemical investigations. MIPs present significant as well as outstanding use today in target compound separation, membrane separation, sensors, and environmental evaluation [17]. This is particularly true for the separation of toxic chemicals. The use of MIPs in the industrial, pharmaceutical, agricultural, and environmental field relies on membrane manufacturing techniques, catalysts, drugs, membranes, cell culture, crystallization, sensors, and analytical studies involving solid-phase extraction, MIP-based solid-phase extraction, as well as high-performance liquid chromatography [18]. The purification of chemical as well as biological reagents, targeted medication administration, chromatographic separation, and environmental analysis are just a few uses made possible by the advantageous characteristics of MIPs. Samples for culinary, environmental, and biological analyses are pretreated with MIPs for commercialization and application [19–21]. When applied in the screen-printed electrode (SPE) method, MIP materials seems to be ideal in comparison to immune-extraction sorbents, as they possess greater adsorption capacity and are highly stable, and thus are better suited for the extraction and enrichment of small amounts of analytes, and might be produced inexpensively as well [22]. However, the MIP production process has a significant flaw that prevents targets from being easily removed from the highly cross-linked polymer, causing accuracy and other performance metrics to be compromised. A few studies attempt to resolve such issues and thus completely remove the template using washing techniques. Instead of the target analyte with a structurally desired molecule, 'dummy templates' are also utilized for the basis of imprint production [23]. The memory effect does not interfere with this method's typical performance as the polymer may still bind to the template molecule even though the leaching template does not co-elute along with the target throughout the chromatographic analysis [24].

12.2 MIPs—artificial antibodies

In last few years, the area of molecular imprinting has been growing rapidly. They are known as artificial templates that bind a specific analyte. MIPs work on the principle of a 'lock and key' mechanism and thus shows specific binding [25]. The first work was reported by D H Dickey, where he primed specific adsorbents [26]. In terms of effective bioimprinting, there has since been ongoing research concerning the preparation of synthetic materials at the nanoscale level. The term 'imprints' was first used in 1955 by Haldeman and Emmett as they observed the indentations left by dye molecules when they were removed from a silica matrix. The footprints or imprints left by the dyes displayed geometry as well as properties parallel to the dyes [27]. As silica possesses the capability to generate imprints, research based on silica was continued by Beckett and co-workers in 1957, where silica was engaged for producing imprints of organic molecules [28]. In addition, imprints based on silica were also utilized for biological receptors [29]. During the 1970s work based on covalent imprinting was published, followed by the non-covalent approach in the 1980s [30–32]. The term 'imprinting polymer' was introduced in 1984 and since then many studies have been published that use the term 'molecularly imprinted polymers' [33]. MIPs possess the ability to bind specifically to molecules such as amino acids, metal ions, antibiotics, drug additives, food additives, pesticides, as well as several environmental pollutants. Additionally, MIPs complement natural antibodies. Natural antibodies are known to be functional polypeptide chains and biological materials. They possess numerous disadvantages such as instability if exposed to extreme conditions, thus it is very challenging to obtain them in large amounts, and the synthesis procedure is quite expensive. On the other hand, MIPs are synthetic materials and have the capability to overcome the limitations of antibodies as they are low-priced, physically as well as chemically stable in nature, extremely specific and selective toward target analytes, and also have excellent longevity [34, 35]. MIPs are mostly applied in immune affinity separation, biosensors, bioimaging, food analysis, explosive detection, pathogen detection, chiral molecule detection, pharmaceutics detection, as well as in clinical treatment as drug delivery agents [36]. The choice of the components applied for the molecular imprinting process depend on the material of the template. In addition to the choice of a well-suited template, the initiators, configuration of the solvent, functional monomers, as well as cross-linkers should be chosen carefully. All of these factors shows a substantial impact on the performance of MIPs [37]. A schematic description of the synthesis procedure for MIPs is provided in figure 12.2 [38].

12.3 MIPs—life-saving tools

Each year, thousands of tonnes of pharmacologically active substances are used to cure diseases, avoid unintended pregnancies, and cope with the strains of modern life. Drug use is even more prevalent in animal farming, where hormones, para-siticides, and antibiotics are used as feed additives. Some of these compounds are excreted unmetabolized or as active metabolites; they avoid breakdown in waste treatment plants and enter the environment. This contamination might also be a

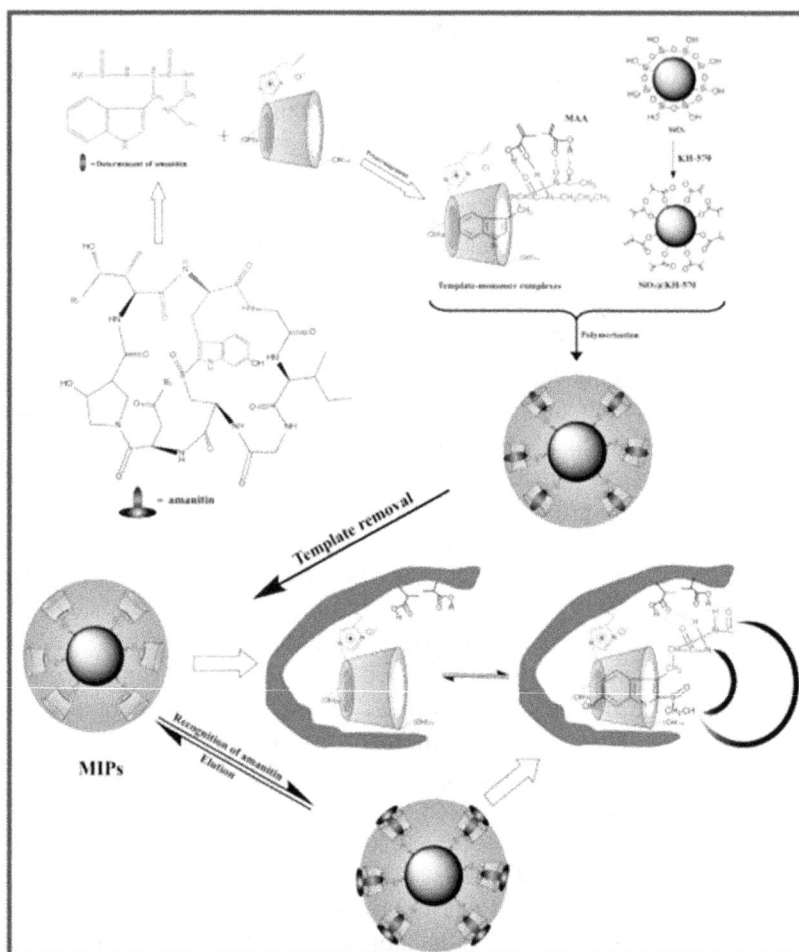

Figure 12.2. Schematic illustration of the synthesis procedure for MIPs. (Reproduced with permission from [38]. Copyright 2020 Elsevier.)

result of improper disposal of expired medication [39]. Manufacturing plants can be significant local point sources. Pharmaceutical drugs can have extended half-lives, so they build up until they are at detectable and biologically active levels. Many routinely prescribed medications, including erythromycin, cyclophosphamide, naproxen, sulphamethoxazole, and sulphasalazine, have a longer environmental persistence than a year [40]. A variety of life-saving medications have been discharged into environmental water during the past few decades from a number of sources and pharmaceutical residues have significantly contaminated both the surface water and groundwater. Due to their high toxicity and bioaccumulation, all of these residues have a detrimental impact on aquatic ecosystems and human health [41]. As a result, sophisticated wastewater treatment must be planned to eliminate pharmaceutical residues. To remove contaminants from wastewater, a number of treatments have

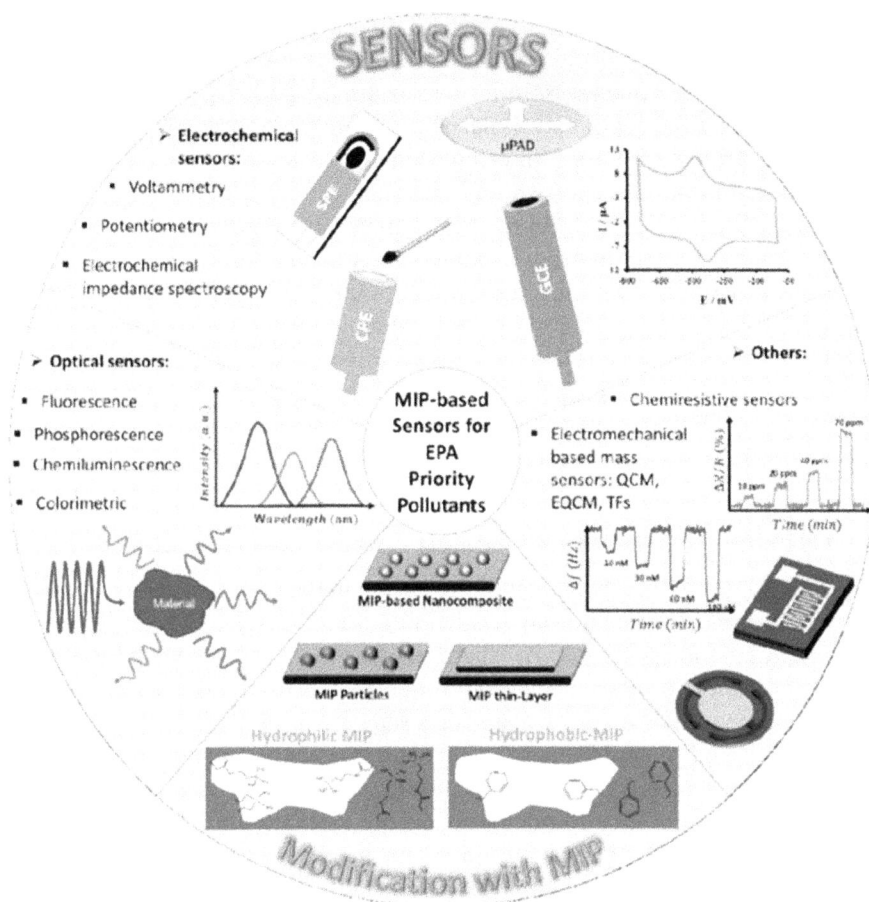

Figure 12.3. MIP-modified sensors for EPA priority pollutants. (Reproduced from [39]. Copyright 2021 the authors. CC BY 4.0.)

been developed, including physical adsorption, oxidation, and bioremediation. The majority of these treatments can only partially eliminate these unmetabolized medications because of their extremely low concentrations (usually at levels of parts per billion or parts per trillion) [42]. Additionally, in the presence of multiple simultaneous pollutants, these procedures are unable to successfully remove certain pollutants from wastewater. These factors highlight the urgent need for novel, selective, and efficient techniques to remove contaminants. Recently, MIPs have gained considerable attention because of their strong selectivity [43]. MIP-modified sensors for EPA priority pollutants are shown in figure 12.3 [44]. It is commonly accepted that samples utilized in the detection of drug residues and additives are typically organic small molecules in an aqueous solution. The majority of MIPs, which use organic small molecules as their template molecules, only exhibit excellent molecular identification capabilities in organic solutions. MIPs recognize target molecules through hydrogen bonds in organic solutions, but in aqueous solutions hydrogen bonds are significantly weaker due to a high hydration action, which has an impact on the MIPs' molecular identification ability [45]. It is crucial to carry out research on MIPs in aqueous solutions in order to address the shortcomings of current MIPs, whose template molecule is aimed at organic small molecule identification strategies. Other intermolecular interactions such as electrostatic contact and metal ion chelation should be taken into account for weak hydrogen bonding [46]. Water molecules do not affect metal ion chelation, even though they are more powerful than hydrogen bonds in aqueous solutions. In an aqueous solution, stable, specific binding sites between a metal ion and a template molecule are created for selective recognition [47]. Metal ion chelation, which is a bond interaction, makes it easier for metal ions to interact with imprinted molecules in aqueous solutions by mildly generating and cleaving them. Molecular imprinted technology using all the necessary elements has the potential to save the environment from the slow poison of pharmaceutical residues [48].

12.4 MIP-based sensors for the detection of pharmaceutical residues

The MIP technique gained considerable attention in the area of sensor development. In the near future, MIP technology has the potential to overtake techniques such as chromatography and spectroscopy. There are two main factors which make sensor technology superior to other technologies, i.e. the sensitivity and selectivity toward the target compound. The performance of a sensor can be enhanced by the efficient modification of the electrode surface through the immobilization of recognition elements. In molecular imprinting technology this is achieved by selective and sensitive binding of the target compounds to the fabricated electrochemical sensor. Molecular imprinting based sensors give the flexibility to analyse a component onsite as well as online. These techniques are cost effective and uncomplicated. MIP-based sensors have a wide variety of applications, one of which is the detection of environmental pollutants, such as pesticides, herbicides, pathogens, drug residues, heavy metals, industrial waste, etc [39, 49–52].

Environmental health is a broad and complex subject area, that can be examined through human health effects from exposure to harmful agents in the environment. Pharmaceutical drug residues are silently playing a supporting role in deteriorating environmental health. This concern has been brought o light by the widespread detection of pharmaceuticals in all aquatic compartments [53]. Antibiotics are a class of drugs that has been used extensively in both human and veterinary medicine. Recently, a lot of antibiotics have been found in the environment due to their widespread use and large production, in addition to the improper disposal of these drugs [54]. Pharmaceuticals are specifically created to exert an impact on particular organs, tissues, or cells in living systems and many of them remain persistent in the body. This is known as having an explicit mechanism of action. As a result, when medications and their unmodified metabolites are released into the environment through a variety of sources and channels, they can have an impact on both humans and animals. From a biological perspective, even potentially 'safe' medications could have detrimental consequences [55]. Nanostructured electrochemical platforms made of carbonaceous nanoparticles, metal nanoparticles, magnetic nanoparticles, metal–organic platforms, and quantum dots have been offered as a solution to this issue for the purpose of detecting pharmaceutical residues [56]. Cefixime is a typical representative of the commonly used cephalosporin antibiotics. Although it represents a beta-lactam antibiotic, which is often rendered inactive due to the action of the beta-lactamase enzyme, cefixime is more stable in the presence of beta-lactamase enzymes than other antibiotics in this family. This stability enhances its antibacterial potential but poses difficulties for its removal from water [57, 58]. Such pharmaceuticals have drawn attention from scientists all around the world due to their possible negative effects on organisms other than their intended targets. In addition to this they have played a vital role in the emergence of antibiotic resistance by promoting the growth of bacteria and genes that are resistant to antibiotics. They are categorized as a priority risk group because of this [59]. Researchers are showing tremendous interest the remediation of pharmaceutical residues as they have become a major element of concern. Several analytical methods have been developed to determine trace level antibiotics in biological or real samples. Analysis of pharmaceutical residues has been done using various analytical methods, including high-performance liquid chromatography, liquid chromatography with mass spectrometry and spectrophotometry, capillary electrophoresis fluorometry, etc [60]. Even though chromatographic and spectrophotometric procedures have been systematically employed for many analyses with high sensitivity and good selectivity, these methods suffer from disadvantages such as expensive instrumentation, a requirement of hazardous organic solvents in large quantities, and time-consuming pretreatment. Several steps are also required for the purification and pre-concentration of pharmaceutical residue when employing these methods [61].

The MIP technique combined with electrochemical sensors or other sensor technology has proven advantages such as low cost, simplicity of operation, high sensitivity and selectivity, and a fast response time. In addition, only a small amount of an analyte is needed, in comparison with other techniques. These qualities have made them superior to other techniques. These key basic features made it possible for MIP-based sensors to satisfy the requirements to overcome false detection in real sample analysis. MIP electrochemical sensors may be made using a variety of approaches consisting of bulk polymerization, the sol–gel method, electropolymerization, and layer-by-layer deposition [62]. Among these, electropolymerization is a simple, safe, and convenient technique of fabricating MIP films directly on the surface of an electrode. The thickness and morphology of the electropolymerized layer can be managed through adjusting the voltage and the quantity of cycles. Selectivity is one of the primary characteristics of MIP-based sensors. Sensitivity is another crucial aspect in an electrochemical sensor for the detection of trace levels of analyte. In addition, many authors have recognized that modifying the sensors with different nanoparticles or adding them to the structure of the electrodes are effective procedures for enhancing the sensitivity and selectivity of the sensors, with the specific binding sites forming predominantly at the material's surface as well as regular morphology, large surface, and good stability [63]. Different types of nanomaterial based electrodes have emerged over the last few decades, considerably altering the scope and sensitivity of voltammetric procedures for the measurement of diverse analytes. When chosen carefully, these nanomaterials have the ability to generate noteworthy improvements in all of the classical analytical methods. Recently, nanomaterials have attracted the attention of scientists worldwide due to their high specific surface area and large pore volume. These characteristics are highly desirable in development of nanoelectrodes [64]. The use of these materials as modifiers in the fabrication of sensors in electro-analytical studies has been increasing lately. Some of the nanoparticles have provided significant breakthroughs in contemporary research [65]. These electrode modifiers have proved to be advantageous because of their properties such as high thermal stability, fast electron transfer, and excellent electrocatalytic activity. This clearly suggests that mesoporous carbon as an electron modifier will prove to be a promising candidate for electroanalytical studies in the future [66]. Dehghani and co-workers fabricated a glass carbon electrode using a drop casting method with graphene oxide and graphene nanowires of 2 μl each. After drying a thin conductive film of nanomaterial was formed which was further was dipped in HNO_3/H_2SO_4 solution containing 0.1 M aniline monomers in the presence of 5.0 mM cefexime. The aniline monomers and cefixime were electropolymerized via cyclic voltammetry. After washing, specific binding sites were created and had the ability to detect cefixime up to a detection limit of 7.1 nM. This method is very specific and accurate in detecting cefexime in tap water and serum samples [67].

The matter of increasing pollutants, even at low concentrations, threatens aquatic environments. Normally called micropollutants, they are introduced directly or indirectly from homes, health centres, commercial wastewater, and runoff water. Among the micropollutants, those of pharmaceutical origin constitute a significant concern because they are the most common and pose risks for aquatic life [68]. Among the different types of antibiotics the most commonly prescribed ones are amoxicillins. Amoxicillin belongs to the amino glycoside class of drugs [69]. Amoxicillin can be degraded by biotic and abiotic factors, but produces distinct intermediate products that are more resistant to breakdown and more toxic than the mother compound [70–72]. Through committed research, Ayankojo and co-workers achieved a breakthrough in developing amoxicillin (AMO)-MIP-modified surface plasmon resonance (SPR) sensors for the detection of amoxillin in tap water. They successfully designed a hybrid organic–inorganic MIP film selective towards amoxicillin and integrated with an SPR sensor. The sensor fabrication step used methacrylamide as the organic functional monomer, tetraethoxysilane as the inorganic precursor, and vinyltrimethoxysilane as the coupling agent. Characterized by SPR, this sensor showed 16 times higher binding capacity for amoxycillin than non-imprinted polymers. This sensor had ability to detect up to 73 pM amoxicillin as pharmaceutical residue in tap water [73]. Vancomycin is a wide-spectrum, glycopeptide antibiotic, used in particular within the prevention and treatment of significant or extreme infections caused by vancomycin inclined lines of methicillin resistant *Staphylococcus aureus*. It is often reserved as the 'drug of last resort' when different antibiotics are not effective. However, high vancomycin blood concentrations can result in serious unfavourable reactions, consisting of ear and renal toxicity, in addition to thrombotic phlebitis. To prevent over- or under-dosing, drug monitoring of vancomycin is usually recommended. An article by Cetinkaya *et al* reported a green synthesis route to develop molecularly imprinted electrochemical sensors for selective detection of vancomycin from aqueous and serum samples. Their study showed how green approaches can be good alternatives for chemically modified sensors. A schematic diagram describing the green synthesis of alginate on a TiO_2/MIP-GCE sensor is shown in figure 12.4 [74]. With the help of green chemistry, they tried to create materials suitable for recognizing and binding targets with high sensitivity and selectivity. This has opened a door for researchers to work with eco-friendly approaches so we can find a better and cost effective option for the well-known and not environmentally friendly organic functional monomers and cross-linkers [75]. The fabricated alginate@TiO_2/MIP-GCE sensor was characterized using scanning electron microscopy and Raman spectroscopy methods. Additionally, cyclic voltammetry (CV) and electrochemical impedance spectroscopy (EIS) methods were used for the electrochemical characterization. This newly developed green approach sensor offers high sensitivity, a very low limit of detection (LOD), i.e. 2.808 pM. In addition to this Cetinkaya and co-workers have tried to minimize the generation of waste. However, the use of expensive monomers and the large amounts of solvents required still remain challenges [76].

After cardiovascular diseases, cancer is reported as the second leading cause of death worldwide. In the last few years, several methods such as surgical procedures, radiation treatments, targeted treatments, interventional radiology, and immunotherapy have been developed and utilized in most cancer treatments. At present, one of the most commonly used cancer treatment techniques is chemotherapy—the use

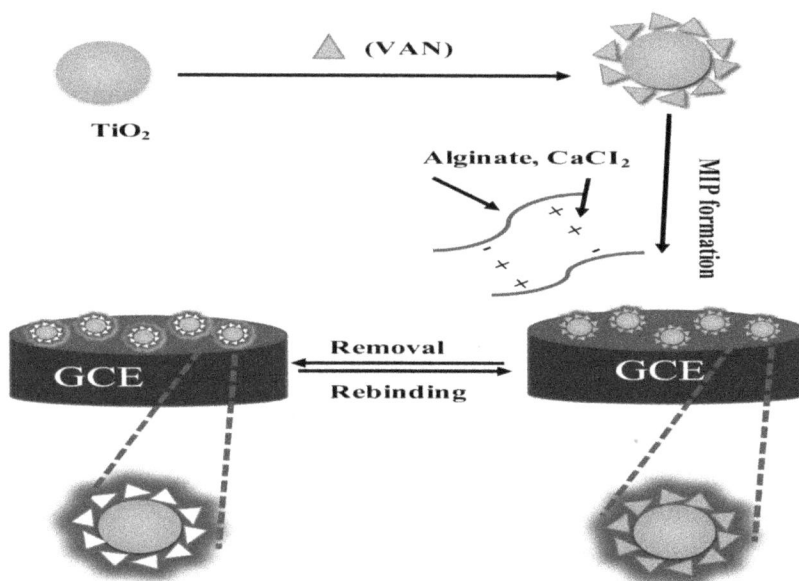

Figure 12.4. A schematic representation of an alginate@TiO_2/MIP-GCE sensor. A step towards green synthesis. (Reproduced with permission from [74]. Copyright 2022 Elsevier. CC BY 4.0.)

of medication to destroy, kill, reduce, or slow the growth of cancer cells. However the electrochemical detection of hydroxyurea (HU) was attempted to achieve a fast response, ease of handling, good stability, portability, and low cost [77]. Thus, a significant surge in research is required for the development of selective and sensitive MIP-based electrochemical sensors for HU analysis in real samples. Pathak *et al* for the first time developed a nanocomposite consisting of acrylated nitrogen doped with quantum dot bimetallic Au/Ag–MIPs, fabricated on an screen-printed carbon electrode (SPCE) surface, using diethyl dithiocarbamate functionalized nitrogen doped graphene quantum dots (N-GQDs) for surface polymerization. This fabricated sensor had an excellent diffusion coefficient and electron transfer kinetics for anticancerous HU, enabling a detection sensitivity as high as 0.05 ng ml^{-1} in real samples without any cross-reactivity or false-positives [78].

Oxytetracycline (OTC) is very commonly used in terrestrial cattle and also in aquaculture because of its affordability, availability, and efficiency. Due to its low bioavailability, it is very poorly absorbed in the animal gut and most OCT used in livestock production is excreted in the form of the parent compounds. In a study by Y Yang and co-workers, the first of its kind, a metal–organic framework was introduced in the construction of a molecularly imprinted photoelectrochemical sensor, and a novel molecularly imprinted photoelectrochemical (MIP-PEC) sensor was fabricated using a NH$_2$-MIL-125(Ti)–TiO$_2$ composite [79]. A simple one-step solvo-thermal method was used for the synthesis of the NH$_2$-MIL-125(Ti)–TiO$_2$ composite. The fabricated MIP-PEC based sensor has had a great impact in the field of environmental and food safety. However, this method has certain drawbacks, as the linear range in this work was still unsatisfactory compared to previously reported methods for the detection of oxtetracycline, and the sensor fabrication is a time-consuming and laborious process. This method has shown new insight into the design and application of metal–organic framework based MIP-PEC sensors. In the future more work should focus on achieving portable sensing platforms [80]. Many MIP-based electrochemical sensors have been designed and there are many more to come for the active detection of antibiotics as pharmaceutical residues in the environment (see table 12.1 for some examples). An MIP electrochemical sensor for the detection of the antidepressant drug trazodone was fabricated for the first time by Seguro and co-workers. They used an eletropolymerization technique using SPCE for the fabrication of electrode [81]. They used this sensor successfully for the determination of trazodone in real samples such as water and human serum. This MIP-based sensor seems to be very suitable for mass production and can be integrated with portable devices for the detection of trazodone as a pharmaceutical residue in the environment [82].

12.5 Conclusion and future perspectives

The emergent pharmaceutical interest in molecularly imprinted polymers is perhaps a direct result of its major advantages over other analytical approaches, namely the enhanced selectivity as well as sensitivity of the method. In order to improve the response of the signal, increase the sensitivity, and reduce the detection limit of the

Table 12.1. Detection of pharmaceutical residues using MIP-based sensors.

Name of drug	Type of drug	Sensor	Range	LOD	Sample
Cefixime	Antibiotic	MIP/GNW/GO	20–950 nM	7.1 nM	Serum and urine
Vancomycin	Antibiotic	AMO-MIP	0.1–2.6 nM	73 pM	Tap water/buffer
Hydroxyurea	Anticancerous	Au/AgNPs@N-GQDS-MIP/SPE	0.62–102.33	0.07 ng ml^{-1}	Blood/plasma
Oxtetracycline	Antibiotic	NH$_2$-MIL-125 (Ti)–TiO$_2$	0.1 nM–10 μM	60 pM	
Primaquine	Antimalaria	C$_{60}$ monoadduct-MIP/PGE	2.7–848.5 nM	0.80 nM	Aqueous environment
Pemetrexed	Anticancerous	MIP/CQD/SPCE	5–10 nM l^{-1}	1.61 nM l^{-1}	Synthetic serum sample
Kanamycin	Antibiotic	MIP/AuNPs-rGO-CS/GCE	10–500 nM	1.87 nM	Water, food, and biological samples
Flouxetine	Antidepressant	MIP/GCE	4.99×10^{-7} to 3.38×10^{-5} mol l^{-1}	3.33×10^{-7}	Blood serum
Clenbuterol	Bronchodilator	MIP-DLLME-IPS	5–40 μg l^{-1}	2 μg l^{-1}	Urine
Diethylsilbestrol	Antidepressant	MIP-QDS	2×10^{-4} to 10 mg l^{-1}	5.9×10^{-5} mg l^{-1}	Water
Trazodone	Antidepressant	MIP/SPCE	5–80 μM	1.6 μM	Tap water and human serum
Cetirizine	Antiallergic	CD-MIP	0.5–500 ng ml^{-1}	0.4 ng ml^{-1}	Human saliva and urine

sensors, due to the complexity of biological samples and the low levels of drugs in biological samples, molecularly imprinted polymers have been utilized. To generate MIP materials, a unique approach has been utilized in combination with electrochemical transduction in order to develop sensors for the analysis of various pharmaceutically active compounds. The electrochemical sensors offer various advantages such as low cost, short analysis time, simple design, portability, miniaturization, ease-of-use, and customization by means of an easy method for specific applications. Moreover, the performance of the sensors could be enhanced through integrating a few conductive nanomaterials such as Au nanoparticles, carbon nanotubes, graphene, nanowires, and magnetic nanoparticles in the polymeric matrix of MIP-based sensors. The application of novel electrochemical sensing supports new multifunctional MIPs and is anticipated to be developed and utilized extensively in the future.

Acknowledgements

The authors express their sincere thanks to the director of CSIR-AMPRI for his support and encouragement in this work. Raju Khan would like to acknowledge SERB for providing funds in the form of the IPA/2020/000130 project.

References

[1] Wulff G and Sarhan A 1972 The use of polymers with enzyme-analogous structures for the resolution of racemates *Angew. Chem. Int. Ed.* **11** 341–4

[2] Puoci F, Lemma F, Cirillo G, Picci N, Matricardi P and Alhaique F 2007 Molecularly imprinted polymers for 5-fluorouracil release in biological fluids *Molecules* **12** 805–14

[3] Lee S H and Doong R A 2016 Design of size-tunable molecularly imprinted polymer for selective adsorption of pharmaceuticals and biomolecules *Biosens. Bioelectron.* **7** 1000228

[4] Oyedeji A O, Msagati T A M M, Williams A B and Benson N U 2021 Detection and quantification of multiclass antibiotic residues in poultry products using solid-phase extraction and high-performance liquid chromatography with diode array detection *Heliyon* **7** 8469

[5] Ranjan N, Singh P K and Maurya N S 2022 Pharmaceuticals in water as emerging pollutants for river health: a critical review under Indian conditions *Ecotoxicol. Environ. Saf.* **247** 114220

[6] Graumans M H F, Hoeben W F L M, van Dael M F P, Anzion R B M, Russel F G M and Scheepers P T J 2021 Thermal plasma activation and UV/H$_2$O$_2$ oxidative degradation of pharmaceutical residues *Environ. Res.* **195** 110884

[7] Abd El-Monaem E M, Eltaweil A S, Elshishini H M, Hosny M, AbouAlsoaud M M, Attia N F, El-Subruiti G M and Omer A M 2022 Sustainable adsorptive removal of antibiotic residues by chitosan composites: an insight into current developments and future recommendations *Arabian J. Chem.* **15** 103743

[8] Barocio M E, Hidalgo-Vazquez E, Kim Y, Isabel Rodas-Zuluaga L, Chen W-N, Barcelo D, Iqbal H N M, Parra-Saldívar R and Castillo-Zacarías C 2021 Portable microfluidic devices for in-field detection of pharmaceutical residues in water: recent outcomes and current technological situation—a short review *Case Stud. Chem. Environ. Eng.* **3** 100069

[9] Ding R, Chen Y, Wang Q, Wu Z, Zhang X, Li B and Lin L 2022 Recent advances in quantum dots-based biosensors for antibiotics detection *J. Pharm. Anal.* **12** 355–64

[10] Samal K, Mahapatra S and Ali M H 2022 Pharmaceutical wastewater as emerging contaminants (EC): treatment technologies, impact on environment and human health *Energy Nexus* **6** 100076

[11] Kayode-Afolayan S D, Ahuekwe E F and Nwinyi O C 2022 Impacts of pharmaceutical effluents on aquatic ecosystems *Sci. Afr.* **17** e01288

[12] Graumans M H F, van Hove H, Schirris T, Wilfred F L M H, van Dael M F P, Anzion R B M, Russel F G M and Scheepers P T J 2022 Determination of cytotoxicity following oxidative treatment of pharmaceutical residues in wastewater *Chemosphere* **303** 135022

[13] Preda D, David I G, Popa D-E, Buleandra M and Radu G L 2022 Recent trends in the development of carbon-based electrodes modified with molecularly imprinted polymers for antibiotic electroanalysis *Chemosensors* **10** 243

[14] Feroz M, Lopes I C, Rehman H, Ata S and Vadgama P 2020 A novel molecular imprinted polymer layer electrode for enhanced sensitivity electrochemical determination of the antidepressant fluoxetine *J. Electroanal. Chem.* **878** 114693

[15] Ansari S and Karimi M 2017 Novel developments and trends of analytical methods for drug analysis in biological and environmental samples by molecularly imprinted polymers *TrAC Trends Anal. Chem.* **89** 146–62

[16] Barati A, Kazemi E, Dadfarnia S and Shabani A M H 2016 Synthesis/characterization of molecular imprinted polymer based on magnetic chitosan/graphene oxide for selective separation/preconcentration of fluoxetine from environmental and biological samples *J. Ind. Eng. Chem.* **46** 212–21

[17] Cai D, Zhu J, Li Y, Li L, Zhang M, Wang Z, Yang H, Li J, Yang Z and Chen S 2020 Systematic engineering of branch chain amino acid supply modules for the enhanced production of bacitracin from *Bacillus licheniformis Metab. Eng. Commun.* **11** 136

[18] Maciuca A M, Munteanu A C and Uivarosi V 2020 Quinolone complexes with lanthanide ions: an insight into their analytical applications and biological activity *Molecules* **25** 1347

[19] Wang X, Ding H, Yu X, Shi X, Sun A, Li D and Zhao J 2019 Characterization and application of molecularly imprinted polymer-coated quantum dots for sensitive fluorescent determination of diethylstilbestrol in water samples *Talanta* **197** 98–104

[20] Zhaoa H, Wanga H, Quana X and Tan F 2013 Amperometric sensor for tetracycline determination based on molecularly imprinted technique *Procedia Environ. Sci.* **18** 249–57

[21] Hu R, Tang R, Xu J and Lu F 2018 Chemical nanosensors based on molecularly-imprinted polymers doped with silver nanoparticles for the rapid detection of caffeine in wastewater *Anal. Chim. Acta.* **1034** 176–83

[22] Mzukisi L, Nikita M, Tavengwa T and Chimuka L 2017 Applications of molecularly imprinted polymers for solid-phase extraction of non-steroidal anti-inflammatory drugs and analgesics from environmental waters and biological samples *J. Pharm. Biomed. Anal.* **147** 624–33

[23] Kolera A, Gornikb T, Kosjekb T, Jerabekd K and Krajnc P 2018 Preparation of molecularly imprinted copoly(acrylic acid-divinylbenzene) for extraction of environmentally relevant sertraline residues *React. Funct. Polym.* **131** 378–83

[24] MassimaMouele E S *et al* 2021 Removal of pharmaceutical residues from water and wastewater using dielectric barrier discharge methods—a review *Int. J. Environ. Res. Public Health* **18** 1683

[25] Bhalla N, Jolly P, Formisano N and Estrela P 2016 Introduction to biosensors *Essays Biochem.* **60** 1–8

[26] Koedrith P, Thasiphu T, Weon J I, Boonprasert R, Tuitemwong K and Tuitemwong P 2015 Recent Trends in Rapid Environmental Monitoring of Pathogens and Toxicants: Potential of Nanoparticle-Based Biosensor and Applications *Sci. World J.* **2015** 1–12

[27] Yılmaz E, Ozgur E, Bereli N, Türkmen D and Denizli A 2017 Plastic antibody based surface plasmon resonance nanosensors for selective atrazine detection *Mater. Sci. Eng.* C **73** 603–10

[28] Rodriguez-Mozaz S, de Alda M J L and Barcelo D 2006 Biosensors as useful tools for environmental analysis and monitoring *Anal. Bioanal. Chem.* **386** 1025–41

[29] Afshar M G, Crespo G A and Bakker E 2016 Flow chronopotentiometry with ion-selective membranes for cation, anion, and polyion detection *Anal. Chem.* **88** 3945–52

[30] Bakker E, Bühlmann P and Pretsch E 1997 Carrier-based ion-selective electrodes and bulk optodes. 1. General characteristics *Chem. Rev.* **97** 3083–132

[31] Bobacka J, Ivaska A and Lewenstam A 2008 Potentiometric ion sensors *Chem. Rev.* **108** 329–51

[32] Chen L D, Mandal D, Pozzi G, Gladysz J A and Bühlmann P 2011 Potentiometric sensors based on fluorous membranes doped with highly selective ionophores for carbonate *J. Am. Chem. Soc.* **133** 20869–77

[33] Meyerhoff M E, Fu B, Bakker E, Yun J H and Yang V C 1996 Polyion-sensitive. membrane electrodes for biomedical analysis *Anal. Chem.* **68** 168A–75A

[34] Bühlmann P, Badertscher M and Simon W 1993 Molecular recognition of creatinine *Tetrahedron* **49** 595–8

[35] Erdőssy J, Horvath V and Gyurcsanyi R E 2016 Electrosynthesized molecularly imprinted polymers for protein recognition *TrAC Trends Anal. Chem.* **79** 179–90

[36] Banerjee S and Konig B 2013 Molecular imprinting of luminescent vesicles *J. Am. Chem. Soc.* **135** 2967–70

[37] Fuchs Y, Soppera O, Mayes A G and Haupt K 2013 Holographic molecularly imprinted polymers for label-free chemical sensing *Adv. Mater.* **25** 566–70

[38] Tana L, Li Y, Denga F, Pana X, Yua H, Marina M L and Jiang Z 2020 Highly sensitive determination of amanita toxins in biological samples using β-cyclodextrin collaborated molecularly imprinted polymers coupled with ultra-high performance liquid chromatography tandem mass spectrometry *J. Chromatogr.* A **1630** 461514

[39] Huang D-L *et al* 2015 Application of molecularly imprinted polymers in wastewater treatment: a review *Environ. Sci. Pollut. Res.* **22** 963–77

[40] Zuccato E, Calamari D, Natangelo M and Fanelli R 2000 Presence of therapeutic drugs in the environment *Res. Lett. Lancet* **355** 1789–90

[41] Ayankojo A G, Reut J, Nguyen V B C, Boroznjak R and Syritski V 2022 Advances in detection of antibiotic pollutants in aqueous media using molecular imprinting technique—a review *Biosensors* **12** 441

[42] Sarpong K A, Xu W, Huang W and Yang W 2019 The development of molecularly imprinted polymers in the clean-up of water pollutants: a review *Am. J. Anal. Chem.* **10** 202–26

[43] Xiao D, Jiang Y and Bi Y 2018 Molecularly imprinted polymers for the detection of illegal drugs and additives: a review *Microchim. Acta.* **185** 247

[44] Zarejousheghani M, Rahimi P, Borsdorf H, Zimmermann S and Joseph Y 2021 Molecularly imprinted polymer-based sensors for priority pollutants *Sensors* **21** 2406

[45] Abd F N, Joda B A and Al-Bayati Y K 2020 Synthesis of molecularly imprinted polymers for estimation of anticoagulation drugs by using different functional monomers *AIP Conf. Proc.* **2290** 030042

[46] Sfoog A A A, Bakar N A, Rahim N A, Mahamod W R W, Hashim N and Che Soh S K 2022 Recent Advances and Future Prospects of Molecular Imprinting Polymers as a Recognition Sensing System for Food Analysis: A Review *Indones. J. Chem.* **22** 1737–58

[47] Ayerdurai V, Lach P, Lis-Cieplak A and Cieplak M 2022 An advantageous application of molecularly imprinted polymers in food processing and quality control *Crit. Rev. Food Sci. Nutr.* **27** 1–34

[48] Guc M and Schroeder G 2017 The molecularly imprinted polymers influence of monomers on the properties of polymers—a review *World J. Res. Rev.* **5** 36–47

[49] Vu V-P, Tran Q-T, Pham D-T, Tran P-D, Thierry B, Chu T-X and Mai A-T 2019 Possible detection of antibiotic residue using molecularly imprinted polyaniline-based sensor *Vietnam J. Chem.* **57** 328–33

[50] Penuela-Pinto O, Armenta S, Esteve Turrillas F A and de la Guardia M 2016 Selective determination of clenbuterol residues in urine by molecular imprinted polymer—Ion mobility spectrometry *Microchem. J.* **134** 62–7

[51] Tarannum N, Khatoon S and Dzantiev B B 2020 Perspective and application of molecular imprinting approach for antibiotic detection in food and environmental samples: A critical review *Food Control* **118** 107381

[52] Viveiros R, Karim K, Piletsky S A, Heggie W and Casimiro T 2017 Development of a molecularly imprinted polymer for a pharmaceutical impurity in supercritical CO_2: Rational design using computational approach *J. Clean. Prod.* **168** 1025–31

[53] Pereira A, Silva L, Laranjeiro C, Lino C and Pena A 2020 Selected pharmaceuticals in different aquatic compartments: part II—toxicity and environmental risk assessment *Molecules* **25** 1796

[54] Yang Q, Gao Y, Ke J, Show P L, Ge Y, Liu Y, Guo R and Chen J 2021 Antibiotics: an overview on the environmental occurrence, toxicity, degradation, and removal methods *Bioengineered* **12** 7376–416

[55] Kar S, Roy K, Leszczynski J and Nicolotti O (ed) 2018 *Impact of Pharmaceuticals on the Environment: Risk Assessment Using QSAR Modeling Approach Comput. Toxicol* 1800 395–443

[56] Joshi A and Kim K-H 2020 Recent advances in nanomaterial-based electrochemical detection of antibiotics: Challenges and future perspectives *Biosens. Bioelectron.* **153** 112046

[57] Zhang T, Zhou R, Wang P, Mai-Prochnow A, McConchie R, LiW, Zhou R, Thompson EW, Ostrikov K K and Cullen P J 2020 Degradation of cefixime antibiotic in water by atmospheric plasma bubbles: Performance, degradation pathways and toxicity evaluation *Chem. Eng. J.* **421** 127730

[58] Imrana M, RazaShahb M, Ullaha F, Ullaha S, Elhissic A M A, Nawazd W, Ahmadb F, Sadiqa A and Al I 2016 Glycoside-based niosomal nanocarrier for enhanced *in-vivo* performance of Cefixime *Int. J. Pharm.* **505** 122–32

[59] Mirzaei R, Yunesian M, Nasseri S, Gholami M, Shoeibi E J S and Mesdaghinia A 2018 Occurrence and fate of most prescribed antibiotics in different water environments of Tehran, Iran *Sci. Total Environ.* **619–620** 446–59

[60] Masoudyfar Z and Elhami S 2019 Surface plasmon resonance of gold nanoparticles as a colorimetric sensor for indirect detection of Cefixime Spectrochim *Acta A Mol. Biomol. Spectrosc.* **211** 234–8

[61] Qin G, Wang J, Li L, Yuan F, Zha Q, Bai W and Ni Y 2021 Highly water-stable Cd-MOF/ Tb3+ ultrathin fluorescence nanosheets for ultrasensitive and selective detection of Cefixime *Talanta* **221** 121421

[62] Bompart M, Haupt K and Ayela C 2012 Micro and nanofabrication of molecularly imprinted polymers *Top. Curr. Chem.* **325** 83–110

[63] Yoshikawa M 2002 Molecularly imprinted polymeric membranes *Bioseparation* **10** 277–86

[64] El-Schich Z, Zhang Y, Feith M, Beyer S, Sternbæk L, Ohlsson L, Stollenwerk M and Wingren A G 2020 Molecularly imprinted polymers in biological applications *Biotechniques* **69** 406–19

[65] Chen L, Wang X, Lu W, Wua X and Li J 2016 Molecular imprinting: perspectives and applications *Chem. Soc. Rev.* **45** 2137–211

[66] Elugoke S E, Adekunle A S, Fayemi O E, Akpan E D, Mamba B B, Sherif E-S M and Ebenso E E 2021 Molecularly imprinted polymers (MIPs) based electrochemical sensors for the determination of catecholamine neurotransmitters–Review *Electrochem. Sci. Adv.* **1** 1–43

[67] Dehghani M, Nasirizadeh N and Yazdanshenas M E 2019 Determination of cefixime using a novel electrochemical sensor produced with gold nanowires/graphene oxide/electropolymerized molecular imprinted polymer *Mater. Sci. Eng.* C **96** 654–60

[68] AremouDaouda M M, OnesimeAkowanou A V, ReineMahunon S E, Adjinda C K, PepinAina M and Drogui P 2021 *S. Afr. J. Chem. Eng.* **38** 78–89

[69] Sharma G, Pahade P, Durgbanshi A, Carda-Broch S, Peris-Vicente J and Bose D 2022 Application of micellar liquid chromatographic method for rapid screening of ceftriaxone, metronidazole, amoxicillin, amikacin and ciprofloxacin in hospital wastewater from Sagar District, India *Total Environ. Res. Themes* **1–2** 100003

[70] Chowdhury J, Mandal T K and Mondal S 2020 Genotoxic impact of emerging contaminant amoxicillin residue on zebra fish (*Danio rerio*) embryos *Heliyon* **6** e05379

[71] Debalke A, Kassa A, Asmellash T, Beyene Y, Amare M, Tigineh G T and Abebe A 2022 Synthesis of a novel diaquabis(1,10-phenanthroline)copper(II)chloride complex and its voltammetric application for detection of amoxicillin in pharmaceutical and biological samples *Heliyon* **8** e11199

[72] Fabregat-Safont D, Pitarch E, Bijlsma L, Matei I and Hernandez F 2021 Rapid and sensitive analytical method for the determination of amoxicillin and related compounds in water meeting the requirements of the European Union watchlist *J. Chromatogr.* A **1658** 462605

[73] Ayankojo A G, Reut J, Opik A, Furchner A and Syritski V 2018 Hybrid molecularly imprinted polymer for amoxicillin detection *Biosens. Bioelectron.* **118** 102–7

[74] Cetinkaya A, Yıldız E, Irem Kayaa S, Emin Çormana M, Uzunb L and Ozkan S A 2022 A green synthesis route to develop molecularly imprinted electrochemical sensor for selective detection of vancomycin from aqueous and serum samples *Green Anal. Chem.* **2** 100017

[75] Alzahrani A M, Naeem A, Alwadie A F, Albogami K, Alzhrani R M, Basudan S S and Alzahrani Y A 2021 Causes of vancomycin dosing error; problem detection and practical solutions; a retrospective, single-center, cross-sectional study *Saudi Pharm. J.* **29** 616–24

[76] Mu F, Zhou X, Fan F, Chen Z and Shi G 2021 A fluorescence biosensor for therapeutic drug monitoring of vancomycin using *in vivo* microdialysis *Anal. Chim. Acta.* **1151** 338250

[77] Saban N *et al* 2013 Antitumor mechanisms of amino acid hydroxyurea derivatives in the metastatic colon cancer model *Int. J. Mol. Sci.* **14** 23654–71

[78] Pathak P K, Kumar A and Prasad B B 2018 Functionalized nitrogen doped graphene quantum dots and bimetallic Au/Ag core-shell decorated imprinted polymer for electrochemical sensing of anticancerous hydroxyurea *Biosens. Bioelectron.* **127** 10–8

[79] Li Z-j, Qi W-n, Feng Y, Liu Y-w, Shehata E and Long J 2019 Degradation mechanisms of oxytetracycline in the environment *J. Integr. Agric.* **18** 1953–60

[80] Yang Y, Yan W, Wang X, Yu L, Zhang J, Bai B, Guo C and Fan S 2021 Development of a molecularly imprinted photoelectrochemical sensing platform based on NH_2-MIL-125(Ti)–TiO_2 composite for the sensitive and selective determination of oxtetracycline *Biosens. Bioelectron.* **177** 113000

[81] Akbaria V, Ghobadia S, Mohammadib S and Khodarahmi R 2020 The antidepressant drug; trazodone inhibits Tau amyloidogenesis: prospects for prophylaxis and treatment of AD *Arch. Biochem. Biophys.* **679** 108218

[82] Seguro I, Rebelo P, Pacheco J G and Delerue-Matos C 2022 Electropolymerized, molecularly imprinted polymer on a screen-printed electrode—a simple, fast, and disposable voltammetric sensor for trazodone *Sensors* **22** 2819

IOP Publishing

Molecularly Imprinted Polymers for Environmental Monitoring
Fundamentals and applications
Raju Khan and Ayushi Singhal

Chapter 13

Molecularly imprinted polymers for the sensing of microorganisms

Ayushi Singhal, Apoorva Shrivastava and Raju Khan

Molecularly imprinted polymers (MIPs) have attracted a lot of interest for applications in portable sensors as biomimetic recognition components capable of replacing antibodies because of their high affinity, low production cost, thermal durability, and superior chemical composition. Microorganisms pose a serious risk to human health as well as to plant life, therefore the sensing of microbes is very much needed. The excellent selectivity, ease-of-use, speed, and outstanding stability of microorganism imprinting on MIPs, as well as their affordability and environmental friendliness, have all drawn considerable interest for this purpose. In this chapter we discuss various detection techniques, including traditional and emerging methods. We further highlight electrochemical, fluorescence based, surface plasmon resonance (SPR) based, and quartz crystal microbalance (QCM) based detection and their applications in microbe sensing. At the end, the main obstacles and future possibilities for the design of microorganism-imprinted polymers are also outlined.

13.1 Introduction

The types of biological cells known as 'microorganisms' consist of bacteria, fungi, viruses, and microalgae, among other prokaryotes. Microorganisms are microscopic in size and come in a variety of shapes, including spirals, spheres, and rods. The great majority of ecosystems, including water, soil, radioactive waste, hot springs, and the biosphere of the Earth's crust, are inhabited by microorganisms, which were the first life to emerge on the planet. Many naturally occurring species have not yet been characterised, and only a few can be produced in a lab [1]. Microorganisms are closely linked to humans and may exist as parasites, symbionts, or irritants on the skin or in the gut. Despite the fact that the majority of them are benign, some species have been identified as harmful in clinical trials and are capable of spreading deadly diseases such as leprosy, cholera, bubonic plague, anthrax, syphilis, and syphilis.

In some situations, these infections can be fatal. For instance, TB alone claims the lives of 2 million people annually [2]. Enteric microorganisms such as bacteria, viruses, and protozoa, which mostly spread via the faecal–oral pathway, are often the source of waterborne infections. Among food-borne microbes, *Salmonella*, *Listeria monocytogenes*, and enterohemorrhagic *Escherichia coli* are responsible for several million cases of illnesses worldwide each year [3]. Microorganisms pose a serious risk to human health particularly considering the emergence of antibiotic resistance and the departure of many pharmaceutical companies whose core business was the discovery of antibiotics [4]. Every year, food-borne bacteria cause millions of illnesses throughout the world. Some bacteria may represent significant economic risks since they may result in agricultural losses from bacterially induced plant diseases [3, 5]. Plant diseases brought on by microorganisms are thought to cause the loss of 10% of crops annually around the globe, which can result in significant financial losses for farmers and even societal issues, particularly in emerging nations [6]. In most fields, pathogenic microorganism contamination is a serious and ongoing problem, making early and effective identification of these organisms crucial to halting the spread of catastrophic diseases [6]. However, bacteria work as a type of cell factory for the creation of chemical substances [7–11]. In a wide range of industries around the world, including food and beverage, sports, medical, and even the construction industry, microbiological contamination is a major concern. Particularly with regard to food and beverage items, contamination raises serious issues for both economic and health-related reasons. Food-borne disease is prevalent, with an estimated one-third of the population experiencing it each year, even in wealthy nations. Delaying the growth of particular bacteria or limiting contamination by bacterial pathogens are the best ways to maintain food quality and safety, so it is crucial to quickly identify and discard contaminated food [12]. Researchers must identify and detect pathogens as soon as feasible in order to limit the spread of pathogenic diseases, as pathogen contamination is a persistent issue in a variety of sectors [13]. The bulk of currently used microbiological techniques, analytical antibody assays, and nucleic acid-based assays, including polymerase chain reaction (PCR) tests are used to detect microorganisms. The analytical community is very interested in the development of a quick, dependable, on-site sensor platform for the detection of microbes because all of these approaches require time-consuming preparation, processes, and/or measurement protocols [14]. Due to the precise recognition between the antibody and the target antigen, which might be released by (e.g. virulence factors) or is present on the surface of the bacterium, antibodies are frequently utilised as a recognition element in sensor development. The drawbacks of antibody-based sensors, however, are their high cost, significant batch-to-batch fluctuation, low stability, and the fact that some antibodies still require animal testing for their manufacture [15]. As a result, numerous artificial receptors that are identical to the corresponding natural bioreceptors in that they can selectively bind target chemicals with high affinity are being designed and manufactured [16]. Due to their distinct qualities of structure predictability, recognition specificity, and application universality, molecularly imprinted polymers (MIPs) produced via molecular imprinting technology (MIT) have garnered considerable interest. MIPs have attracted a lot of interest for application in portable sensors as biomimetic recognition components capable of replacing antibodies because of their high affinity, low production cost, thermal durability, and superior chemical properties [17]. They can be made utilising a wide variety of monomers and synthesis techniques, which can result in a huge variety of sensitive and selective recognition components [18, 19]. It has been demonstrated that when combined with optical, electrochemical, and thermal detection techniques, they can generate highly specialised sensing platforms. MIPs are the height of adaptability when it comes to target sensing. They can be used for the detection of a wide variety of targets, from tiny molecules to proteins and cells, with little optimisation [20, 21]. In this chapter, we discuss various microbe detection methods using molecular imprinting techniques, their applications and future prospects. The key MIP characteristics, different MIP templates, and published literature are all shown in figure 13.1.

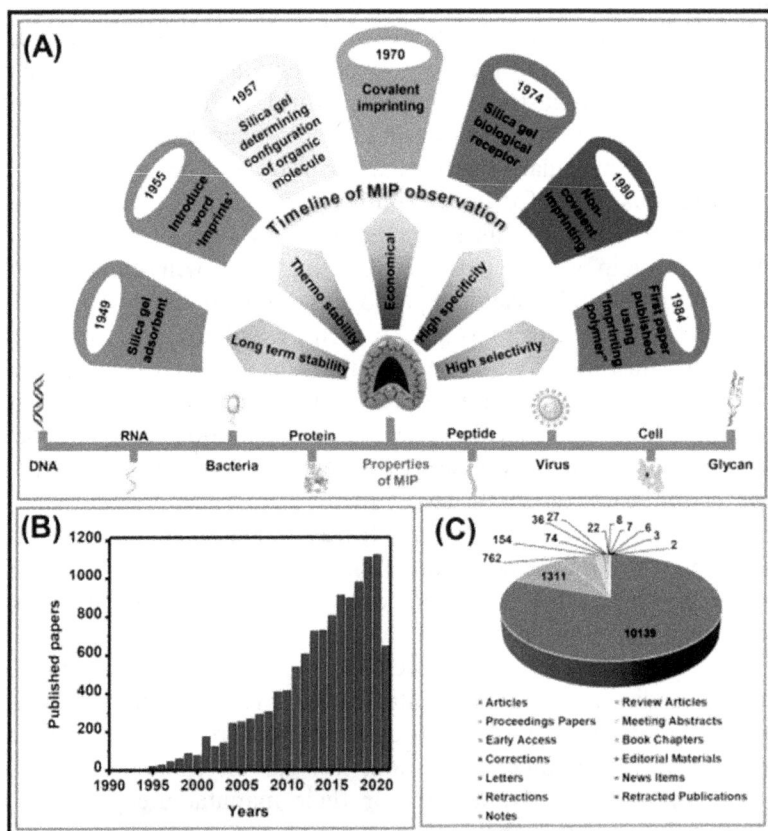

Figure 13.1. Representation of (A) various templates and properties along with the early history of MIPs. (B) and (C) The amount of year-wise published literature and its types. (Data were obtained from the Web of Science with 'Molecularly Imprinted Polymer' entered as the subject in the search box (last date accessed: 10.08.2021). (Reproduced with permission from [19]. Copyright 2022 Elsevier.)

13.2 Detection of microorganisms

Governments and societies all around the world are very concerned about the direct and indirect consequences that microbes and their products have on human health. Since many of the microorganisms found in the air, water, soil, food, plants, and animals are helpful or even necessary for human survival, it is important to distinguish between dangerous and safe bacteria and to ascertain their quantity in various types of samples [22]. The development of new techniques and more effective tools for this aim is currently a focus of study in several domains, including environment, food safety, and healthcare. Among them, designing and manufacturing more affordable biosensors with improved selectivity, sensibility, and stability is crucial [22, 23].

13.3 Traditional methods of detection and analysis

Traditional techniques for finding and analysing microorganisms often involve culture techniques, microscopy, antibody-based assays, polymer chain reaction (PCR), and genomic analysis, which can take up to a week [24, 25]. The nucleic acid-based tests, which specifically rely on the detection of nucleotide sequences inside the genome of a given microbe strain, require high levels of expertise as well as sample preparation techniques, such as cleaning and biomolecular purification [25, 26]. Although the culturing process takes a lot longer, it can produce findings that are more accurate [27, 28]. An immunoassay is a different strategy that makes use of antibodies and the chemical data on microorganisms' cell surfaces. Despite the benefits of high specificity and selectivity, antibodies nevertheless have several fundamental constraints, including high production costs, laborious processes, and low stability [29].

13.4 Emerging methods for detection and selective methods of recognition

The proper operation of biological systems depends on molecular recognition, which is frequently seen between receptor–ligand, antigen–antibody, enzyme–inhibitor, RNA–ribosome, DNA–protein, sugar–lectin, etc. Cell division, cell migration, gene expression, secretion, and differentiation are only a few of the specific actions that are brought on by the manipulation of molecular recognition in living systems [31]. Non-covalent interactions, which include hydrophobic and electrostatic contacts, hydrogen bonding, and the van der Waals force, are the main forces behind the development of receptor–ligand complexes. High affinity is made possible by multivalent bonding that may occur at several locations. When bonding to their corresponding ligand, the natural receptors display complementarity in shape, size, and activities, which is essential to achieving excellent specificity [32, 33]. Primary recognition, orientation, and physical binding are all aspects of the dynamic process of receptor–ligand recognition. Designing artificial receptors that can imitate the natural recognition qualities requires a fundamental understanding of the molecular recognition mechanism between the ligand and natural receptor. The synthetic receptors, including MIPs, provide a more affordable option to natural receptors

[34–37]. A method called template-assisted synthesis is used to create MIPs, which have receptor-like features, by creating specific recognition sites inside a polymeric network. The polymerisation of monomers and cross-linkers while a target molecule serves as a template is part of the molecular imprinting process. The target template is complimentary to the imprinting cavities created when the template is extracted from the cross-linked polymer network in terms of size, shape, and functionality. These chemically created artificial receptors have various benefits, including affordability, high stability, increased robustness and endurance, and worldwide applicability and designability [38, 39].

Covalent imprinting and non-covalent imprinting are the two distinct techniques that have been established in molecular imprinting since the fundamental work pioneered by Wulff in the early 1970s [40]. MIPs were initially used to separate enantiomers chirally. However, a significant barrier in the typical broad peak of a stronger retained enantiomer, which was not competitive with the better-performing chiral stationary phases, prevented this application from being well-developed. Later, the potential of MIPs as selective sorbents for solid-phase extraction of trace analytes from complicated samples was investigated. The unique selectivity of MIPs to their templates makes this use seem more promising. Numerous research has been conducted using MIPs to target ions, small molecules, peptides, and proteins due to the benefits of MIPs [41–43]. More encouragingly, cells and microbes have been added to the list of things that MIPs can detect [44, 45].

13.5 Electrochemical based detection

Biomolecules and the detecting electrode surface interact to produce changes that electrochemical sensors measure. These sensors have a sensitive surface that can recognise a target and translate interactions into electrical impulses [46, 47]. There are two different kinds of sensors: potential-type sensors and current-type sensors. The detection of biological recognition or chemically reactive compounds in a chemical reaction serves as the driving power for current-type sensors, which track variations in current over time by using the fixed electrode potential of an active electron transfer reaction [48]. The biometric reaction is transformed into an electrical signal by the potential-type sensor, and this electrical signal is proportional to the logarithm of the concentration of the active substance consumed or created during the biometric reaction. Voltammetry, amperometry, potentiometry, conductivity, and capacitance variations are often measured parameters in electrochemical sensors [49]. Biosensors integrate minute amounts of biological components such nucleic acids, functioning proteins, cell organelles, and even whole living cells [50]. The biological samples are immobilised on a physicochemical transducer's surface, and the transducer may convert the specific interactions between the immobilised biological samples and their associated binding analytes into quantifiable, concentration-dependent electrical signals [51]. For practical availability, electrochemical biosensors combine specialised biochemical recognition with high sensitivity electrochemical detection. Electrochemical sensors based on MIP are thought to be promising methods for detecting and measuring a variety of analytes [20, 29, 52, 53]. Electropolymerization of imprinted polymer film offers more benefits than conventional polymerisation techniques, including the ability to regulate the film's thickness and adherence to the surface of the transducer. High sensitivity, simplicity of downsizing, and affordability are advantages of electrochemical biosensors. As a result, the field of biosensors uses them extensively [20, 54].

Using magnetic nanoparticles and MIP technology, a novel and affordable electrochemical sensor was created to quantify the Gram-negative bacterial quorum signalling molecules N-acyl-homoserine lactones (AHLs), with the successful surface polymerisation of magnetic MIPs (MMIPs) that can selectively absorb AHLs. Electrochemical tests were used to characterise the particles after they had been put onto a magnetic carbon paste electrode (MGCE) surface. The oxidative current signal indicative of AHL was recorded using differential pulse voltammetry (DPV). The linear detection range for this assay was found to be 2.5×10^{-9} mol l^{-1} to 1.0×10^{-7} mol l^{-1}, with an 8×10^{-10} mol l^{-1} detection limit. A useful new instrument that enables the quantitative detection of Gram-negative bacterial quorum signalling molecules is this $Fe_3O_4@SiO_2$–MIP-based electrochemical sensor. It has potential applications in the fields of clinical diagnosis or food analysis with real-time detection capability, high specificity, excellent reproducibility, and good stability [55]. An MIP-based electrochemical biosensor for *Klebsiella pneumoniae* detection was developed. *K. pneumoniae* concentrations ranging from 10 to 10^5 CFU ml^{-1} were detected quantitatively using an electrochemical method to evaluate the changes in electrical signals, as shown in figure 13.2. With an R2 value of 0.9919, the MIP-based *K. pneumoniae* sensor was discovered to provide a high linear response.

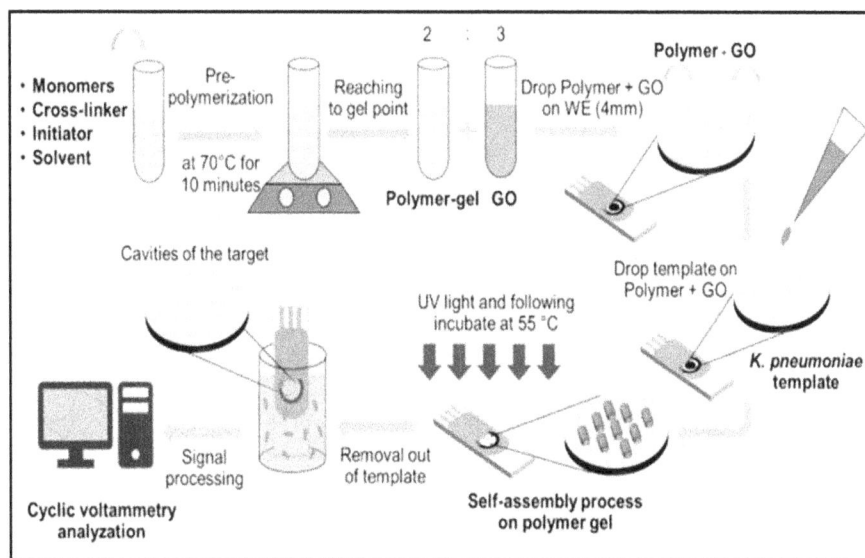

Figure 13.2. Schematic representation of the preparation of polymer–GO on a gold electrode for *K. pneumoniae* detection. (Reproduced with permission from [30]. Copyright 2022 the authors. CC BY 4.0.)

A sensitivity test on bacteria with a structure resembling that of *K. pneumoniae* was also carried out. When compared to MIP sensors used with *Pseudomonas aeruginosa* and *Enterococcus faecalis*, whose sensitivity was 2.634 and 2.226, respectively, the sensitivity results show that the MIP-based *K. pneumoniae* biosensor with a gold electrode was the most sensitive, with a 7.51 (% relative current/log concentration). The sensor was also able to achieve a limit of detection (LOD) of 0.012 CFU ml^{-1} and limit of quantitation (LOQ) of 1.61 CFU ml^{-1} [30]. By utilising the covalent interaction between the 1,2-diols of the highly glycosylated protein and the boronic acid group of 3-aminophenyl boronic acid, an electrochemical sensor based on a molecularly imprinted polymer synthetic receptor for the quantitative detection of SARS-CoV-2 spike protein subunit S1 (ncovS1) was created (APBA). The sensor performs satisfactorily with a 15 min reaction time and is able to identify ncovS1 in samples of patient's nasopharyngeal fluid and phosphate buffered saline with LOD values of 15 fM and 64 fM, respectively. Figure 13.3A(a) shows the Calibration plot of ncovS1 sensor at the low concentration range (26.7–194 fM) of ncovS1 in PBS, figure 13.3A(b) Selectivity of ncovS1 sensor against increasing concentrations (40, 60, 80, 100, and 120 fM) of different proteins (ncovNP, HSA, E2, IgG, and S1) in

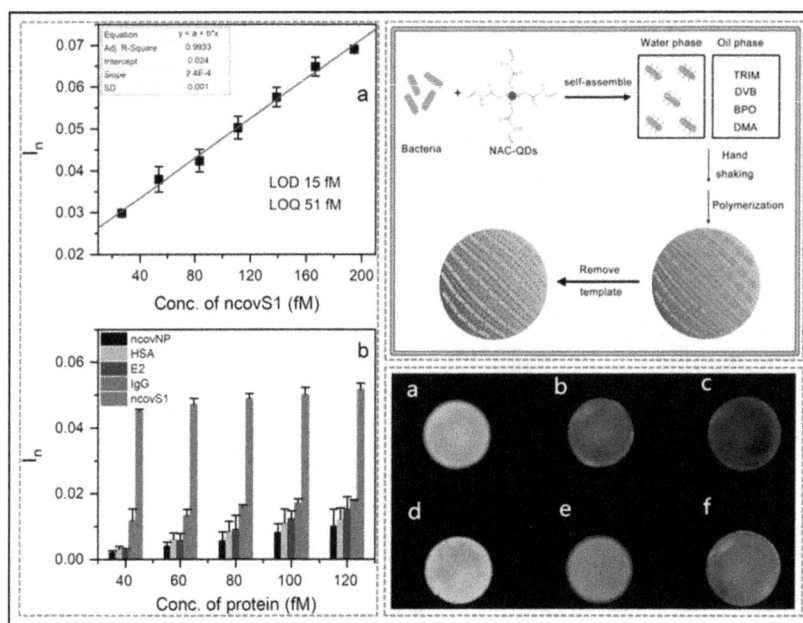

Figure 13.3. (A) (a) Calibration plot of the ncovS1 sensor of ncovS1 at a low concentration in PBS. (b) Selectivity of ncovS1 sensor against increasing concentrations of different proteins (ncovNP, HSA, E2, IgG, and S1) in PBS. (Reproduced with permission from [56]. Copyright 2022 Elsevier.) (B) MIP synthesis process (NAC, TRIM, DVB, BPO, and DMA) related to *N*-acrylchitosan, trimethylolpropane trimethacrylate, divinylbenzene, benzoyl peroxide, and *N,N*-dimethylaniline. (C) Images of MIPs (for (a)–(c), the concentrations of bacteria were 0, 103, and 105 CFU ml^{-1}) and NIPs (for (d)–(f), the concentrations of bacteria were 0, 103, and 105 CFU ml^{-1}) adsorbing different concentrations of bacteria. (Reproduced from [57]. Copyright 2019 the authors. CC BY 4.0.)

PBS. The sensor has a lot of potential as a point-of-care testing tool for quick and accurate diagnosis of COVID-19 patients because it is also compatible with portable potentiostats, allowing on-site readings. Each step of the sensor preparation was characterised by cyclic voltammetry (CV) in the potential range of −0.2 to 0.2 V at a scan rate of 100 mV s^{-1} and square wave voltammetry (SWV) at a potential range of −0.2 to 0.2 V, pulse amplitude of 12.5 mV, frequency of 10 Hz [56].

13.6 Fluorescence based detection

Due to its nondestructive operation method and high signal output, fluorescent detection has emerged as one of the most widely used optical techniques [58, 59]. Fluorescence biosensors are an analytical tool that employs fluorescence signals as a signalling unit and mounted biosensor materials such as enzymes, antigens/antibodies, aptamers, nucleic acids, liposomes, cells, and microorganisms as recognition components. The molecularly imprinted fluorescent biosensors typically have fluorescence acting as the signalling unit and MIP acting as the recognition unit [60, 61]. Fluorescence detection technology has emerged as the most widely employed technique in the field of bioanalysis, in particular in the diagnosis of tumours, due to this type of biosensor's nondestructive mode of operation, high signal production, and reading speed [60]. The inclusion of fluorescence in molecularly imprinted sensors has received a great deal of investigative attention since these biosensors can quickly identify bacteria based on fluorescence [62]. Traditional fluorescent labels, such as organic dyes, frequently exhibit narrow absorption and broad emission spectra with long tails and are easily photobleached, all of which led to low detection sensitivity [63]. Traditional fluorescent labels have some drawbacks, but fluorescent nanoparticles can create strong fluorescence signals, considerably increase sensitivity, and work in a variety of applications [62]. Despite all the benefits listed above, the fluorescence sensing technique is more complicated than electrochemical, QCM, SPR, and HTM detection methods [64]. Due to their huge dimensions, microbes present unique challenges for the precise quantification utilising fluorescent probes. However, the combination of MIP/fluorescent detection with microfluidic flow cell cytometry can address these issues [65].

Oil-in-water Pickering emulsion polymerisation was used to create a brand-new MIP with water-soluble CdTe quantum dots (QDs) utilising entire *L. monocytogenes* as the template. Figure 13.3(B) shows the synthesis process of MIPs related to *N*-acrylchitosan, trimethylolpropane trimethacrylate, divinylbenzene, benzoyl peroxide, and *N,N*-dimethylaniline. The *L. monocytogenes* was initially treated with chitosan that had been acryloyl functionalized using QDs to create a bacterial-chitosan network in the aqueous phase. Afterwards, it was stabilised in an oil-in-water emulsion made of a cross-linker, a monomer, and an initiator, which led to the appearance of recognition sites on the surface of the CdTe QD-embedded micro-spheres. Through recognition cavities, the generated MIP microspheres allowed for the selective capture of the target microorganisms. *L. monocytogenes*, the intended bacteria, was discovered. The analysis of the MIPs using scanning electron microscopy (SEM) revealed that they were approximately spherical in shape. *L. monocytogenes* may be qualitatively detected in milk and ham sample using visual fluorescence detection by quenching in the presence of the target molecule. The MIPs and non-imprinted polymers (NIPs) were initially incubated with 10^3 and 10^5 CFU ml^{-1} concentrations of bacteria, and the fluorescence intensity of the resultant polymer beads was directly seen using a fluorescence microscope, to demonstrate the usefulness of these materials. Figure 13.3(C) shows the images of MIPs and NIPs adsorbing different concentrations of bacteria. The fluorescent MIPs sensor was used for quick and qualitative detection of *L. monocytogenes* with the qualitative detection of 10^3 since microbial imprinting has not yet reached the level of accuracy that can be attained by imprinting small molecules. The created technology eliminated the need for sample preparation and streamlined the analytical process. In addition, the fluorescence sensor provided an effective, fast, and convenient method for *L. monocytogenes* detection in food samples [57]. Utilising molecular imprinting, a unique technique was created to isolate *Staphylococcus aureus* from complicated (food) samples. A functional monomer of dopamine was employed, and fluorescence microscopy was used to detect it. Investigations were conducted into the conditions for MIP production, adsorption performance, adsorption kinetics, and selectivity of the polymeric layers. The imprinted layer on the surface of the magnetic particles was extracted using a single extraction technique that incorporated the multiple procedures (magnetic MIPs). MIPs were then employed to remove *S. aureus* from rice and milk. Furthermore, tests on raw milk from cows with mastitis were successful. It was possible to detect bacteria in milk at 1×10^3 CFU ml^{-1} using this new MIP-based technique, which is the legal limit for microbiological control of food in the European Union. Measurements were made of *S. aureus*'s binding characteristics at different concentrations ($1 \times 10^6 - 1 \times 10^2$ CFU ml^{-1}), as well as the rate of adsorption [66]. In order to detect and quantify the bacterial quorum signalling molecules AHLs, a class of autoinducers from Gram-negative bacteria, a unique class-specific artificial receptor based on MIP-coated quantum dots (QDs@MIP) was created. The CdSe/ZnS QDs were used as the signal transducing material to prepare the QDs@MIP employing the surface imprinting process under regulated conditions. Transmission electron microscopy, scanning electron microscopy, Fourier transform infra-red spectroscopy, x-ray diffraction analysis, and fluorescence spectroscopy were used to characterise the synthesis of the QDs@MIP. After template elution, the obtained cavities recognised the target AHLs of interest with sensitivity and selectivity. When exposed to various concentrations of AHL, the fluorescence intensity of the QDs@MIP was noticeably reduced in comparison to the control non-imprinted polymer (QDs@NIP). Along with a detection limit of 0.66, 0.54, 0.88, 0.72, and 0.68 nM for DMHF, C4-HSL, C6-HSL, C8-HSL, and *N*-3oxo-C6-HSL, respectively, it also demonstrated good linearity in the range of 2–18 nM. The proposed sensor most intriguingly displayed great sensitivity, good stability, and a quick reaction time (30 s) towards the target molecules due to efficient surface imprint generation. The analysis of bacterial supernatant samples, with acceptable recoveries ranging from 89% to 103%, further demonstrated the viability of the designed sensor in real samples. According to these results, the as-prepared QDs@MIP can be used as a new potential supporting technique for the rapid and real-time detection of bacterial pathogens in food safety and healthcare facilities [67].

13.7 Surface plasmon resonance based detection

Due to its compact size, low cost, label-free sensing, etc, surface plasmon resonance (SPR) technology for sensing applications has garnered great interest during the last 20 years [68]. SPR is a non-invasive and label-free detection method based on the surface plasmons of metal–dielectric waveguides. The ease-of-use, high sensitivity, short response time, and automation adaptability of SPR sensors are all appealing qualities. Surface plasmon waves, electromagnetic waves that can be induced by light on gold sensor surfaces, are utilised by the SPR sensor technology [68]. A portion of the incident light's energy is transferred to the surface plasmons for their excitation, which causes the surface plasmon resonance phenomenon, when p-polarised light is incident on a metal–dielectric interface in the Kretschmann configuration at an angle greater than the critical angle and the wave vectors of the surface plasmon wave and evanescent wave match [69]. Real-time monitoring of the minute variations in refractive index at a metal–dielectric interface caused by interactions between biomolecules and SPR sensors is possible [70]. SPR sensors can analyse the binding of target molecules to be determined on an SPR sensor chip surface coated with a metal film directly without any labelling. One of the advantages of SPR sensors is, thanks to their sensitive refractive index changing during the surface binding, giving the results immediately and label free. SPR sensors have been used extensively for detecting various analytes in conjunction with MIPs [71–73].

To undertake simultaneous quantitative analysis of *Pseudomonas* spp., an MIP-based SPR sensor was developed. Using the microcontact printing technique, polymeric nanofilm was imprinted on the surface of the SPR sensor. SEM examination, ellipsometer measurement, and contact angle NIP and MIP-SPR chips were used to describe the nanofilm. The thickness of the MIP and NIP nanofilms were determined to be 150.4 nm and 72.4 nm, respectively. Various *Pseudomonas* spp. were produced and subjected to the SPR system at concentrations ranging from 1×10^2 to 1×10^4 CFU ml^{-1}. In comparison to the culture method, the total detection time for all procedures was reported to be just 150 s for microbe detection (2 days). The calculation for the detection of the limit value was 0.5×10^2 CFU ml^{-1}. SPR sensors based on nanofilms made of *Pseudomonas* spp. were found to have a selectivity of 8.3 for MIP and 0.25 for NIP. For *S. aureus*, *Salmonella paratyphi*, and *E. coli*, the relative selectivity coefficients (k) of the MIP nanofilm-based SPR sensor were reported to be 25.82, 31.69, and 31.24 times, respectively. The generated *Pseudomonas* spp. suppressed surface plasmon resonance sensor can be employed as a replacement for the present culturing method and has effective suppression quality, selectivity, and affinity, according to the findings collected in [74]. Tailor-made *E. coli* receptors were made using the microcontact imprinting approach, and binding activities of *E. coli* were monitored in real time and label-free by an SPR sensor in urine mimic and aqueous solution. Figure 13.4(A) shows the entire process of synthesis of *E. coli* imprinted polymeric film via the microcontact imprinting technique. The functional compound *N*-methacryloyl-(L)-histidine methyl ester (MAH) was chosen. Ag nanoparticles (AgNPs) were entrapped into

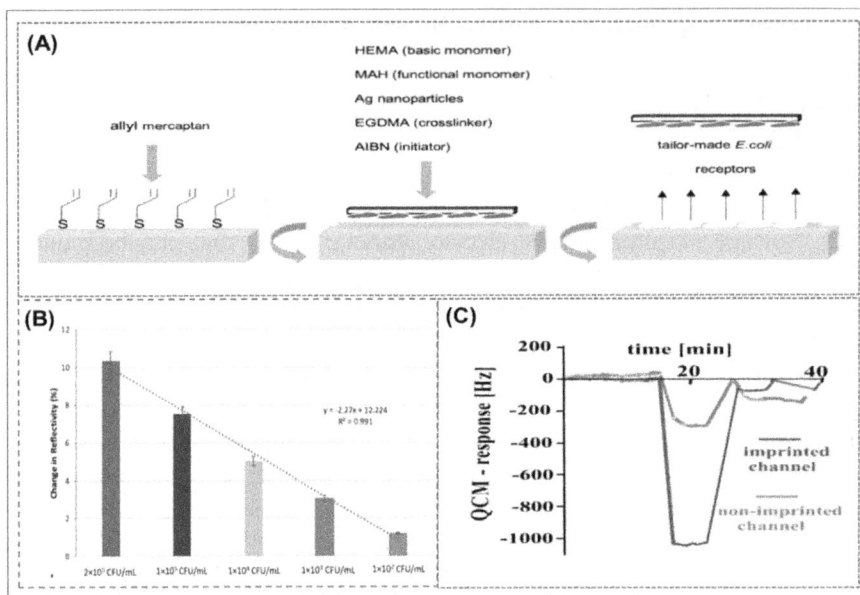

Figure 13.4. (A) Schematic representation of *E. coli* imprinted polymeric film synthesis via a microcontact imprinting technique. (Reproduced with permission from [75]. Copyright 2020 Elsevier.) (B) The calibration curve of *S. aureus* obtained in a range of 1.0×10^2–2.0×10^5 CFU ml^{-1} bacterial concentrations. (Reproduced from [76]. Copyright 2021 the authors. CC BY 4.0.) (C) Mass-sensitive sensor responses to *Bacteria subtilis* spores to imprinted and non-imprinted dual electrodes as a function of time. (Reproduced from [77]. Copyright 2020 the authors. CC BY 4.0.)

the polymer mixture during the production of the polymeric film on a chip surface in order to lower the detection limit of the biomimetic SPR based sensor. Atomic force microscopy (AFM), SEM, an ellipsometer, and contact angle measurements were used to characterise the polymeric layer. The biomimetic sensor's viability was examined using a urine mimic and the LOD was discovered to be 0.57 CFU ml^{-1} [75]. *S. aureus* was detected using an SPR sensor with microcontact imprinted sensor chips. MAH was used for whole-cell imprinting during UV polymerisation. The concentration range used for the sensing studies was 1.0×10^2–2.0×10^5 CFU ml^{-1}. Microcontact imprinting and optical sensor technologies were utilised, with a detection limit of 1.5×10^3 CFU ml^{-1}, to successfully identify *S. aureus*. Figure 13.4(B) shows the calibration curve of *S. aureus* obtained in a range of 1.0×10^2–2.0×10^5 CFU ml^{-1} bacterial concentrations. Through the injection of competing bacterial strains, the resulting sensor's selectivity was assessed. The reactions of the various strains were contrasted with those of *S. aureus*. Additionally, real studies were conducted using milk samples that had been tainted with *S. aureus*, and it was shown that the ready-made sensor platform could be used with actual samples [76]. For the purpose of detecting and measuring a secreted bacterial factor (RoxP) from skin, an SPR biosensor was created. Sensor chips were created using a molecular imprinting technique, and five distinct monomer cross-linker compositions were tested for their sensitivity, selectivity, affinity, and kinetics. SEM, AFM, and cyclic voltammetry were used to characterise the MIPs that showed the greatest promise. For the promising MIPs, the LOD value was calculated to be 0.23 nM with an affinity constant of 3.3×10^9 M. The estimated selectivity coefficients demonstrated that the devised system was not only exceedingly sensitive but also very selective for the template protein RoxP. Finally, the created MIP-SPR biosensor was used to assess the absolute RoxP concentrations in several skin swabs and the results compared to an ELISA competitor. The created system provides a very effective tool for the detection and quantification of RoxP as an early signal for several oxidative skin illnesses, particularly when they are present in low abundance levels (e.g. skin samples) [73].

13.8 Quartz crystal microbalance-based detection

Due to its high sensitivity and on-line collection capacity, the quartz crystal microbalance (QCM) is a simple, affordable, high-resolution, and label-free mass sensing device that has been widely utilised for analytical purposes [78]. By tracking the change in resonance frequency in real time, it may identify surface mass changes on quartz crystals. A quartz disc with mounted electrodes makes up QCM. Based on the Sauerbrey equation, the mass variations caused by the analyte adsorption on the disc surface lead to a change in frequency in a quartz crystal resonator. Resonance energy variation is a clear indicator of biomolecular interactions. For almost all analytes, the combination of MIPs and QCM offers a focused and sensitive detection technique. QCM has also been used frequently in a variety of pharmaceutical, biomedical, and environmental assays [42, 79, 80].

Clinical and commercial applications are very interested in small molecule identification. However, its availability is still limited, due to reliability, cost, and complexity issues with system integration and miniaturisation. For label-free, real-time detection of N-hexanoyl-L-homoserine lactone (199 Da), a gram-negative bacterial infection indicator, the authors paired a 13.3 MHz quartz crystal resonator (QCR) with a nanomolecular imprinted polymer. Without any optimization, the lowest concentration found (1 M) was similar to the expensive laboratory gold standard BIAcore SPR system, with a notable improvement in sensitivity and specificity over the most advanced QCR. Potentially enabling a multiplexed 'QCR-on-chip' technology, the analytical formula-based fixed frequency drive (FFD) approach could lead to a revolution in the speed, availability, and cost of tiny molecule detection [46]. Mass-sensitive transducers were paired with a molecular imprinting method to create dependable biomimetic sensor devices for the detection of bioparticles. The authors were able to use a QCM to detect the bacteria *E. coli* thanks to the patterning of polymers with bioanalytes. An appreciable sensor response is obtained by an imprinted channel when it selectively recognises the templated analyte in comparison to non-imprinted material, as shown in figure 13.4(C). A QCM with a fundamental frequency of 10 MHz was used to identify *E. coli* germs. Direct AFM measurements were made in order to compare the QCM sensor results with the number of bacteria cells that adhered to the sensor coatings. The recognition sites produced by *Bacillus subtilis* spores allowed for quick sensing and the successful and reversible recognition of the template analyte. Cross-sensitive tests amply demonstrated the benefit of the molecular imprinting technique, which made it simple to distinguish and selectively detect the spores of *Bacillus* species (*B. subtilis* and *B. thuringiensis*). Over the course of 15 h at 42 °C in the proper nutritional solution of glucose and ammonium sulphate, *B. subtilis* development from its spores was seen. Additionally, the development of *B. subtilis* bacteria from their individual spores was examined by raising the glucose concentration until the sensor reached saturation. In order to investigate osmotic effects according to a frequency response of 400 Hz by varying the ionic strength of 0.1 M, the polymeric sensor coatings were patterned to fix the *B. subtilis* [77]. For the purpose of detecting staphylococcal enterotoxins (SEs) in actual samples, a QCM with dissipation biosensor based on a two-dimensional (2D) molecularly imprinted film-coated by organo silanes and the template protein staphylococcal SEs synthesising *in situ* on the surface of the PQC as transducer was developed. Frequency shifts (FSs) and dissipation shifts (DSs), which were integrated with the sensor, were recorded and observed in real time for the evaluation of the capability of the sol–gel imprinted films for target identification. The lowest detection limits for Staphylococcus aureus A (SEA) and staphylococcal enterotoxin B (SEB) were 7.97 and 2.25 ng ml^{-1}, respectively, for both working ranges of 0.1–1000 g ml^{-1}. AFM was used to observe and confirm the performances and surface structures of the manufactured 2D sol–gel films. The results showed that the template SEs may be repeatedly used and recognised by the 2D molecularly imprinted film-coated QCM. After pre-treatment, the intelligentized biosensor was used to detect SEA and SEB in milk samples. The recoveries ranged from 97.00% to 114.20% and it could be applied ten times, showing that the method was practical and useful for the detection of SEs [81] (table 13.1).

13.9 Conclusion and future perspectives

While the imprinting of small compounds, peptides, or even proteins is a simple and well established process, the creation of MIPs against microbes still faces significant difficulties, such as the following: (i) the binding affinity heterogeneity of the imprinted cavities; (ii) the large dimensions of the microorganisms limit diffusion and transfer within the highly cross-linked polymer networks in bulk imprinting; (iii) live microorganisms are not robust during the imprinting process, and effective approaches are required to maintain the conformational integrity of the cell template; and (iv) live microorganisms feature environmental adaptations. Indirect imprinting, which uses specific molecular components on the cell surface as the template, is one way to address these drawbacks. Investigating successful ways to achieve whole-cell imprinting is another tactic.

As a result of molecular recognition in nature, molecular imprinting has become a promising method for developing synthetic materials that resemble receptors (i.e. MIPs). MIPs have intrinsic advantages over natural receptors, such as low cost, excellent stability, and adaptability. As a result, MIPs are frequently utilised in a variety of applications, such as separation, sample pre-treatment, sensors, catalysis,

Table 13.1. Various recent studies involving the application of molecularly imprinted polymers.

S. No.	Analyte	Detection method	LOD	Linear range	Reference
1	Gram-negative bacterial quorum signalling molecules, N-acyl-homoserine lactones (AHLs)	Electrochemical Based detection	8×10^{-10} mol·l^{-1}	2.5×10^{-9} mol·l^{-1} – 1.0×10^{-7} mol·l^{-1}	[55]
2	Klebsiella pneumoniae	Electrochemical based detection	0.012 CFU ml^{-1}	10–105 CFU ml^{-1}	[30]
3	SARS-CoV-2 spike protein (ncovS1)	Electrochemical based detection	15 fM (in phosphate buffer saline)	0–400 fM	[56]
			64 fM (nasopharyngeal sample)		
4	Staphylococcus aureus	Electrochemical based detection	2 CFU ml^{-1}	10–10^8 CFU ml^{-1}	[82]
5	Listeria monocytogenes	Fluorescent based detection	10^3 CFU ml^{-1}	10^3–10^5 CFU ml^{-1}	[57]
6	Staphylococcus aureus	Fluorescent based detection	1×10^3 CFU ml^{-1}	1×10^6–1×10^2 CFU ml^{-1}	[66]
7	Bacterial quorum signalling molecules, AHLs	Fluorescent based detection	0.66 nM for DMHF, 0.54 nM for C4-HSL, 0.72 nM for C6-HSL, 0.68 for N-3oxo-C6-HSL	2–18 nM	[67]
8	Japanese encephalitis virus	Fluorescent based detection	0.11 pM	2.4–24 pmol ml^{-1}	[83]
9	N-hexanoyl-L-homoserine	Quartz crystal resonator	1 μM–50 μM	1 μM	[46]
10	Escherichia coli	Quartz crystal microbalance sensor	—	—	[77]
11	Staphylococcal enterotoxins	Quartz crystal microbalance sensor	0.1–1000 μg ml^{-1}	7.97 ng ml^{-1}	[81]
12	Escherichia coli	Quartz crystal microbalance sensor	Staphylococcal enterotoxins	1.6×10^8 cells/ml	[84]
13	Picornaviruses	Quartz crystal microbalance sensor	—	—	[85]
14	Bacillus cereus	Quartz crystal microbalance sensor	—	—	[86]
15	Classical swine fever virus	Quartz crystal microbalance sensor	4–21 μg ml^{-1}	1.7 μg ml^{-1}	[87]
16.	Neisseria meningitidis	Piezoelectric sensor		15.71 ng ml^{-1}	[88]
17.	Pseudomonas spp.	Surface plasmon resonance sensor	1×10^2 to 1×10^4 CFU ml^{-1}	0.5×10^2 CFU ml^{-1}	[74]
18.	Escherichia coli	Surface plasmon resonance sensor	1.5×10^1–1.5×10^6 CFU ml^{-1}	0.57 CFU ml^{-1}	[75]
19.	Staphylococcus aureus	Surface plasmon resonance sensor	1.0×10^2–2.0×10^5 CFU ml^{-1}	1.5×10^3 CFU ml^{-1}	[76]
20.	Bacterial factor (RoxP)	Surface plasmon resonance sensor	5.2×10^{-4} μM and 68 μM	0.23 nM	[73]

drug transfer, etc. In addition to ions and tiny molecules, proteins and peptides are now included in the imprinting templates. However, early research on the potential of these natural receptor mimics in the selective identification of microbes was limited. MIPs have developed rapidly during the past 10 years, with promising applications in the targeted inactivation and sensing of microbes as well as microbial fuel cells. We have provided a thorough summary of the existing MIP synthesis techniques that target microorganisms in this review. In addition, we outlined three key specialised uses of MIPs, including microbial inactivation, microbial fuel cells, and microorganism detection and sensing. The reported imprinting techniques range from whole-cell imprinting to focusing on certain elements on cell surfaces. The former tactic uses cell membrane components as the targeted templates and implicitly utilises the conventional molecular imprinting concepts. The specificity and selectivity of MIPs in this situation depend heavily on the screening of appropriate molecules. Only a small number of proteins, peptides, and saccharides have been studied in MIP production up to this point, greatly limiting the application of this technology. Therefore, there is a strong need for the creation of new indicators that contain a wealth of chemical information about microorganisms. Tools from molecular biology and cell biology can be used for this. Since these cell membrane molecules' partial configurations are bound to the cell membrane, the molecular imprinting procedure should also take this into account. Due of live microorganisms' vast dimensions, complexity, non-robustness, and environmental adaptability, the whole-cell imprinting approach is both simpler and more difficult. The combination of size- and shape-dependent physical recognition and chemical recognition is essential for the technique to be successful. In addition, novel imprinting techniques are needed to develop imprinting cavities that can preserve the integrity of the form and size of microorganisms. Under the assumption that inactivation does not alter a cell's characteristics, using inactivated microorganisms at the imprinting templates can be one way to address problems that arise during the operation of living cells. Despite several obstacles, MIPs that target microbes have gained significant attention in a variety of applications. We anticipate significant investments in more interdisciplinary fields, including quorum sensing, biofilm development, and microbial engineering for chemical product manufacturing. Success in this area will result in a promising paradigm with lots of room for future expansion.

Acknowledgements

The authors express their sincere thanks to Director, CSIR-AMPRI for his support and encouragement in this work. Raju Khan would like to acknowledge SERB for providing funds in the form of the IPA/2020/000130 project.

References

[1] Fredrickson J K *et al* 2004 Geomicrobiology of high-level nuclear waste-contaminated vadose sediments at the Hanford Site, Washington state *Appl. Environ. Microbiol.* **70** 4230–41

[2] Belland R J, Ouellette S P, Gieffers J and Byrne G I 2004 *Chlamydia pneumoniae* and atherosclerosis *Cell Microbiol.* **6** 117–27

[3] Lopez-Roldan R, Tusell P, Cortina J L and Courtois S 2013 On-line bacteriological detection in water *TrAC Trends Anal. Chem.* **44** 46–57

[4] Speight R E and Cooper M A 2012 A survey of the 2010 quartz crystal microbalance literature *J. Mol. Recognit.* **25** 451–73

[5] Newell D G *et al* 2010 Food-borne diseases—the challenges of 20 years ago still persist while new ones continue to emerge *Int. J. Food Microbiol.* **139** S3–15

[6] Scott P 2005 Plant disease: a threat to global food security *Annu. Rev. Phytopathol.* **43** 83–116

[7] Ko Y S *et al* 2020 Tools and strategies of systems metabolic engineering for the development of microbial cell factories for chemical production *Chem. Soc. Rev.* **49** 4615–36

[8] Luo X *et al* 2019 Complete biosynthesis of cannabinoids and their unnatural analogues in yeast *Nature* **567** 123–6

[9] Nielsen J and Keasling J D 2016 Engineering cellular metabolism *Cell* **164** 1185–97

[10] Zhao X R, Choi K R and Lee S Y 2018 Metabolic engineering of *Escherichia coli* for secretory production of free haem *Nat. Catal.* **1** 720–8

[11] Yu T *et al* 2018 Reprogramming yeast metabolism from alcoholic fermentation to lipogenesis *Cell* **174** 1549–58

[12] Jamieson O *et al* 2021 Electropolymerised molecularly imprinted polymers for the heat-transfer based detection of microorganisms: a proof-of-concept study using yeast *Therm. Sci. Eng. Prog.* **24** 100956

[13] Buchanan R L, Gorris L G M, Hayman M M, Jackson T C and Whiting R C 2017 A review of *Listeria monocytogenes*: an update on outbreaks, virulence, dose–response, ecology, and risk assessments *Food Control* **75** 1–13

[14] Mothershed E A and Whitney A M 2006 Nucleic acid-based methods for the detection of bacterial pathogens: present and future considerations for the clinical laboratory *Clin. Chim. Acta* **363** 206–20

[15] Choi S Y *et al* 2016 One-step fermentative production of poly(lactate-co-glycolate) from carbohydrates in *Escherichia coli Nat. Biotechnol.* **34** 435–40

[16] Ashley J, Feng X, Halder A, Zhou T and Sun Y 2018 Dispersive solid-phase imprinting of proteins for the production of plastic antibodies *Chem. Commun.* **54** 3355–8

[17] Chen L, Xu S and Li J 2011 Recent advances in molecular imprinting technology: current status, challenges and highlighted applications *Chem. Soc. Rev.* **40** 2922–42

[18] Dar K K, Shao S, Tan T and Lv Y 2020 Molecularly imprinted polymers for the selective recognition of microorganisms *Biotechnol. Adv.* **45** 107640

[19] Singhal A, Parihar A, Kumar N and Khan R 2022 High throughput molecularly imprinted polymers based electrochemical nanosensors for point-of-care diagnostics of COVID-19 *Mater. Lett.* **306** 130898

[20] Jia M, Zhang Z, Li J, Ma X, Chen L and Yang X 2018 Molecular imprinting technology for microorganism analysis *TrAC Trends Anal. Chem.* **106** 190–201

[21] Singhal A *et al* 2022 Molecularly imprinted polymers-based nanobiosensors for environmental monitoring and analysis *Nanobiosensors for Environmental Monitoring* (Cham: Springer International) pp 263–78

[22] Pourmadadi M, Yazdian F, Hojjati S and Khosravi-Darani K 2021 Detection of microorganisms using graphene-based nanobiosensors *Food Technol. Biotechnol.* **59** 496–506

[23] Lv L, Jin Y, Kang X, Zhao Y, Cui C and Guo Z 2018 PVP-coated gold nanoparticles for the selective determination of ochratoxin A via quenching fluorescence of the free aptamer *Food Chem.* **249** 45–50

[24] Bej A K, Mahbubani M H, Dicesare J L and Atlas R M 1991 Polymerase chain reaction-gene probe detection of microorganisms by using filter-concentrated samples *Appl. Environ. Microbiol.* **57** 3529–34

[25] Skottrup P D, Nicolaisen M and Justesen A F 2008 Towards on-site pathogen detection using antibody-based sensors *Biosens. Bioelectron.* **24** 339–48

[26] Parra E, Segura F, Tijero J, Pons I and Nogueras M M 2017 Development of a real-time PCR for *Bartonella* spp. detection, a current emerging microorganism *Mol. Cell. Probes* **32** 55–9

[27] Leoni E and Legnani P P 2001 Comparison of selective procedures for isolation and enumeration of *Legionella* species from hot water systems *J. Appl. Microbiol.* **90** 27–33

[28] Ku S *et al* 2017 Protein particulate retention and microorganism recovery for rapid detection of *Salmonella Biotechnol. Prog.* **33** 687–95

[29] Yang Y *et al* 2020 Magnetic molecularly imprinted electrochemical sensors: a review *Anal. Chim. Acta* **1106** 1–21

[30] Pintavirooj C, Vongmanee N, Sukjee W, Sangma C and Visitsattapongse S 2022 Biosensors for *Klebsiella pneumoniae* with molecularly imprinted polymer (MIP) technique *Sensors* **22** 4638

[31] Persch E, Dumele O and Diederich F 2015 Molecular recognition in chemical and biological systems *Angew. Chem. Int. Ed.* **54** 3290–327

[32] Guryanov I, Fiorucci S and Tennikova T 2016 Receptor–ligand interactions: advanced biomedical applications *Mater. Sci. Eng.* C **68** 890–903

[33] Pandey A, Chauhan P and Singhal A 2022 Potential electrochemical biosensors for early detection of viral infection *Advanced Biosensors for Virus Detection: Smart Diagnostics to Combat SARS-CoV-2* (Cambridge, MA: Academic Press) pp 133–54

[34] Andersson L I, Müller R, Vlatakis G and Mosbach K 1995 Mimics of the binding sites of opioid receptors obtained by molecular imprinting of enkephalin and morphine *Proc. Natl Acad. Sci.* **92** 4788–92

[35] Arshady R and Mosbach K 1981 Synthesis of substrate-selective polymers by host–guest polymerization *Makromol. Chem.* **182** 687–92

[36] Haupt K and Mosbach K 1998 Plastic antibodies: developments and applications *Trends Biotechnol.* **16** 468–75

[37] Sellergren B 2000 Imprinted polymers with memory for small molecules, proteins, or crystals *Angew Chem. Int. Ed. Engl.* **39** 1031–7

[38] Lv Y, Qin Y, Svec F and Tan T 2016 Molecularly imprinted plasmonic nanosensor for selective SERS detection of protein biomarkers *Biosens. Bioelectron.* **80** 433–41

[39] Lv Y, Tan T and Svec F 2013 Molecular imprinting of proteins in polymers attached to the surface of nanomaterials for selective recognition of biomacromolecules *Biotechnol. Adv.* **31** 1172–86

[40] Wulff G, Sarhan A and Zabrocki K 1973 Enzyme-analogue built polymers and their use for the resolution of racemates *Tetrahedron Lett.* **14** 4329–32

[41] Beltran A, Borrull F, Marcé R M and Cormack P A G 2010 Molecularly-imprinted polymers: useful sorbents for selective extractions *TrAC Trends Anal. Chem.* **29** 1363–75

[42] Chen J Y, Penn L S and Xi J 2018 Quartz crystal microbalance: sensing cell-substrate adhesion and beyond *Biosens. Bioelectron.* **99** 593–602

[43] Muldoon M T and Stanker L H 1997 Molecularly imprinted solid phase extraction of atrazine from beef liver extracts *Anal. Chem.* **69** 803–8

[44] Daoud Attieh M, Zhao Y, Elkak A, Falcimaigne-Cordin A and Haupt K 2017 Enzyme-initiated free-radical polymerization of molecularly imprinted polymer nanogels on a solid phase with an immobilized radical source *Angew. Chem.* **129** 3387–91

[45] Singhal A *et al* 2022 Multifunctional carbon nanomaterials decorated molecularly imprinted hybrid polymers for efficient electrochemical antibiotics sensing *J. Environ. Chem. Eng.* **10** 107703

[46] Guha A *et al* 2020 Direct detection of small molecules using a nano-molecular imprinted polymer receptor and a quartz crystal resonator driven at a fixed frequency and amplitude *Biosens. Bioelectron.* **158** 112176

[47] Parihar A, Singhal A, Kumar N, Khan R, Khan M A and Srivastava A K 2022 Next-generation intelligent MXene-based electrochemical aptasensors for point-of-care cancer diagnostics *Nano-Micro Lett.* **14** 100

[48] Shahzad F, Zaidi S A and Koo C M 2017 Highly sensitive electrochemical sensor based on environmentally friendly biomass-derived sulfur-doped graphene for cancer biomarker detection *Sensors Actuators* B **241** 716–24

[49] Suginta W, Khunkaewla P and Schulte A 2013 Electrochemical biosensor applications of polysaccharides chitin and chitosan *Chem. Rev.* **113** 5458–79

[50] Zhang Z and Liu J 2016 Molecularly imprinted polymers with DNA aptamer fragments as macromonomers *ACS Appl. Mater. Interfaces* **8** 6371–8

[51] Namvar A and Warriner K 2007 Microbial imprinted polypyrrole/poly(3-methylthiophene) composite films for the detection of *Bacillus* endospores *Biosens. Bioelectron.* **22** 2018–24

[52] Beluomini M A, da Silva J L, de Sá A C, Buffon E, Pereira T C and Stradiotto N R 2019 Electrochemical sensors based on molecularly imprinted polymer on nanostructured carbon materials: a review *J. Electroanal. Chem.* **840** 343–66

[53] Scheller F W, Zhang X, Yarman A, Wollenberger U and Gyurcsányi R E 2019 Molecularly imprinted polymer-based electrochemical sensors for biopolymers *Curr. Opin. Electrochem.* **14** 53–9

[54] Singhal A *et al* 2022 MXene-modified molecularly imprinted polymers as an artificial bio-recognition platform for efficient electrochemical sensing: progress and perspectives *Phys. Chem. Chem. Phys.* **24** 19164–76

[55] Jiang H, Jiang D, Shao J and Sun X 2016 Magnetic molecularly imprinted polymer nanoparticles based electrochemical sensor for the measurement of gram-negative bacterial quorum signaling molecules (N-acyl-homoserine-lactones) *Biosens. Bioelectron.* **75** 411–9

[56] Ayankojo A G, Boroznjak R, Reut J, Öpik A and Syritski V 2022 Molecularly imprinted polymer based electrochemical sensor for quantitative detection of SARS-CoV-2 spike protein *Sensors Actuators* B **353** 131160

[57] Zhao X, Cui Y, Wang J and Wang J 2019 Preparation of fluorescent molecularly imprinted polymers via pickering emulsion interfaces and the application for visual sensing analysis of *Listeria monocytogenes Polymers* **11** 984

[58] Bhardwaj N, Bhardwaj S K, Nayak M K, Mehta J, Kim K H and Deep A 2017 Fluorescent nanobiosensors for the targeted detection of foodborne bacteria *TrAC Trends Anal. Chem.* **97** 120–35

[59] Martynenko I V *et al* 2019 Magneto-fluorescent microbeads for bacteria detection constructed from superparamagnetic Fe_3O_4 nanoparticles and AIS/ZnS quantum dots *Anal. Chem.* **91** 12661–9

[60] Panagiotopoulou M *et al* 2016 Molecularly imprinted polymer coated quantum dots for multiplexed cell targeting and imaging *Angew. Chem.* **128** 8384–8

[61] Ivanova-Mitseva P K *et al* 2012 Cubic molecularly imprinted polymer nanoparticles with a fluorescent core *Angew. Chem. Int. Ed.* **51** 5196–9

[62] Qian H S, Guo H C, Ho P C L, Mahendran R and Zhang Y 2009 Mesoporous-silica-coated up-conversion fluorescent nanoparticles for photodynamic therapy *Small* **5** 2285–90

[63] Chao M R, Hu C W and Chen J L 2014 Fluorescent turn-on detection of cysteine using a molecularly imprinted polyacrylate linked to allylthiol-capped CdTe quantum dots *Microchim. Acta* **181** 1085–91

[64] Zhu L, Wu W, Zhu M Q, Han J J, Hurst J K and Li A D Q 2007 Reversibly photoswitchable dual-color fluorescent nanoparticles as new tools for live-cell imaging *J. Am. Chem. Soc.* **129** 3524–6

[65] Chatterjee D K, Gnanasammandhan M K and Zhang Y 2010 Small upconverting fluorescent nanoparticles for biomedical applications *Small* **6** 2781–95

[66] Bezdekova J *et al* 2020 Magnetic molecularly imprinted polymers used for selective isolation and detection of *Staphylococcus aureus Food Chem.* **321** 126673

[67] Habimana J *et al* 2018 A class-specific artificial receptor-based on molecularly imprinted polymer-coated quantum dot centers for the detection of signaling molecules, *N*-acyl-homoserine lactones present in gram-negative bacteria *Anal. Chim. Acta* **1031** 134–44

[68] Agrawal H, Shrivastav A M and Gupta B D 2016 Surface plasmon resonance based optical fiber sensor for atrazine detection using molecular imprinting technique *Sensors Actuators* B **227** 204–11

[69] Usha S P, Mishra S K and Gupta B D 2015 Fabrication and characterization of a SPR based fiber optic sensor for the detection of chlorine gas using silver and zinc oxide *Materials* **8** 2204–16

[70] Šípová H and Homola J 2013 Surface plasmon resonance sensing of nucleic acids: a review *Anal. Chim. Acta* **773** 9–23

[71] Baldoneschi V, Palladino P, Banchini M, Minunni M and Scarano S 2020 Norepinephrine as new functional monomer for molecular imprinting: an applicative study for the optical sensing of cardiac biomarkers *Biosens. Bioelectron.* **157** 112161

[72] Lach P, Cieplak M, Majewska M, Noworyta K R, Sharma P S and Kutner W 2019 'Gate effect' in *p*-synephrine electrochemical sensing with a molecularly imprinted polymer and redox probes *Anal. Chem.* **91** 7546–53

[73] Ertürk Bergdahl G, Andersson T, Allhorn M, Yngman S, Timm R and Lood R 2019 *In vivo* detection and absolute quantification of a secreted bacterial factor from skin using molecularly imprinted polymers in a surface plasmon resonance biosensor for improved diagnostic abilities *ACS Sens.* **4** 717–25

[74] Turkmen D, Yilmaz T, Bakhshpour M and Denizli A 2022 An alternative approach for bacterial growth control: *Pseudomonas* spp. imprinted polymer-based surface plasmon resonance sensor *IEEE Sens. J.* **22** 3001–8

[75] Özgür E, Topçu A A, Yılmaz E and Denizli A 2020 Surface plasmon resonance based biomimetic sensor for urinary tract infections *Talanta* **212** 120778

[76] Idil N, Bakhshpour M, Perçin I and Mattiasson B 2021 Whole cell recognition of *Staphylococcus aureus* using biomimetic SPR sensors *Biosensors* **11** 140

[77] Latif U, Can S, Sussitz H F and Dickert F L 2020 Molecular imprinted based quartz crystal microbalance sensors for bacteria and spores *Chemosensors* **8** 64

[78] Kurosawa S, Park J W, Aizawa H, Wakida S I, Tao H and Ishihara K 2006 Quartz crystal microbalance immunosensors for environmental monitoring *Biosens. Bioelectron.* **22** 473–81

[79] Arif S, Qudsia S, Urooj S, Chaudry N, Arshad A and Andleeb S 2015 Blueprint of quartz crystal microbalance biosensor for early detection of breast cancer through salivary autoantibodies against ATP6AP1 *Biosens. Bioelectron.* **65** 62–70

[80] Marx K A 2003 Quartz crystal microbalance: a useful tool for studying thin polymer films and complex biomolecular systems at the solution–surface interface *Biomacromolecules* **4** 1099–120

[81] Liu N, Li X, Ma X, Ou G and Gao Z 2014 Rapid and multiple detections of staphylococcal enterotoxins by two-dimensional molecularly imprinted film-coated QCM sensor *Sensors Actuators* B **191** 326–31

[82] Wang R *et al* 2021 Rapid, sensitive and label-free detection of pathogenic bacteria using a bacteria-imprinted conducting polymer film-based electrochemical sensor *Talanta* **226** 122135

[83] Feng W, Liang C, Gong H and Cai C 2018 Sensitive detection of Japanese encephalitis virus by surface molecularly imprinted technique based on fluorescent method *New J. Chem.* **42** 3503–8

[84] Samardzic R, Sussitz H F, Jongkon N and Lieberzeit P A 2014 Quartz crystal microbalance in-line sensing of *Escherichia coli* in a bioreactor using molecularly imprinted polymers *Sens. Lett.* **12** 1152–5

[85] Jenik M *et al* 2009 Sensing picornaviruses using molecular imprinting techniques on a quartz crystal microbalance *Anal. Chem.* **81** 5320–6

[86] Spieker E and Lieberzeit P A 2016 Molecular imprinting studies for developing QCM-sensors for *Bacillus cereus Procedia Eng.* **168** 561–4

[87] Klangprapan S, Choke-arpornchai B, Lieberzeit P A and Choowongkomon K 2020 Sensing the classical swine fever virus with molecularly imprinted polymer on quartz crystal microbalance *Heliyon* **6** e04137

[88] Gupta N, Shah K and Singh M 2016 An epitope-imprinted piezoelectric diagnostic tool for *Neisseria meningitidis* detection *J. Mol. Recognit.* **29** 572–9

IOP Publishing

Molecularly Imprinted Polymers for Environmental Monitoring
Fundamentals and applications
Raju Khan and Ayushi Singhal

Chapter 14

Future perspectives on molecularly imprinted polymers for environmental monitoring

Sadhna Chaturvedi

Molecular imprinting technology (MIT) is an exceptional procedure for preparing artificial receptors with a predetermined selectivity and specificity for a given analyte, and this strategy can be utilized as ideal materials in different application fields. Molecularly imprinted polymers (MIPs), the engraving innovation brings about polymeric matrices, are strong atomic recognition components that are able to mimic natural recognition particles, such as antibodies and cell receptors, and are capable of extracting and measuring analytes in complex matrices such as biological liquids and natural samples. Molecular imprinting provides a significant addition to the utilization of macromolecules in the development of sensors for the investigation of target substances because of their exceptionally specific recognition qualities. In addition, their excellent properties make them ideal for use in applications that are not usually suitable for the utilization of biomolecules. In this chapter the ideas behind the method and its utilization in various research designs appropriate for use in environmental monitoring are introduced. The aim is to give an overview of the field of MIPs by first examining first broad viewpoints in MIP preparation and various utilizations of MIPs, and then present a summary of the future prospects of MIPs, focusing on environmental monitoring, green MIP innovations, electrochemical biosensors in disease diagnosis, MIP nanoparticles, drug conveyance, and catalysis. A few critical perspectives on the future uses of MIPs from recent publications will also be examined.

Thus, in this chapter, we will provide an overview of the reasoning behind using MIPs as recognition elements in various fields of environmental research, and discuss the future perspectives this promising and innovative technology.

14.1 Introduction

The need for improved sensitivity and robustness are the two main considerations influencing the development of new research strategies. Nature's ability to deliver highly specific recognition frameworks (e.g. antigen–antibody, receptor–ligand, enzyme–substrate) has allowed biomolecules to be utilized as recognition components in scientific techniques, specifically for the identification of target analytes in complex samples, in which they are generally effective. However, the effectiveness of biomolecules in terms of operation tolerance (e.g. temperature, pH, ionic strength, and natural solvents) is quite limited. Molecularly imprinted polymers (MIPs) can overcome these limitations and find use in a variety of contexts. In this section we will examine their use in environmental monitoring, green aspects of molecular imprinting technology innovation, electrochemical biosensors in the diagnosis of disease, and MIP nanoparticles.

14.2 Molecular imprinting for environmental monitoring

Molecular imprinting is becoming a significant method for creating exceptionally specific artificial receptors for a constantly expanding range of biochemical and other compounds [1–4]. This method involves the development of cavities in an engineered polymer matrix that are complementary in functional and structural character to a template molecule.

MIPs are able to specifically recognize and bind to target structures in the presence of closely related compounds and species, and they are utilized in an extensive variety of biotechnological and biomedical applications. MIPs as antibody binding site mimics have demonstrated binding affinities and cross-reactivity profiles comparable to their biological counterparts and have even been employed as substitutes for biological antibodies in different environmental evaluation studies and for medical diagnosis as well. MIPs have also been used as highly selective chiral chromatographic stationary phases. Over time, the original practical methodology has been developed to use the exceptionally specific recognition ability of MIPs for the creation of enzyme mimics [5–7], following a strategy such as that utilized for delivering synergistic immunoglobulins.

The main advantages of molecular imprinting are the robust framework it provide for creating antibody mimics with specific molecular attributes, and it also provides other critical advantages, as follows: no need for laboratory animals, speed-of-use, low production costs, no requirement for the haptane formation protocol, and no issues with non-immunogenic substances. In addition, these types of materials a stable in extremes of temperature, pH, and natural or organic solvents, and offer additional great benefits over the utilization of macromolecules with regard to the development of vigorous scientific frameworks for execution in the field.

The adaptability of this method in terms of the choice of template, polymer creation, and polymerization design allow their inclusion with various configurations of sensor. MIPs have been used in various sensor designs for trace examination. Currently, the attention is on optical sensing and it has been the focus point of much work, as exemplified by the fluoroimmuno study on the identification of 2,4-D and related substances [3] and in the treatment of asthma, in which for the recognition of medications a surface plasmon resonance (SPR) strategy is utilized [8]. In the detection of

2-methylisobomeol, a substance that causes the smell of stagnant water, MIP-based quartz crystal microbalances (QCMs) have been developed [9]. Different kinds of electrochemical sensors have been studied to date, the most pivotal being one for the testing of mixtures arising because of contamination by sarin, a nerve gas [10]. For the identification of triazine herbicides, competitive immunoassay techniques have been utilized [11]. In the continued improvement of these MIPs, a strong argument can be introduced for their use in sensor advancement, particularly under the conditions expected for robust analysis examination strategies such as in the field and in industrial applications.

The primary benefits of MIPs are their high selectivity and specificity for the target molecule utilized in the imprinting technique. In contrast to natural frameworks such as proteins and nucleic acids, MIPs have higher actual robustness, strength, and resistance to temperature, pressure, acids, bases, metal particles, and natural or organic solvents. Furthermore, they are cheaper to synthesize and the lifetime of the polymers can be extremely high, retaining their recognition ability for several years at room temperature [12].

14.3 Pre-polymerization studies

Good interaction between the monomer and template is a preliminary necessary condition to obtain MIP networks with potential recognition sites. Currently, two main approaches are employed for MIP technology depending on the nature of the repolymerization interactions between the monomer and template. The first is the self-assembly approach, similar to biological recognition systems, that uses non-covalent forces, such as ionic interactions, hydrophobic interactions, van der Waals forces, hydrogen bonds, and van der Waals forces. The second is the preorganized approach [13], which is based on covalent reversible bonds and provides a rather homogeneous population of binding sites and reduces non-specific sites. However, in this approach it is necessary to cleave the covalent bonds to remove the template from the polymer matrix. Self-assembly is currently the most regularly adopted strategy for the production of MIPs due to the simplicity of complex formation and dissociation and the flexibility in terms of available functional monomers that can relate to almost any form of template. However, the self-assembled MIPs have lower binding affinity than those synthesized using covalent interactions. Hwang and Lee prepared cholesterol-imprinted polymers using a bulk technique, with either a covalent or a non-covalent approach, and they were able to differentiate the MIPs acting as stationary phases in high performance liquid chromatography (HPLC) columns [14]. They compared the covalently and non-covalently imprinted polymers and found less peak broadening, higher adsorption capacity, and about five-fold higher chromatographic efficiency for the covalently imprinted polymers compared to the non-covalently imprinted polymers. There are various reports in the literature showing that pre-polymerization studies on self-assembling systems can be helpful for the selection of acceptable effective monomers and solvents for specific template molecules. The formation of useful monomer–template complexes has been investigated by spectroscopic and theoretical approaches, that include nuclear magnetic resonance (NMR) and UV–vis spectroscopies and theoretical models as well [15–21].

Titration curves and binding isotherms using Job's method are currently applied to determine the nature of the interactions, the coordination number of the monomer–template complex, and the association constant by using spectroscopy techniques. For example, with the help of NMR spectroscopy the formation of the complex between nitrofurantoin with 2,6-bis(methaacrylamido) pyridine (BMP) was studied. This study revealed a hydrogen bonding interaction between the imide moiety of nitrofurantoin and the protons of the pyridine moiety of BMP and also permitted the calculation of the association constant [22].

In 2010 researchers analysed the binding affinity and selectivity of a new phthalocyanine, as potential monomer towards nucleoside derivatives, using UV–vis titration experiments. The experiment allowed the calculation of the association [89] constant Ka, determined by the modified Benesi–Hildebrand equation, of a zinc phthalocyanine with tri-O-acetyladenosine (TOAA).

14.4 Green aspects in molecularly imprinted microspheres

In developing an economical, robust, and sustainable approach that is harmless to the ecosystem, the reusability and safety of the imprinted material assumes a significant role. It is notable that polymer degradation products can also contaminate a sample during their application.

There are four principal factors that influence the reusability and reliability of MIPs, specifically the crosslinker, crosslinking degree, condition template extraction, and functional monomer. These variables were explored by Kupai's group in 2017 utilizing 11 distinct L-phenylalanine methyl ester (ME)-imprinted polymers in different structures. Their study was directed at the assessment of long-term robustness and reusability through performing adsorption–recovery cycles multiple times [23]. As far as the crosslinking degree, the divinylbenzene (DVB)-based polymers showed an extraordinary outcome and could be reused multiple times without losing their designed selectivity under acidic or basic conditions or increased temperature (65 °C). In contrast to DVB-base polymers, the crosslinking level of acrylamide- and methacrylate-based polymers decreased in both acidic and basic circumstances because of their irreversible degradation. Both the acrylamide and methacrylate crosslinker are regularly utilized in the preparation of molecularly imprinted microspheres (MIM). This may pose difficulty for the future exploration toward fostering an extended stable and reusable design for MIMs. Decreasing the utilization of solvents and energies could also have a critical effect. These critical effects on the environmental aspects were discussed at length in several studies [24–26].

As per these studies, elective methodologies that can be considered incorporate the utilization of green templates, green monomers, green solvents (for example, porogens and template expulsion solvents), green crosslinkers and initiators, energy efficiency, the incorporation of ultrasound and microwaves in improving response rates, scaled down procedures, and the utilization of computational devices for enhancing both the polymer and combination processes. Energy efficiency is vital, since higher energy needs cause critical effects for the climate, such as global warming. All controlled radical precipitation polymerization (CRPP) methods can be carried out under gentle temperature conditions, which is valuable with regard to the environmental perspective. MIPs depend on the development of specific interactions between a template (atom, particle, ion, complex or molecular, ionic, or macromolecular assembly, including microorganisms) and a functional monomer, and progressive polymerization in the presence of a large excess of crosslinker.

Imprinted polymers, in contrast to natural frameworks such as proteins and nucleic acids, have higher actual function, strength, and resistance to raised temperatures and pressure, and inertness toward acids, bases, metal particles, and natural or organic solvents [27–30]. The MIP community has more recently devoted considerable research and development efforts toward eco-friendly processes. Among other materials, biomass waste, which is a large environmental problem because most of it is discarded, can represent a potential sustainable alternative source for green synthesis, which can be addressed through the production of high-value carbon-based materials for different applications [31]. Molecular imprinting innovation technology (MIT) is a multidisciplinary innovation mimicking the specific limited selectivity of enzymes to substrates or antigens to antibodies; along with its quick advancement and wide applications, MIT faces the test of conforming to the need for sustainable green improvements. With the identification of the ecological dangers related to unsustainable MIT, another part of MIT, called green MIT, has emerged.

Up to this point, no single definition has been provided for green MIT. In this chapter, the execution cycle of green science in MIT is shown. The term 'green-ification' is proposed for the introduction of green MIT standards. The whole greenificated imprinting process is reviewed in [32], including component choice, polymerization execution, energy input, imprinting procedures, waste treatment and recovery, as well as the effects of these processes on operator wellbeing and the climate. In green chemistry, the basic properties of MIPs should incorporate reusability, and chemical and mechanical stability. In most cases, hazardous reagents are utilized in the formation of MIPs. Occasionally, the manufacturing process for the generation of MIPs needs to be repeated a few times before the ideal product is achieved. Such repeated procedures cause massive solid and fluid waste generation, which end-up in logical inconsistency with perspective of green chemistry. The utilization of computational devices for the choice of reagents that are reasonable for the combination of MIPs is being assessed. The reason for such a study is obtain knowledge for a rational design of MIPs with the aim to attend to the ecological effects brought about by the consequences of ineffective imprinting. In addition, it has been seen that in recent years, MIPs are being synthesized with monomers that are harmless to the ecosystem, and wider use of such reagents is expected in the coming years. In addition, greener uses of MIPs, for example, miniaturized strategies are also being examined.

The simplicity of production, minimal cost, capacity to tailor the recognition component for analyte molecules, and stability under harsh conditions make MIPs promising candidates in recognition techniques for biosensing. In contrast to natural frameworks, molecular imprinting methods enjoy a few benefits, including high recognition capacity, long-term durability, minimal expense, and robustness, permitting MIPs to be utilized in drug delivery, biosensor innovation technology, and nanotechnology. However, MIP-based sensors have specific weaknesses in the determination of biomolecules (nucleic acid, protein, lipids, and carbohydrates), taking into account the immense volume of the most recent writing on bio-micromolecules. Despite their potential, MIP materials need to address some weaknesses before earning their position with respect to biomacromolecules. This is expected to be a feature of the ongoing advancements in MIP-based sensors for the determination of DNA and other nucleobases [33].

14.5 The detection of contaminants using MIPs

Environmental pollution causing harm, including death, has become a major issue around the world, and includes chemical toxins, such as agricultural pesticides [34], human and veterinary medications [35], persistent organic toxins [36], dyes [37], and heavy metals [38]. These contaminants can enter the body through the food chain and cause skin, breathing, gastrointestinal tract, and systemic responses, and that can prompt lethal anaphylactic shock, seriously threatening the wellbeing and health of individuals [31, 39, 40]. At present, the conventional strategies for the identification of pollutants include gas chromatography, HPLC, and different kinds of procedures [41]. These chromatography methods have costly lab set-ups and this is the greatest weakness of this kind of traditional procedure, along with its long processing time and the necessity for sample preparation, which restricts its applications [42]. Hence there is a need for improvements and new systems for the detection of chemical toxins which require less time to carry out and provide extraordinary usefulness through quick recognition, precise evaluation of synthetic contaminants, precise detection of significant contamination levels, and low test costs [43–46].

Researchers have proposed MIPs using conventional detection strategies and have worked to further develop them for exceptionally selective and sensitive methods for the detection of toxins since the first synthesis of MIPs in 1977, using three steps:

A. Combination of the template molecule and monomers to form a complex by covalent and non-covalent bonding [47, 48].
B. The composite is immobilized with the augmentation of crosslinking and pore shaping agents [49].
C. The template molecules with the desired shape and design for the target molecules are eluted, leaving the cavities of the templates in the polymer [50].

MIPs have different exceptional qualities of specific recognition [51–57]. MIPs have been created from single templates to composite templates, and the preparation process has been consistently advanced to further develop the application range, adsorption execution, and specific selectivity [58]. They have been utilized in different fields such as environmental pollutant examination, food quality and security, and organic sample separation and remediation. There remain a few issues to be investigated and settled [59]. MIPs generally show the best performance in hydrophobic natural solvents, which means that the presence of polar solvents (particularly water) in functional utilizations of MIP can seriously interfere with the development of pre-polymerized complexes in the imprinting system and disrupt the collaboration between the monomer and the template, which can also be utilized in future [60].

The remarkable properties of nanostructured MIP materials depend on their size, however, due to their fast equilibrium with the substance to be estimated, and because of trouble eliminating the template totally after the preparation phase of the highly cross-linked polymer, MIP materials frequently suffer from the effects of issues such as molecular leakage, causing the arrangement of molecularly imprinted nanomaterials with irregular particle shapes, different molecule sizes, non-uniform recognition sites, and low affinity [61]. In the synthesis of MIPs, there are limited varieties of functional monomers and crosslinking agents available, and the chemical reagents used are not only toxic, but also face problems such as high capital expenditure and low conversion efficiency, making it difficult to achieve mass production from the laboratory to the factory and to maximize commercial conversion [62]. The efforts to tackle these issues never cease, and in the future the combination of MIPs with other permeable or nanostructured materials might provide new ways to deal with compound impurities for use in food and, specifically, MIP/permeable polymers and carbon nanomaterials will be a significant forward leap in the field of biotech sensors [63]. Likewise, the utilization of MIPs in combination with various logical instruments to simulate discovery frameworks is additionally an ideal objective that is continually being sought after and might be achieved soon [64, 65].

14.6 MIPs for microbial contaminants

Currently, nanotechnology-based MIP detection systems are receiving a great deal of attention in the microbial research community. Increasingly, nanosensors are being introduced as promising tools in order to overcome infection problems through the identification of contaminants. Nanosensors can provide monitoring of microorganisms, viruses, toxins, spores, signaling molecules, cell wall components, etc [66]. Nanosensors make it possible to remove contaminants, decrease the incidence of infections, control the spread of infection agents and in this respect decrease outbreaks, and improve suitable control measures [67]. The increasing incidence of different kinds of infections supports the requirements for developing novel techniques. It is necessary to detect microbial causative agents rapidly, accurately, and cost-effectively. Traditional laboratory techniques applied for the detection of microorganisms have some disadvantages such as being laborious, time-consuming, and expensive. In light of new sensing approaches, nanomaterial-based technologies have some desirable properties, providing rapid, specific, sensitive, cheap, and reliable detection [68].

14.7 MIPs for water contaminants

Water contamination is a general threat brought about by a number of dangerous toxins. Because of the adverse consequences of these contaminants on the wellbeing of humans and other living beings in different ecosystems, various detection and remediation systems have been created [69]. MIPs have been reported for the extraction of water toxins. Such MIPs are tailored adsorbent polymers, created through bonding between a template particle and monomer, that are then polymerized by a crosslinker. Because of their predetermined target molecules, these MIPs can be utilized as adsorbent for the treatment of water [70]. MIPs with tailor-made recognition sites are utilized in water testing for pre-treatment before evaluation. In [71], recent advances in MIP innovation, including the design conventions, applications, and improvements, are examined, and, furthermore, the demonstration of MIPs which can be improved by streamlining the polymeric organization is examined in regard to the attributes of the rebinding medium. The limitations of MIPs for water investigation, particularly their restricted selectivity for water dissolvable analytes, material wettability, and MIP inhomogeneity are discussed in [72], providing potential solutions. Finally, a few novel applications and possibilities for on-line, fast, and direct investigation of environmental samples utilizing MIPs are included [72].

14.8 Electrochemical biosensors

Electrochemical biosensing for clinical biomarker recognition has, in particular, received critical consideration because of the fast, low cost, and versatile nature of electrochemical sensing. The quick identification of clinical biomarkers for diagnosing and monitoring illness is particularly significant in numerous clinical fields such as malignant growth, sepsis, and cardiovascular disease (CVD) [73], with a wide range of evidence demonstrating better patient outcomes following early diagnosis [74].

Imprinting for the identification of proteins has perhaps been one of the most difficult areas being developed, and there are huge difficulties that should be defeated to match the specificity and responsiveness of current innovations and accomplish their broader use as a clinical diagnostics tool [75]. There is a huge range of malignant growths, requiring different biomarkers for certain aspects of their diagnosis and monitoring. For colorectal and breast disease, the cost for health services of treatment is decreased substantially with early diagnosis (stages 1 and 2), underlining the interest in quick point-of-care testing (POCT) to help with fast analysis and treatment [76]. The significance of fast diagnosis is clearly shown by diseases that have high death rates without quick treatment, for example, for each hour that sepsis is not recognized and treatment has not been started, there is a 7.6% decrease in survival rate [77].

MIPs offer numerous potential benefits over the utilization of organic recognition components as they offer superior compound and temperature stability, the capability to bond to the target particle, and are cheaper to produce [78]. Most reports using a combination of MIPs and electrochemical detection involve electropolymerization as their chosen strategy for MIP development. This synergizes well with the general stage because of the possibility of delivering both conductive and non-conductive polymers and the immediate development of the MIP layer on the outer layer of the transducer, successfully eliminating the time-consuming immobilization step. This philosophy depends on the development of polymers through the use of potential, causing an oxidation or reduction in the monomers, and has been applied to the identification of a range of biological targets [79].

Although advantageous, the arrangement of MIPs through electropolymerization has huge difficulties to conquer, for example, a more restricted supply of appropriate monomers, reduced homogeneity of the recognition sites delivered, and the overall adaptability of the process [80]. Different strategies for sensors for clinical bio-markers, including careful optimization of the synthesis, template evacuation, and recognition parameters, should be applied.

14.9 MIP nanoparticles

MIPs are also called artificial antibodies or plastic antibodies. These artificial antibodies are chemically synthesized affinity materials with tailor-made binding cavities which are complementary to the template or target molecules in size, shape, and performance [81]. MIP-based nanoparticles (nanoMIPs) are straightforward combinations of the biomimetic recognition of MIPs with the special characteristics of nanoparticles (optical, magnetic, thermal, acoustic and electric properties, nano-scale size, and high surface-to-volume ratio) that have received great attention due to their unique properties [82]. NanoMIPs have shown significant potential for cancer therapy [83]. The unique properties of nanoMIPs bring desired and multi-faceted functions to the nanomedicine used in cancer treatment. Currently, a number of outstanding studies on the fundamentals of preparation strategies, chemical design, and applications of nanoMIPs in the biomedical field (e.g. disease diagnosis, bioimaging, and drug delivery) have been reported [84]. It is also highlighted that MIPs are used as artificial antibodies in cancer therapy.

MIPs have been used in various devices (e.g. coating, patches, and bulk hydro-gels) for ocular, intravenous, dermal, and oral drug delivery. The molecular imprinting technology elicits new polymer formats and shifts the material size to the nanoscale, providing a versatile tool for cancer nanomedicines with controlled drug release [85]. For systemically administered cancer nanomedicines, side effects caused by premature drug release during circulation in the blood and reduction of the drug amount arriving at tumors is possible, but nanomedicines are expected to fullly release the loaded drug only upon reaching the target site, to achieve a functional therapeutic concentration [86]. Despite the fact that they have faced many different challenges at the primary level, nanoMIPs have opened a new research approach and show eminent potential in cancer nanomedicine. With the progress of MIT, nanoMIPs and their excellent properties will provide a remarkable push forward in cancer therapy [87].

14.10 Conclusions

In this chapter, several applications of MIPs have been discussed. However, we cannot conclude which technique is better because of the absence of studies. We believe studies based on improving MIM properties and applications will continue to grow rapidly. Future studies may focus on the following areas: (i) comparison studies, as future investigations to conclude which technique is better still need to be performed; (ii) applicability of MIMs as drug delivery systems, as MIMs have excellent potential as drug delivery systems because of their selective binding characteristics and their ability to release the template from the matrix, and MIMs can also be used as targeting systems for the recognition of large molecules in gene therapy; (iii) MIP nanoparticles; (iv) electrochemical biosensors based on MIPs; and (v) green aspects of MIMs.

The protability, result comprehension, and real-world applicability can all be improved with the use of smartphones and other analogous readers. Also, researchers must look into methods for manufacturing POCT devices with multiplexed detection in order to decrease cost and test time while enhancing assay productivity. The growing movement toward the fusion of various important detection technologies into a single system will allow for the simultaneous detection of numerous analytes. The field needs additional research as it is still in its early phases. Environmental safety analysis will become simpler, quicker, less expensive, and more efficient with the advent of such developments in POCT equipment. When combined, multidisciplinary methods for the creation of MIPs electrochemical nanosensors for any analyte detection have the potential to not only monitor antibiotic levels but also open the door to modifying their use to combat antibiotic resistance. By incorporating smart technologies such as artificial intelligence (AI) and the Internet of Medical Things (IoMT), the potential uses of MIP electrochemical nanosensors for the efficient detection of antibiotics will certainly grow, as shown in figure 14.1.

The fusion of various important detection technologies into a single system allows for the simultaneous detection of numerous analytes. The field needs additional research as it is still in its early phases.

For MIPs used as recognition components in electrochemical biosensors for fundamental biomarkers, we need to examine the issues around synergizing MIPs and electrochemical read-out methodologies and experiences into the future

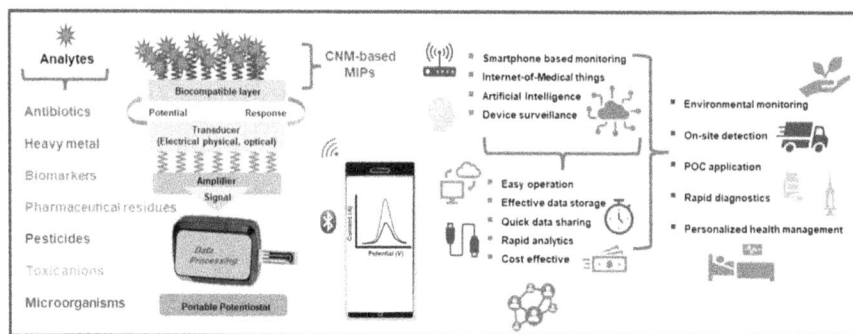

Figure 14.1. Futuristic applications of MIPs for the electrochemical sensing of different analyte. (Reproduced with permission from [88]. Copyright 2022 Elsevier.)

perspectives of this promising and imaginative innovation. In recent years, specific consideration has been focused on the improvement of MIPs as materials for sensors and biosensors. However, MIP-based biomimetic sensors are still not at standard compared to biosensors on the basis that further enhancement of the MIPs and the transducers is required. Hence, although stimulating work continues in this field, the marketability of molecular imprinting sensors is still in its earliest stages.

References

[1] Sellergren B 2000 Imprinted polymers with memory for small molecules, proteins, or crystals *Angew. Chem. Int. Ed.* **39** 1031–7

[2] Andersson H S and Nicholls I A 1997 Molecular imprinting: recent innovations in antibody and enzyme mimicking synthetic polymers *Recent Res. Develop. Pure Appl. Chem.* **1** 133–57

[3] Haupt K, Mayes A G and Mosbach K 1998 Herbicide assay using an imprinted polymer based system analogous to competitive fluoroimmunoassays *Anal. Chem. Anal. Chem.* **70** 3936–9

[4] Asanuma H, Hishiya T and Komiyama M 2000 Tailor-made receptors by molecular imprinting *Adv. Mater.* **12** 1019–30

[5] Matsui J, Nicholls I A, Karube I and Mosbach K 1996 Carbon–carbon bond formation using substrate selective catalytic polymers prepared by molecular imprinting: an artificial class II aldolase *J. Org. Chem.* **61** 5414–7

[6] Strikovsky A G, Kasper D, Grun M, Green B S, Hradil J and Wulff G 2000 Catalytic molecularly imprinted polymers using conventional bulk polymerization or suspension polymerization: selective hydrolysis of diphenyl carbonate and diphenyl carbamate *J. Am. Chem. Soc.* **122** 6295–6

[7] Sellergren B, Karmalkar R N and Shea K J 2000 Enantioselective ester hydrolysis catalyzed by imprinted polymers *J. Org. Chem.* **65** 4009–27

[8] Lai E P C, Fafra A, Vandernoot V A, Kono M and Polsky B 1998 Surface plasmon resonance sensors using molecularly imprinted polymers for sorbent assay of theophylline, caffeine, and xanthine *Can. J. Chem.* **76** 265–73

[9] Ji H-S, McNivan S, Ikebukuro K and Karube I 1999 Selective piezoelectric odor sensing using molecularly imprinted polymers *Anal. Chim. Acta* **390** 93–100

[10] Jenkins A L, Uy O M and Murray G M 1999 Polymer-based lanthanide luminescent sensor for detection of the hydrolysis product of the nerve agent Soman in water *Anal. Chem.* **70** 373–8

[11] Piletsky S A, Piletskaya E V, El'skaya A, Levi R, Yano K and Karube I 1997 Optical detection system for triazine based on molecularly-imprinted polymers *Anal. Lett.* **30** 445–55

[12] Cirillo P F G, Curcio M, Parisi O I, Iemma F and Picci N 2011 Molecularly imprinted polymers in drug delivery: state of art and future perspectives *Expert Opin. Drug Deliv.* **8** 1379–93

[13] Wulff G and Sarhan A 1972 Use of polymers with enzyme-analogous structures for the resolution of racemates *Angew. Chem. Int. Ed.* **11** 341–2

[14] Hwang C C and Lee W C 2002 Chromatographic characteristic of cholesterol-imprinted polymers prepared by covalent and non-covalent imprinting methods *J. Chromatogr.* A **962** 69–78

[15] O'Mahony J, Molinelli A, Nolan K, Smyth M R and Mizaikoff B 2005 Towards the rational development of molecularly imprinted polymers: ^{1}H NMR studies on hydrophobicity and ion-pair interactions as driving forces for selectivity *Biosens. Bioelectron.* **20** 1884–93

[16] Wei S, Jakusch M and Mizaikoff B 2007 Investigating the mechanisms of 17β-estradiol imprinting by computational prediction and spectroscopic analysis *Anal. Bioanal. Chem.* **389** 423–31

[17] Karlsson B C G, O'Mahony J, Karlsson J G, Bengtsson H, Eriksson L A and Nicholls I A 2009 Structure and dynamics of monomer–template complexation: an explanation for molecularly imprinted polymer recognition site heterogeneity *J. Am. Chem. Soc.* **131** 13297–304

[18] Yánez-Sedeno P, Pingarrón J M, Riu J and Rius F X 2010 Electrochemical sensing based on carbon nanotubes *TrAC Trends Anal. Chem.* **29** 939–53

[19] Vasapollo G, Del Sole R, Mergola L, Lazzoi M R, Scardino A, Scorrano S and Mele G 2011 Molecularly imprinted polymers: present and future perspective *Int. J. Mol. Sci.* **12** 5908–45

[20] Pietrzyk A, Kutner W, Chitta R, Zandler M E, D'Souza F, Sannicolò F and Mussini P R 2009 Melamine acoustic chemosensor based on molecularly imprinted polymer film *Anal. Chem.* **81** 10061–70

[21] O'Mahony J, Karlsson B C G, Mizaikoff B and Nicholls I A 2007 Correlated theoretical, spectroscopic and x-ray crystallographic studies of a non-covalent molecularly imprinted polymerisation system *Analyst* **132** 1161–8

[22] Ecclesia O T, Pari P, Ago A, Sombo A, Rahayu D and Nur Hasanah A 2020 Microsphere polymers in molecular imprinting: current and future perspectives *Molecules* **25** 3256

[23] Athikomrattanakul U, Katterle M, Gajovic-Eichelmann N and Scheller F W 2009 Development of molecularly imprinted polymers for the binding of nitrofurantoin *Biosens. Bioelectron.* **25** 82–7

[24] Kupai J, Razali M, Buyuktiryaki S, Kecili R and Szekely G 2017 Long-term stability and reusability of molecularly imprinted polymers *Polym. Chem.* **8** 666–673

[25] Madikizela L M, Tavengwa N T, Tutu H and Chimuka L 2018 Green aspects in molecular imprinting technology: from design to environmental applications *Trends Environ. Anal. Chem.* **17** 14–22

[26] Keçili R, Büyüktiryaki S, Dolak İ and Hussain C M 2020 The use of magnetic nanoparticles in sample preparation devices and tools *Handbook of Nanomaterials in Analytical Chemistry* (Amsterdam: Elsevier) pp 75–95

[27] Viveiros R, Rebocho S and Casimiro T 2018 Green strategies for molecularly imprinted polymer development *Polymers* **10** 306

[28] Vasapollo G, Del Sole R, Mergola L, Lazzoi M R, Scardino A, Scorrano S and Mele G 2011 Molecularly imprinted polymers: present and future perspective *Int. J. Mol. Sci.* **12** 5908–45

[29] Chen L, Wang X, Lu W, Wu X and Li J 2016 Molecular imprinting: perspectives and applications *Chem. Soc. Rev.* **45** 2137–211
Schirhagl R 2014 Bioapplications of Molecularly Imprinted Polymers *Anal. Chem.* **86** 250–61
Dmitrienko E V, Pyshnaya I A, Martyanov O N and Pyshnyi D V 2016 Molecularly imprinted polymers for biomedical and biotechnological applications *Russ. Chem. Rev.* **85** 513–36

[30] Culver H A, Steichen S D and Peppas N A 2016 A closer look at the impact of molecular imprinting on adsorption capacity and selectivity for protein templates *Biomacromolecules* **17** 4045–53

[31] Singhal A, Parihar A, Kumar N and Khan R 2022 High throughput molecularly imprinted polymers based electrochemical nanosensors for point-of-care diagnostics of COVID-19 *Mater. Lett.* **306** 130898

[32] Ding S, Lyu Z, Niu X, Zhou Y, Liu D, Falahati M, Du D and Lin Y 2020 Integrating ionic liquids with molecular imprinting technology for biorecognition and biosensing: a review *Biosens. Bioelectron.* **149** 111830

[33] Janczura M, Lulínski P and Sobiech M 2021 Imprinting technology for effective sorbent fabrication: current state-of-art and future prospects *Materials* **14** 1850

[34] Cai L, Zhang Z, Xiao H, Chen S and Fu J 2019 An eco-friendly imprinted polymer based on graphene quantum dots for fluorescent detection of *p*-nitroaniline *RSC Adv.* **9** 41383–91

[35] Zhou J W, Zou X M, Song S H and Chen G H 2018 Quantum dots applied to methodology on detection of pesticide and veterinary drug residues *J. Agric. Food Chem.* **66** 1307–19

[36] Ren B, Shen W, Li L, Wu S and Wang W 2018 3D CoFe$_2$O$_4$ nanorod/flower-like MoS$_2$ nanosheet heterojunctions as recyclable visible light-driven photocatalysts for the degradation of organic dyes *Appl. Surf. Sci.* **447** 711–23

[37] Horak E, Hranjec M, Vianello R and Steinberg I M 2017 Reversible pH switchable aggregation-induced emission of self-assembled benzimidazole-based acrylonitrile dye in aqueous solution *Dyes Pigm.* **142** 108–15

[38] Rudd N D, Wang H, Fuentes-Fernandez E M, Teat S J, Chen F, Hall G, Chabal Y J and Li J 2016 Highly efficient luminescent metal–organic framework for the simultaneous detection and removal of heavy metals from water *ACS Appl. Mater. Interfaces* **8** 30294–303

[39] Ashley J, Shahbazi M A, Kant K, Chidambara V A, Wolff A, Bang D D and Sun Y 2017 Molecularly imprinted polymers for sample preparation and biosensing in food analysis: progress and perspectives *Biosens. Bioelectron.* **91** 606–15

[40] Pakchin P S, Nakhjavani S A, Saber R, Ghanbari H and Omidi Y 2017 Recent advances in simultaneous electrochemical multi-analyte sensing platforms *TrAC Trends Anal. Chem.* **92** 32–41

[41] Ebrahimnejad P, Dinarvand R, Sajadi A, Jafari M R, Movaghari F and Atyabi F 2009 Development and validation of an ion-pair HPLC chromatography for simultaneous determination of lactone and carboxylate forms of SN-38 in nanoparticles *J. Food Drug Anal.* **17** 8

[42] Mazouz Z *et al* 2020 Computational approach and electrochemical measurements for protein detection with MIP-based sensor *Biosens. Bioelectron.* **151** 111978

[43] Lu Y *et al* 2015 Impacts of soil and water pollution on food safety and health risks in China *Environ. Int.* **77** 5–15

[44] Carvalho F P 2017 Pesticides, environment, and food safety *Food Energy Secur.* **6** 48–60

[45] Piletsky S, Canfarotta F, Poma A, Bossi A M and Piletsky S 2020 Molecularly imprinted polymers for cell recognition *Trends Biotechnol.* **38** 368–87

[46] Yadav S, Sadique M A, Ranjan P, Kumar N, Singhal A, Srivastava A K and Khan R 2021 SERS based lateral flow immunoassay for point-of-care detection of SARS-CoV-2 in clinical samples *ACS Appl. Bio Mater.* **4** 2974–95

[47] Figueiredo L, Erny G L, Santos L and Alves A 2016 Applications of molecularly imprinted polymers to the analysis and removal of personal care products: a review *Talanta* **146** 754–65

[48] Singhal A, Yadav S, Sadique M A, Khan R, Kaushik A, Sathish N and Srivastava A K 2022 MXene-modified molecularly imprinted polymer as an artificial bio-recognition platform for

efficient electrochemical sensing: progress and perspectives *Phys. Chem. Chem. Phys* **24** 19164–76

[49] Kupai J, Razali M, Buyuktiryaki S, Kecili R and Szekely G 2017 Long-term stability and reusability of molecularly imprinted polymers *Polym. Chem.* **8** 666–73

[50] Ramström O, Ye L and Mosbach K 1998 Screening of a combinatorial steroid library using molecularly imprinted polymers *Anal. Commun.* **35** 9–11

[51] Huang D L, Wang R Z, Liu Y G, Zeng G M, Lai C, Xu P, Lu B A, Xu J J, Wang C and Huang C 2015 Application of molecularly imprinted polymers in wastewater treatment: a review *Environ. Sci. Pollut. Res.* **22** 963–77

[52] Iskierko Z, Sosnowska M, Sharma P S, Benincori T, D'Souza F, Kaminska I, Fronc K and Noworyta K 2015 Extended-gate field-effect transistor (EG-FET) with molecularly imprinted polymer (MIP) film for selective inosine determination *Biosens. Bioelectron.* **74** 526–33

[53] Song X, Xu S, Chen L, Wei Y and Xiong H 2014 Recent advances in molecularly imprinted polymers in food analysis *J. Appl. Polym. Sci.* **131** 16

[54] He J, Song L, Chen S, Li Y, Wei H, Zhao D, Gu K and Zhang S 2015 Novel restricted access materials combined to molecularly imprinted polymers for selective solid-phase extraction of organophosphorus pesticides from honey *Food Chem.* **187** 331–7

[55] Ren X, Cheshari E C, Qi J and Li X 2018 Silver microspheres coated with a molecularly imprinted polymer as a SERS substrate for sensitive detection of bisphenol A *Microchim. Acta* **185** 1–8

[56] Andersson T, Bläckberg A, Lood R and Ertürk Bergdahl G 2020 Development of a molecular imprinting-based surface plasmon resonance biosensor for rapid and sensitive detection of *Staphylococcus aureus* alpha hemolysin from human serum *Front. Cell. Infect. Microbiol.* **10** 571578

[57] Singhal A, Singh A, Shrivastava A and Khan R 2023 Epitope imprinted polymeric materials: application in electrochemical detection of disease biomarkers *J. Mater. Chem.* B **11** 936–54

[58] Singhal A, Ranjan P, Sadique M A, Kumar N, Yadav S, Parihar A and Khan R 2022 Molecularly imprinted polymers-based nanobiosensors for environmental monitoring and analysis *Nanobiosensors for Environmental Monitoring: Fundamentals and Application* (Cham: Springer International Publishing) pp 263–78

[59] Scialabba N and Hattam C (ed) 2002 *Organic Agriculture, Environment and Food Security* (Rome: Food and Agriculture Organization)

[60] Cormack P A and Elorza A Z 2004 Molecularly imprinted polymers: synthesis and characterisation *J. Chromatogr.* B **804** 173–82

[61] Spivak D A 2005 Optimization, evaluation, and characterization of molecularly imprinted polymers *Adv. Drug Delivery Rev.* **57** 1779–94

[62] Bossi A, Bonini F, Turner A P and Piletsky S A 2007 Molecularly imprinted polymers for the recognition of proteins: the state of the art *Biosens. Bioelectron.* **22** 1131–7

[63] Uzun L and Turner A P 2016 Molecularly-imprinted polymer sensors: realising their potential *Biosens. Bioelectron.* **76** 131–44

[64] Dong C, Shi H, Han Y, Yang Y, Wang R and Men J 2021 Molecularly imprinted polymers by the surface imprinting technique *Eur. Polym. J.* **145** 110231

[65] Guoning C, Hua S, Wang L, Qianqian H, Xia C, Hongge Z, Zhimin L, Chun C and Qiang F 2020 A surfactant-mediated sol–gel method for the preparation of molecularly imprinted

polymers and its application in a biomimetic immunoassay for the detection of protein *J. Pharm. Biomed. Anal.* **190** 113511

[66] Bompart M, De Wilde Y and Haupt K 2010 Chemical nanosensors based on composite molecularly imprinted polymer particles and surface-enhanced Raman scattering *Adv. Mater.* **22** 2343–8

[67] Shojaei S, Nasirizadeh N, Entezam M, Koosha M and Azimzadeh M 2016 An electro-chemical nanosensor based on molecularly imprinted polymer (MIP) for detection of gallic acid in fruit juices *Food Anal. Methods* **9** 2721–31

[68] Akgönüllü S, Bakhshpour M, Neslihan İ D, Andaç M, Yavuz H and Denizli A 2020 Versatile polymeric cryogels and their biomedical applications *Hacettepe J. Biol. Chem.* **48** 99–118

[69] Murray A and Örmeci B 2012 Application of molecularly imprinted and non-imprinted polymers for removal of emerging contaminants in water and wastewater treatment: a review *Environ. Sci. Pollut. Res.* **19** 3820–30

[70] Parlapiano M, Akyol Ç, Foglia A, Pisani M, Astolfi P, Eusebi A L and Fatone F 2021 Selective removal of contaminants of emerging concern (CECs) from urban water cycle via molecularly imprinted polymers (MIPs): potential of upscaling and enabling reclaimed water reuse *J. Environ. Chem. Eng.* **9** 105051

[71] Lowdon J W, Diliën H, Singla P, Peeters M, Cleij T J, van Grinsven B and Eersels K 2020 MIPs for commercial application in low-cost sensors and assays—an overview of the current status quo *Sensors Actuators* B **325** 128973

[72] Pardeshi S and Dhodapkar R 2022 Advances in fabrication of molecularly imprinted electrochemical sensors for detection of contaminants and toxicants *Environ. Res.* **212** 113359

[73] Fleischmann-Struzek C, Goldfarb D M, Schlattmann P, Schlapbach L J, Reinhart K and Kissoon N 2018 The global burden of paediatric and neonatal sepsis: a systematic review *Lancet Respir. Med.* **6** 223–30

[74] Apple F S, Smith S W, Pearce L A and Murakami M M 2009 Assessment of the multiple-biomarker approach for diagnosis of myocardial infarction in patients presenting with symptoms suggestive of acute coronary syndrome *Clin. Chem.* **55** 93–100

[75] Weller D *et al* 2012 The Aarhus statement: improving design and reporting of studies on early cancer diagnosis *Br. J. Cancer* **106** 1262–7

[76] Laudicella R, Burger I A, Panasiti F, Longo C, Scalisi S, Minutoli F, Baldari S, Grimaldi L M and Alongi P 2021 Subcutaneous uptake on [18F] florbetaben PET/CT: a case report of possible amyloid-beta immune-reactivity after COVID-19 vaccination *SN Compr. Clin. Med.* **3** 2626–8

[77] Park R, Jeon S, Jeong J, Park S Y, Han D W and Hong S W 2022 Recent advances of point-of-care devices integrated with molecularly imprinted polymers-based biosensors: from biomolecule sensing design to intraoral fluid testing *Biosensors* **12** 136

[78] Moczko E, Poma A, Guerreiro A, de Vargas Sansalvador I P, Caygill S, Canfarotta F, Whitcombe M J and Piletsky S 2013 Surface-modified multifunctional MIP nanoparticles *Nanoscale.* **5** 3733–41

[79] Erdőssy J, Horváth V, Yarman A, Scheller F W and Gyurcsányi R E 2016 Electrosynthesized molecularly imprinted polymers for protein recognition *TrAC Trends Anal. Chem.* **79** 179–90

[80] Crapnell R D, Hudson A, Foster C W, Eersels K, Grinsven B V, Cleij T J, Banks C E and Peeters M 2019 Recent advances in electrosynthesized molecularly imprinted polymer sensing platforms for bioanalyte detection *Sensors* **19** 1204

[81] Zhu X, Wei D, Chen O, Zhang Z, Xue J, Huang S, Zhu W and Wang Y 2016 Upregulation of CCL3/MIP-1alpha regulated by MAPKs and NF-kappaB mediates microglial inflammatory response in LPS-induced brain injury *Acta Neurobiol. Exp* **76** 304–17

[82] Zhong C, Yang B, Jiang X and Li J 2018 Current progress of nanomaterials in molecularly imprinted electrochemical sensing *Crit. Rev. Anal. Chem.* **48** 15–32

[83] Cecchini A, Raffa V, Canfarotta F, Signore G, Piletsky S, MacDonald M P and Cuschieri A 2017 *In vivo* recognition of human vascular endothelial growth factor by molecularly imprinted polymers *Nano Lett.* **17** 2307–12

[84] El-Schich Z, Zhang Y, Feith M, Beyer S, Sternbæk L, Ohlsson L, Stollenwerk M and Wingren A G 2020 Molecularly imprinted polymers in biological applications *Biotechniques* **69** 406–19

[85] Tuwahatu C A, Yeung C C, Lam Y W and Roy V A 2018 The molecularly imprinted polymer essentials: curation of anticancer, ophthalmic, and projected gene therapy drug delivery systems *J. Controlled Release* **287** 24–34

[86] Wicki A, Witzigmann D, Balasubramanian V and Huwyler J 2015 Nanomedicine in cancer therapy: challenges, opportunities, and clinical applications *J. Controlled Release* **200** 138–57

[87] Shevchenko K G, Garkushina I S, Canfarotta F, Piletsky S A and Barlev N A 2022 Nano-molecularly imprinted polymers (nanoMIPs) as a novel approach to targeted drug delivery in nanomedicine *RSC Adv.* **12** 3957–68

[88] Singhal A, Sadique M A, Kumar N, Yadav S, Ranjan P, Parihar A, Khan R and Kaushik A K 2022 Multifunctional carbon nanomaterials decorated molecularly imprinted hybrid polymers for efficient electrochemical antibiotics sensing *J. Environ. Chem. Eng.* **10** 107703

[89] Longo L, Scorrano S and Vasapollo G 2010 RNA nucleoside recognition by phthalocyanine-based molecularly imprinted polymers *J. Polym. Res.* **17** 683–7

www.ingramcontent.com/pod-product-compliance
Lightning Source LLC
Chambersburg PA
CBHW080513220326
41599CB00032B/6069